David S. Jordan

A Manual of the Vertebrate Animals of the Northern United States

including the district north and east of the Ozark Mountains, south of the

Laurentian Hills, north of the southern boundary of Virginia, and east of the

Missouri River. Fifth Edition

David S. Jordan

A Manual of the Vertebrate Animals of the Northern United States
including the district north and east of the Ozark Mountains, south of the Laurentian Hills, north of the southern boundary of Virginia, and east of the Missouri River. Fifth Edition

ISBN/EAN: 9783337287962

Printed in Europe, USA, Canada, Australia, Japan

Cover: Foto ©berggeist007 / pixelio.de

More available books at **www.hansebooks.com**

OF THE

VERTEBRATE ANIMALS

OF THE

NORTHERN UNITED STATES

*INCLUDING THE DISTRICT NORTH AND EAST OF THE
OZARK MOUNTAINS, SOUTH OF THE LAURENTIAN
HILLS, NORTH OF THE SOUTHERN BOUNDARY
OF VIRGINIA, AND EAST OF THE
MISSOURI RIVER*

INCLUSIVE OF MARINE SPECIES

BY

DAVID STARR JORDAN

PRESIDENT OF THE UNIVERSITY OF INDIANA

FIFTH EDITION

ENTIRELY REWRITTEN AND MUCH ENLARGED

CHICAGO
A. C. McCLURG AND COMPANY
1888

PREFACE

———

THIS book is designed to give to students and collectors
a ready means of identifying the Vertebrate fauna of the
region which it covers, and of recognizing the characters
on which the families, genera, and species of these animals are founded.

To these ends, I have made use of a system of analytical keys by which differential characters are brought into
contrast. The usefulness of such keys has long been
recognized by botanists, and in ornithology the recent
works of Coues and Ridgway have proved their value to
the student.

That the book might not reach a size too large for field
or class use, I have made all descriptions very concise,
with as few repetitions as possible. I have confined the
generic characters to the analytical keys, using as a rule
only such characters as are distinctive as well as descrip-
tive. The need of condensation has caused the omission
of synonymy, and of references to authorities except in
special cases.

In the first four editions of this work (1876, 1878, 1880,
1884), large use was made of artificial characters in the
analyses of the genera. The use of such characters is
often a help to quick identification of species, but with
the disadvantage of hiding from the student the real characters on which classification is based. In the present
edition, these artificial keys have been chiefly set aside,

and I have tried, with more or less of success, to set before the student the essential characters of each group.

The present edition is wholly re-written and it is printed from new stereotype plates. The order of arrangement is reversed, the lowest forms being placed first.

The region covered by the Manual has been extended in the present edition so as to include, in addition, Missouri, Iowa, Minnesota, the Provinces of Canada, and the sea-coast from Nova Scotia to Cape Hatteras. The deep-sea fishes of this region are, however, omitted, as well as the tropical and semi-tropical forms which occasionally drift northward in the Gulf Stream, without gaining any permanent place in the northern fauna. Several species of birds which have been once or twice taken in our limits, but which are merely accidental wanderers from the West or South or from Europe, have also been omitted. I have wished to include only those animals which really form a part of the fauna of the region in question.

I have made free use of every available source of information, and I believe that the present state of our knowledge in this field is fairly represented. The arrangement of the fishes is essentially that of Jordan and Gilbert's "Synopsis of the Fishes of North America" (1883), and, almost exactly that of Jordan's "Catalogue of the Fishes of North America" (1885). The manuscript of the fresh water fishes, in the present edition, has been carefully revised by Prof. Charles H. Gilbert.

The arrangement of the Batrachians and Reptiles is essentially that set forth in the various papers of Prof. Edward D. Cope. I have made use of Boulenger's Catalogues of the Reptiles in the British Museum, and of the "Catalogue of North American Batrachia and Reptilia" by N. S. Davis and Frank L. Rice. The manuscripts of the Reptiles and Batrachians have been revised by Prof. O. P. Hay.

In the nomenclature and classification of the Birds, I have followed exactly the "Check List of North American Birds," published by the American Ornithologists' Union. In the preparation of analytical keys to the genera of Birds, I have made large use of Ridgway's "Manual of North American Birds," and of Cones' "Key to North American Birds." In the arrangement of the Mammals, I have been guided primarily by Professor Baird's "History of North American Mammals." In the Rodentia, I have made use of the elaborate monographs of Dr. Elliott Coues and Dr. J. A. Allen; and in the other groups reviewed by Dr. Coues, I have adopted most of his conclusions. In the Cetaceans, I have used chiefly the papers of Mr. Frederick W. True and Prof. E. D. Cope, and both these naturalists have kindly furnished me with unpublished catalogues of the species recognized by them.

In the preparation of the present edition I am also personally indebted for aid in various ways to Prof. Edward D. Cope, Mr. Leonhard Stejneger, Prof. Charles H. Gilbert, Prof. Oliver P. Hay, Mr. Frederick W. True, Mr. Robert Ridgway, Mr. Amos W. Butler, Dr. J. Sterling Kingsley, Mr. Charles H. Bollman, Dr. Stephen A. Forbes, Mr. Barton W. Evermann, and others. I may again refer to the obligations acknowledged in the earlier edition, — especially to my indebtedness to Dr. Elliott Coues, Dr. Theodore Gill, Prof. Edward D. Cope, Prof. Herbert E. Copeland, and Mr. Edward W. Nelson.

DAVID S. JORDAN.

BLOOMINGTON, INDIANA,
June, 1888.

THE VERTEBRATE ANIMALS,

NORTHERN UNITED STATES.

VERTEBRATA. (THE VERTEBRATES.)

THE Vertebrates are, in popular language, "animals with a back-bone." They are distinguished from all other animals, says Professor Huxley, " by the circumstance that a transverse and vertical section of the body exhibits two cavities, completely separated from one another by a partition. The dorsal cavity contains the cerebro-spinal nervous system; the ventral, the alimentary canal, the heart, and, usually, a double chain of ganglia, which passes under the name of the 'sympathetic.' A vertebrated animal may be devoid of articulated limbs, and it never possesses more than two pairs. These are always provided with an internal skeleton, to which the muscles moving the limbs are attached."

Modern researches have shown that, besides the ordinary "backboned animals," certain other creatures, formerly considered as Mollusks or Worms, are really degenerate forms of Vertebrates, and must be considered as members, or at least as associates, of this group. The resemblance to the other Vertebrates on the part of the forms in question is seen in their early or larval development, and scarcely at all in the adult condition. "Many of the species start in life with the promise of reaching a point high in the scale, but after a while they turn around, and, as one might say, pursue a downward course, which results in an adult which displays but few resemblances to the other vertebrates." (*Kingsley.*) These are the Tunicates or Ascidians, forming the Class or Province of "Urochordata." The essential character of the Vertebrata, in the broad sense of the term, is now understood to be this: "The possession of a cellular cord, — the ' notochord,' — which runs underneath the central nervous system, and which in the higher forms is surrounded by the permanent vertebral column and skull, and is largely obliterated by the development of these structures. So the term CHORDATA is frequently employed as synonymous with VERTEBRATA in its wide sense." (*R. R. Wright.*)

Without further discussion of the VERTEBRATA or "CHORDATA" as a whole, we may proceed to the account of the several subordinate groups or classes. The existing forms may first be divided into about six primary groups, which have been called "provinces" by Professor Huxley. These are (I) the *Urochordata*, including the class *Tunicata*; (II) the *Hemichordata* or *Enteropneusta*: (III) the *Cephalochordata*, corresponding to the class *Leptocardii*; (IV) the *Ichthyopsida*, including the classes of *Cyclostomi*, *Pisces*, and *Batrachia*; (V) the *Sauropsida*, including the *Reptilia* and *Aves*; and finally (VI) the *Mammalia*, corresponding to the single class of the same name.

The relations of these provinces and classes are shown in the following analysis taken, in part, from Dr. Gill's "Arrangement of the Families of Fishes." Only the more obvious characters are here mentioned. Others may be found in the more elaborate works on Comparative Anatomy.

Analysis of the Classes of Chordata.

a. Anterior end of the central nervous axis not dilated into a brain, and not surrounded by a protective capsule or skull.

  b. Notochord confined to the tail and usually present only in the tadpole-like larval stage of the animal (UROCHORDATA): adult animal not fish-like nor worm-like, its body invested with a tough envelope or "tunic." . . . . TUNICATA, A.

  bb. Notochord not confined to the tail, but extending forward to the anterior end of the body; sides of body with numerous gill slits which are persistent through life.

    d. Notochord developed in anterior end of body only (HEMICHORDATA): adult animal worm-like, without trace of fins; a long proboscis before the mouth. . . . . . ENTEROPNEUSTA, B.

    dd. Notochord perfect, continued forward to a point before the mouth (CEPHALOCHORDATA): body elongate, lanceolate, somewhat fish-like in form, not worm-like nor enveloped in a "tunic"; middle line of body with rudimentary fins; no proboscis; the mouth slit-like, fringed with cirri. . . . . LEPTOCARDII, C.

aa. Anterior end of the nervous axis dilated into a "brain," which is contained within a protective capsule, the "skull"; notochord not continued forwards beyond the pituitary body; heart developed and divided into at least two parts, an auricle and a ventricle. (CRANIOTA.)

  e. Respiration during part or the whole of life performed by means of gills; blood cold. (ICHTHYOPSIDA.)

f. Skull imperfectly developed and without jaws; paired fins undeveloped, with no shoulder girdle or pelvic elements; a single
median nostril; gills purse-shaped; skin naked; skeleton cartilaginous. . . . . CYCLOSTOMI, D.

ff. Skull well developed, and with jaws; shoulder girdle and pelvic
elements developed; nostrils not median.

 g. Limbs developed as rayed fins (rarely abortive); rayed fins normally present on the median line of the body; respiration
throughout life by means of gills; lungs usually not developed.
               PISCES, E.

 gg. Limbs not developed as rayed fins, but, if present, having the same
skeletal elements as in the higher vertebrates; respiration in the
adult chiefly accomplished by means of lungs, the gills usually
not persistent; skin usually naked. . . . . BATRACHIA, F.

 ee. Respiration performed throughout life by means of lungs, the gill
slits disappearing before birth.

  h. Mammary glands not present; diaphragm incomplete; a single
occipital condyle; oviparous (or sometimes ovoviviparous),
the young hatched from a rather large egg. (SAUROPSIDA.)

   i. Exoskeleton developed as scales or bony plates: blood cold;
heart with three (rarely four) cavities. . . REPTILIA, G.

   ii. Exoskeleton developed as feathers; blood warm; heart with
four cavities. . . . . AVES, H.

  hh. Mammary glands present; the young developed within the body
from a minute egg (except in the *Monotremata*), and nourished
for a time after birth by milk secreted in the mammary glands;
exoskeleton developed as hair; two occipital condyles; diaphragm complete; heart with four cavities; blood warm.
               MAMMALIA, I.

Of these classes, the *Tunicata* (A) and the *Enteropneusta* (B) are
excluded from the plan of the present work. The *Tunicata* are all
marine forms, of small size, the larger species being familiarly
known as "Sea Squirts," "Sea Peaches," and "Sea Pears"; but
the most of them are without common names. A considerable
number of species, representing several families, are found on our
Atlantic coast. The *Enteropneusta* consist of the single genus
Balanoglossus, a worm-like creature, of which two or three species
are found on our coasts. They reach a length of six to twelve
inches. They have been considered as worms having possible
affinities with the Echinoderms, but the recent studies of Mr.
William Bateson seem to show conclusively that their place is
among the *Chordata.*

Leaving these groups aside, we take up

CLASS C. — **LEPTOCARDII.** (THE LANCELETS.)

Skeleton membrano-cartilaginous; no brain ; no skull; the notochord persistent and extending to front of body; no heart, its place
being taken by pulsating sinuses; blood colorless ; respiratory cavity confluent with cavity of abdomen ; gill slits in great number ;
the water expelled from an abdominal pore in front of vent ; no
jaws; the mouth inferior, slit-like, with cirri on each side. (Gr.
λεπτός, thin ; καρδία, heart.)

ORDER I. **CIRROSTOMI.**

The single order of this class contains but a single family. (Lat.,
cirrus, hair ; Gr. στόμα, mouth.)

FAMILY I. **BRANCHIOSTOMATIDÆ.** (THE LANCELETS.)

Body elongate-lanceolate, compressed, naked, colorless, the fins
represented by a low fold which extends along the back around
the tail, past the vent, to the abdominal pore ; eye rudimentary ;
liver a blind sac of the simple intestine.   One genus, with 5 or
6 species ; small, translucent creatures found imbedded in the
sand on warm coasts. These animals are highly interesting to
the anatomist as showing the vertebrate type in its simplest
condition.

1. BRANCHIOSTOMA Costa. (*Amphioxus* Yarrell.)
(βράγχια, gills ; στόμα, mouth.)

1. B. caribæum Sundevall. LANCELET. Muscular bands
(myocommas) 55 to 60 ($37 + 14 + 9 = 60$); tail short; extremities attenuate. (Otherwise as in the European *B. lanceolatum*,
which has 56 to 60 myocommas; $35 + 12 + 13 = 60$). N. Y. to
S. A. buried in soft sand, locally abundant. (Name from Carib-
bean Sea.)

CLASS D. CYCLOSTOMI. (THE MYZONTS.)

Skeleton cartilaginous; skull imperfect, not separate from vertebral column; no jaws; no limbs; no ribs; no shoulder girdle nor pelvic elements; gills in the form of fixed sacs, 6 or more on each side; nostril single, median; mouth subinferior, nearly circular, adapted for sucking; heart without arterial bulb; alimentary canal straight, simple; vertical fins with feeble rays. Naked, eel-shaped animals found in all cool waters. (Gr. κύκλος, circle; στόμα, mouth.)

Orders of Cyclostomi.

a. Nostril tube-like with cartilaginous rings, penetrating the palate; gill openings remote from the head; no eyes. . . . . HYPEROTRETA, 2.

aa. Nostril a blind sac not entering the palate; gill openings close behind the head; eyes well developed in the adult. . . . HYPEROARTIA, 2.

ORDER II. HYPEROTRETA.

Characters as given above. Only one family. (ὑπερῴα, palate; τρητός, perforate.)

FAMILY II. MYXINIDÆ. (THE HAG-FISHES.)

Snout with eight barbels; no lips; a median tooth on the palate and two rows on each side of the tongue, which is a powerful organ with a strong fibrous tendon moving in a muscular sheath; each side of abdomen with a series of mucous sacs; no eyes: intestine without spiral valve; skin thin and loose; eggs large, with a horny case and threads for adhesion: genera 2; species 4 or 5. Lamprey-like animals, burrowing into the flesh of fishes, on which they feed; marine.

a. Gill openings one on each side, this leading by six ducts to six branchial sacs. . . . . MYXINE, 2.

2. MYXINE Linnæus. (Gr. μύξα, slime.)

2. M. glutinosa L. HAG-FISH, BORER. Bluish; head 3½ to 4 in length. N. Atl., S. to Cape Cod. (Eu.)

Order III. HYPEROARTIA.

Characters given above. One family only. (ὑπερῴα, palate; ἄρτιος, complete.)

Family III. PETROMYZONTIDÆ. (The Lampreys.)

Body eel-shaped, naked, compressed behind; mouth subcircular, armed with horny teeth, which rest on papillæ; gill openings 7, arranged in a row along the side of the "chest"; lips present, fringed: nostril on top of head, just in front of eyes; dorsal fin more or less notched; intestine with a spiral valve; eggs small. The lampreys undergo a metamorphosis, the larva of all species being toothless and having the eyes rudimentary. The name Ammocœtes was formerly applied to the larval forms; originally, however, to that of *A. branchialis*. Genera 3 or 4, species about 15, chiefly of the fresh waters of temperate regions. They attach themselves to fishes, and feed by scraping off the flesh with their rasp-like teeth.

a. Second dorsal joined to the caudal.
  b. Supraoral lamina ("maxillary tooth") expanded laterally, forming a crescent-shaped plate, with a cusp at each end, and sometimes a median cusp; anterior lingual teeth serrate. . . . . . . AMMOCŒTES, 3.
  bb. Supraoral lamina contracted, of two or three teeth close together; discal teeth numerous, in concentric series; buccal disk large (in adult, very small in larva). . . . . PETROMYZON, 4.

3. AMMOCŒTES Duméril. (ἄμμος, sand; κοίτη, bed.)

a. Supraoral lamina with a very small median cusp or none; edge of anterior lingual tooth small, crescent-shaped, dentate, the median denticle enlarged; buccal disk small, with few teeth. (AMMOCŒTES.)

3. A. branchialis (L.). Mud Lamprey. Brook Lamprey.

Dorsal continuous, deeply notched, both parts high; about 3 bicuspid teeth on each side of buccal disk; the other teeth simple; infraoral plate with 5 to 9 blunt subequal cusps; head with gills 4¾; myocommas 67, between gills and vent; an anal papilla present in spring. Color bluish black. L. 8. Cayuga L. (*Meek*) to Minn. and Ky., ascending brooks in spring. (*Eu.*) (*P. niger*, Raf., not of Lacepède.) (Lat., having gills.)

4. PETROMYZON (Artedi) Linnæus. (πέτρα, stone; μύζω, to suck.)

a. Anterior lingual tooth divided in two by a median groove; dorsal fin continuous, with a broad notch. (*Ichthyomyzon* Girard.)

4. P. castaneus (Girard). Supraoral lamina (maxillary tooth) tricuspid; some lateral teeth bicuspid; infraoral lamina (mandib-

ulary tooth) with 7 to 12 cusps. Color yellowish. L. 10. Miss. Valley, Minn. to Kans. and La. (*Ichth. hirudo* Girard.) (Lat., chestnut-colored.)

5. **P. concolor** (Kirtland). Supraoral lamina bicuspid; teeth on disk all simple, and placed in about 4 concentric series; infraoral lamina with 7 cusps; head $7\frac{1}{2}$; with gills $4\frac{3}{4}$; 51 muscular impressions between gills and vent. Color bluish silvery, sometimes mottled; a small bluish spot above each gill opening, — this found even in the larva. L. 12. L. Erie to Mo. and N., a common parasite on the Sturgeon and other large fishes. (*P. argenteus* Kirtland, not of Bloch.) (Lat., uniformly colored.)

aa. Anterior lingual tooth with a deep median groove, and extending in an incurved point; dorsal fin divided. (*Petromyzon.*)

6. **P. marinus** L. GREAT SEA LAMPREY. "LAMPER EEL." Supraoral lamina bicuspid; infraoral cusps 7 to 9; first row of lateral teeth on side of mouth bicuspid; the others simple; myocommas, 64 between gills and vent; males in spring usually with an elevated fleshy ridge before the dorsal. Color dark brown, usually mottled with blackish. L. 3 feet. N. Atlantic, S. to Va., ascending rivers to spawn, and permanently land-locked (var. *unicolor*, Dekay) in the lakes of W. and N. N. Y. The larva is blind, toothless, with a contracted mouth, in which the lower lip forms a lobe distinct from the upper. The eyes appear before the mouth is enlarged. (*Eu.*)

In the spring the Lamprey ascends small brooks for the purpose of depositing its spawn. They are then often found clinging to stones and clods of earth. Later in the season they disappear, and are seldom seen except when attached to some unlucky fish. They are rarely seen descending the stream, and "it is thought by fishermen that they never return, but waste away and die, clinging to rocks and stumps of trees for an indefinite period; a tragic feature in the scenery of the river bottoms worthy to be remembered with Shakespeare's description of the sea floor." (*Thoreau.*)

Class E. — PISCES. (The Fishes.)

A " fish " in the popular sense is a member of any one of the
three classes of aquatic or fish-like vertebrates, the groups here
designated as *Leptocardii, Marsipobranchii,* and *Pisces.* But the
Lancelets and the *Lampreys* differ so widely from the other groups
that we must exclude them from consideration as fishes. Many
writers go still further and remove from the *Pisces,* the *Sharks,
Chimæras,* and *Dipnoans,* but for our present purposes all these
may be referred to the same class as the true fishes, or *Teleosts.*
The *Pisces* or " Fishes " may then be defined as cold-blooded vertebrates adapted for life in the water, breathing by means of gills
which are not purse-shaped, but attached to bony or cartilaginous
gill arches ; having the skull well developed and with a lower jaw ;
with the limbs present and developed as fins, or rarely wanting
through atrophy ; with shoulder girdle present, furcula-shaped,
curved forward and with the sides connected below ; with pelvic
bones present ; having the exoskeleton developed as scales or bony
plates or horny appendages, sometimes obsolete, and with the median line of body with one or more fins composed of cartilaginous
rays connected by membrane. The existing representatives of the
class *Pisces* may be conveniently divided into four subclasses :
Selachii or *Elasmobranchii, Holocephali, Teleostomi,* and *Dipnoi.*
The last group (*Ceratodus, Lepidosiren*) has well-developed lungs
and the paired fins flipper-like. It forms a connecting link between the *Ganoidei* and the *Batrachia.* As there are no North
American species of *Dipnoi,* the group needs no further men-
tion in this work.

Subclasses of Pisces.

a. Gills not free, being attached to the skin by the outer margin. Ova few
    and large, impregnated and sometimes developed internally : embryo
    with deciduous external gills ; membrane bones of head undeveloped,
    except sometimes a rudimentary opercle ; skeleton cartilaginous ; skull
    without sutures ; tail heterocercal ; ventral fins abdominal ; male with
    large intromittent organs or claspers attached to ventral fins ; skin
    naked or covered with minute rough scales, sometimes with spines ; no
    air-bladder ; arterial bulb with three series of valves : intestine with a
    spiral valve ; optic nerves united by a chiasma ; cerebral hemispheres
    united.

b. Gill openings slit-like, 5 to 7 in number ; jaws distinct from the skull,
    joined to it by suspensory bones ; no membrane bones ; teeth distinct.
    (*Sharks* and *Skates.*) . . . . SELACHII, page 14.

bb. Gill opening single, leading to four gill clefts; jaws coalescent with the skull; a rudimentary opercle; teeth coalescent forming bony plates. (*Chimæras.*) . . . . HOLOCEPHALI, page 24.

aa. Gills free, attached at base only to the gill arches; gill opening single on each side; eggs comparatively small and numerous; no claspers; membrane bones present on head; cerebral hemispheres not united. (*True Fishes.*) . . . . TELEOSTOMI, page 25

Subclass SELACHII.   (The Selachians.)

This group, sufficiently defined above, includes two orders, the Sharks and the Rays, — marine fishes of large size, abundant in most seas. (Gr. σέλαχος, shark.)

Orders of Selachii.

a. Gill openings lateral. . . . . Squali, 4.
aa. Gill openings ventral. . . . . Ralæ, 5.

Order IV. SQUALI. (The Sharks.)

The typical sharks are elongate in form, quite unlike the skates in appearance. Intermediate forms connect the two groups so closely that the position of the gill openings is the only constant character by which the two orders can be separated. (Lat., a shark, from Gr. γαλεός, allied to γαλέη, a weasel.)

Note. — The Sharks are mostly fishes of the high seas, and any of the larger Atlantic species may stray to our coasts. Besides those here described, the following have been at least once taken within our limits:—
Echinorhinus spinosus (Gmelin), Cape Cod; *Centrocyllium fabricii* (Reinhardt), off Gloucester; *Centroscymnus cœlolepis* (Bocage & Capello), Gloucester; *Pseudotriacis microdon* (Capello), Long Island; *Aprionodon isodon* (Müller & Henle); *Isogomphodon limbatus* (Müller & Henle), Wood's Holl.

Omitting extralimital families, we have the following analysis of

Families of Squali.

a. Pectoral fins moderate, without deep notch at base in front; gill openings 5.
  b. Anal fin wanting.
    c. Dorsal fins each with a stout spine. . . . . . . Squalidæ, 4.
    cc. Dorsal fins without spine. . . . . Somniosidæ, 5.
  bb. Anal fin present; both dorsals without spine, the first inserted before the ventrals.
    d. Caudal fin not lunate, the upper lobe very much longer than the lower, with a notch below, towards its tip; side of tail without keel.
      e. Last gill opening above base of pectoral.
        f. Tail moderately developed, not half length of rest of body; eyes with nictitating membrane.
          g. Head kidney-shaped or hammer-shaped, much wider than long.
                                             Sphyrnidæ, 6.
          gg. Head normally formed. . . . . . Galeorhinidæ, 7.
        ff. Tail very long, as long as rest of body; no nictitating membrane. . . . . Alopiidæ, 8.

ee. Last gill opening before base of pectoral; dorsal fins subequal.

CARCHARIID.E, 9.

dd. Caudal fin lunate, the lower lobe not much shorter than the upper; tail with a keel on each side, last gill opening before pectorals.

h. Gill openings rather large; teeth large. . . . . LAMNID.E, 10.

hh. Gill openings very large. nearly meeting both above and below; teeth small (largest of all fishes). . . . . CETORHINID.E, 11.

aa. Pectoral fins very large, wing-like, expanded at the base in front, this expansion being separated from the neck by a deep notch; no anal fin.

SQUATINID.E, 12.

FAMILY IV. SQUALIDÆ. (THE DOG-FISHES.)

Sharks with two dorsal fins, each armed with a stout spine, and without anal fin; no nictitating membrane; spiracles moderate; gill openings narrow, all before pectorals; ventral fins inserted posteriorly; teeth small, compressed: nostrils inferior, near front of snout. Genera 6; species about 15; small sharks, chiefly of the Atlantic. (*Spinacidæ* Auct.)

a. Teeth in both jaws, simple, subquadrate, each with a nearly horizontal cutting edge, and a point directed outward; dorsal spines strong.

SQUALUS, 5.

5. SQUALUS (Artedi) Linnæus.

7. S. acanthias L. DOG-FISH. Dorsal spines not grooved: slate-color, back with whitish spots fading with age. L. 3 feet. North Atl., S. to Cuba; abundant N., its liver valued for the "Dog-fish oil." (*Eu.*) (Gr. ἀκανθίας, having spines.)

FAMILY V. SOMNIOSIDÆ. (THE SLEEPER SHARKS.)

Sharks with two dorsal fins, both without spine, and no anal fin, the first dorsal much before ventrals, otherwise essentially as in the *Squalidæ*. Genera 5; species 5 or 6, mostly large sharks of the Atlantic.

a. Dorsal fins about equal: upper teeth lancet-shaped, incurved; lower quadrate with a horizontal edge, ending in a point directed outwards; fins very small. . . . . SOMNIOSUS, 6.

6. SOMNIOSUS Le Sueur. (Lat., sleepy.)

8. S. microcephalus (Bloch). SLEEPER. NURSE. Color blackish; caudal blunt. L. 10 to 18 feet. Arctic seas, S. to Cape Cod. (*Eu.*) (μικρός, small; κεφαλή, head.)

FAMILY VI. SPHYRNIDÆ. (THE HAMMER-HEADED SHARKS.)

Characters of the *Galeorhinidæ*, except that the head has a form hammer-shaped or kidney-shaped, its sides being much extended, the eyes borne at the ends of the hammer. One genus, with 4 or 5 species; large sharks of the warm seas.

7. SPHYRNA Rafinesque. (An old name from σφύρα, hammer.)

a. Teeth in both jaws oblique, each with a notch on the outside near the base; no spiracles.

b. Head truly hammer-shaped: a long groove extending forward from nostrils. (*Sphyrna.*)

9. **S. zygæna** (L.). HAMMER-HEADED SHARK. Width of "hammer" twice its length. Gray. L. 15 to 20 feet. All warm seas. N. to Cape Cod. (*Eu.*) (An old name from ζυγόν, a cross-beam.)

bb. Head kidney-shaped, the frontal groove obsolete. (*Reniceps*, Gill.)

10. **S. tiburo** (L.). BONNET-HEAD SHARK. Width of "hammer" not nearly twice its length. Ashy gray. L. 3 to 5 feet. Warm seas, N. to Va. (*Eu.*) (*Tiburo*, an Italian name of some shark.)

FAMILY VII. **GALEORHINIDÆ.** (THE TYPICAL SHARKS.)

Sharks with two dorsals and an anal fin; no spines; tail moderate, not lunate, bent upwards, the fin notched below near the tip ; basal lobe short; no caudal keel; last gill opening above base of pectoral; eye with nictitating membrane; head normally formed. Genera 15, species about 60, found in all seas.

a. Teeth blunt, paved, without cusps or cutting edges; spiracles present; no pit at root of tail; labial folds about mouth. . . . GALEUS, 8.
aa. Teeth more or less compressed, with sharp cutting edges.
  b. Spiracles present; teeth large; serrated.
    c. Root of tail with a pit above; caudal fin with two notches.
                                    GALEOCERDO, 9.
  bb. Spiracles none: teeth sharp; a pit at root of tail.
    d. Teeth all serrate in the adult. . . . . . . CARCHARHINUS, 10.
    dd. Teeth all entire, all except the median ones oblique: their points turned away from the middle so that the inner margins are nearly horizontal, and form a cutting edge. . SCOLIODON, 11.

8. GALEUS (Rafinesque) Leach. (*Mustelus* Cuvier.)
(γαλεός, shark ; γαλέη, weasel.)

a. Embryo not attached to uterus by a placenta; teeth very blunt. (*Galeus.*)

11. **G. canis** (Mitchill). DOG SHARK. HOUND SHARK. BOCA DULCE. First dorsal higher than long, its middle midway between pectorals and ventrals; snout shortish. Pale gray. L. 3 feet. Smallest of our sharks. N. Atl.; common N. (*Eu.*)

9. GALEOCERDO Müller & Henle. (γαλεός, shark ; κερδώ, fox).

12. **G. maculatus** (Ranzani). TIGER SHARK. Brown, with numerous large dark spots. L. 10 feet. Warm seas; rarely N. to N. Y. (Lat., spotted.)

10. CARCHARHINUS Blainville. (*Carcharias* Cuvier.)
(κάρχαρος, rough ; ῥίνη, shark.)

(The largest genus of sharks, represented in most warm seas. It is often divided into several genera, but intergradations make it difficult to maintain these divisions. In young specimens the serration of the teeth is not evident.)

a. First dorsal far behind pectoral, nearer root of ventral than that of pectoral. (*Carcharhinus.*)

13. C. glaucus (L.). GREAT BLUE SHARK. Snout very long; color grayish blue. A large shark, rare on our coast. (*Eu.*) (Lat., grayish blue.)

aa. First dorsal not far behind pectoral.

    b. Upper teeth oblique; deeply notched on outer margin. (*Platypodon* Gill.)

14. C. obscurus (Le Sueur). Pectorals large ; second dorsal evidently smaller than anal; first dorsal large ; head pointed. L. 10 feet. N. Atl. Frequently on our coast.

    bb. Upper teeth sub-erect, triangular, scarcely notched at outer margin. (*Eulamia* Gill.)

15. C. caudatus (Dekay). Snout moderate, its length from mouth forward not less than width of mouth ; pectoral fin not very long. Atlantic coast : a little known species of uncertain synonymy. (Lat., long-tailed.)

11. SCOLIODON Müller & Henle.
(σκολιός, oblique ; ὀδών, tooth.)

16. S. terræ-novæ (Richardson). SHARP-NOSED SHARK. Body slender; snout depressed ; mouth with short labial grooves on both jaws; second dorsal smaller than anal; gray, tail dusky-edged. West Indies, N. to Cape Cod, common S. (erroneously ascribed to Newfoundland). (Lat. *terra*, land ; *nova*, new. Newfoundland.)

FAMILY VIII. ALOPIIDÆ. (THE THRESHER SHARKS.)

Body rather slender; snout short: teeth equal, flat, triangular, entire; gill openings moderate, the last above P.; no nictitating membrane ; spiracles obsolete; first dorsal large, second dorsal and anal very small ; tail about as long as rest of body ; no caudal keel; pectorals falcate, very large. One species, a large shark, found in most warm seas.

12. ALOPIAS Rafinesque. (ἀλώπηξ, a fox.)

17. A. vulpes (Gmelin). THRESHER. SWINGLE-TAIL. FOX SHARK. Color gray. L. about 20 feet. Open sea ; occasionally on our coast. (*Eu.*)

FAMILY IX. **CARCHARIIDÆ.** (THE SAND SHARKS.)

Body elongate, the snout sharp; mouth wide, the teeth large, long, narrow, entire, very sharp, most of the teeth with one or two small cusps at base; gill openings all in front of pectorals : dorsals small, similar to the anal; tail as in *Galeorhinidæ ;* no nictitating membrane; spiracles minute. One genus and 3 species; rather small sharks, of the Atlantic.

13. **CARCHARIAS** Rafinesque. (*Odontaspis* Agassiz.)
(κάρχαρος, jagged.)

a. First and fourth teeth of the upper jaw, and first tooth of the lower without basal cusps. (*Eugomphodus* Gill.)

18. **C.** littoralis (Mitchill). SAND SHARK. Pectoral short. Color gray. L. 6 feet. Cape Cod to S. C., rather common N. A voracious little shark. (Lat., of the shore.)

FAMILY X. **LAMNIDÆ.** (THE PORBEAGLES.)

Body robust, contracted to a rather slender tail, which has a keel on each side ; caudal fin lunate, the lower lobe nearly as large as the upper, and not very different in form ; teeth large; gill openings wide, all in front of pectorals; first dorsal and pectorals large ; second dorsal and anal very small ; a pit at root of caudal ; spiracles obsolete. Large, voracious sharks of the warm seas. Genera 3, species about 6.

a. Teeth slender, sharp, with entire edges; tail very slender.
    b. Teeth very slender, flexuous, without basal cusps . . . ISURUS, 14.
    bb. Teeth broader, most of them with a small cusp on each side at base.
                                                LAMNA, 15.
aa. Teeth broad, compressed, triangular, distinctly serrate; tail rather stout.
                                        CARCHARODON, 16.

14. **ISURUS** Rafinesque. (ἴσος, equal ; οὐρά, tail.)

a. First dorsal entirely behind pectorals, nearly midway between base of P. and V. (*Isuropsis*, Gill.)

19. **I.** dekayi (Gill). MACKEREL SHARK. Color bluish. L. 15 feet. W. I., rarely N. (For James E. Dekay, author of the Fauna of New York.)

15. **LAMNA** Cuvier. (λάμνα, a kind of shark.)

20. **L.** cornubica (Gmelin). PORBEAGLE. MACKEREL SHARK. First dorsal close behind pectorals; snout conical, sharp ; back elevated ; third tooth on each side in upper jaw small. L. 8 feet. Warm seas, frequently N. to Cape Cod. (*Eu.*) (Lat., pertaining to Cornwall.)

16. CARCHARODON Andrew Smith. (κάρχαρος, jagged; ὀδών, tooth.)

21. C. carcharias (L.). MAN-EATER SHARK. GREAT WHITE SHARK. First dorsal somewhat behind pectorals. Color leaden-gray, P. edged with black. L. 25 feet. Most voracious of all sharks, and next in size to *Cetorhinus*, weighing nearly a ton. Warm seas, occasional off our coasts. Linnæus says, "Jonam prophetam ut veteres Herculem, in hujus trinoctem ventriculo tridui spateo, bræsisse verosimile est." The fossil teeth of a far larger extinct species, *Carcharodon megalodon*, are often found in tertiary beds along our South Atlantic coast. (*Eu.*) (καρχ ιρίας, old name of large sharks.)

FAMILY XI. **CETORHINIDÆ.** (THE BASKING SHARKS.)

Largest of all fishes; immense sharks with the gill openings extremely wide, nearly meeting above and below; mouth moderate; teeth very small, numerous, conical, simple; no nictitating membrane; spiracles very small; first dorsal and pectorals large; second and anal small; caudal lunate, the upper lobe the larger: tail keeled on the side. One species, a huge, sluggish creature, found in Northern seas.

17. CETORHINUS Blainville. (κῆτος, whale; ρίνη, a shark.)

22. C. maximus (Gunner). BASKING SHARK. Head small, snout blunt. Gray. L. 35 feet; depth nearly 6 feet. Open sea, S. to Va. (*Eu.*)

FAMILY XII. **SQUATINIDÆ.** (THE ANGEL-FISHES.)

Ray-like sharks, with the body depressed, the pectoral fins very large, expanded in the plane of the body, the anterior margin bearing some resemblance to the bend of the wing in birds; ventrals very large; dorsal fins two, small, subequal, behind ventrals: caudal small; no anal; gill openings wide, subinferior, partly hidden by base of pectoral; spiracles wide, crescent-shaped, behind eyes: mouth and nostrils anterior; teeth small, conical, pointed, distant. A single species, in most seas. The singularly formed pectoral fins give an absurd resemblance to the conventional pictures of angels.

18. SQUATINA Duméril. (*Rhina* Günther.) (Latin name, from *squatus*, skate.)

23. S. squatina (L.). ANGEL-FISH. MONK-FISH. Skin rough, with small, stiff prickles; ashy gray above, usually much mottled. L. 3 or 4 feet. Warm seas, rarely N. (*Eu.*)

ORDER V. **RAIÆ.** (THE RAYS.)

The Rays, as a whole, differ from the sharks in having the gill openings underneath the flat disk formed by the body and the expanded pectoral fins. The tail is comparatively slender, and its fins are small. Spiracles present. The *Rajidæ* produce large eggs, enclosed in leathery cases; most of the other *Raiæ* are ovoviviparous, bringing forth their young alive.

Families of Raiæ.

*. Tail comparatively thick, with two dorsal fins; no serrated caudal spine nor cephalic fins.

   b. Snout much produced, flat, armed with strong teeth on each side, set at right angles to its axis; body somewhat shark-like, the disk gradually passing into the tail. . . . . PRISTIDIDÆ, 13.

   bb. Snout not saw-like; disk ending abruptly at base of tail.

     c. Electric organs wanting; skin not perfectly smooth. . RAJIDÆ, 14.

     cc. Electric organs present ; a structure of honeycomb-like tubes between pectoral fins and head; skin perfectly smooth. . TORPEDINIDÆ, 14.

aa. Tail slender, with but one dorsal fin or none, and usually armed with a serrated spine.

   d. Pectoral fins uninterrupted, confluent about the snout; teeth small.
                          DASYATIDÆ, 15.

   dd. Pectoral fins divided, leaving detached appendages ("cephalic fins") on the snout.

     e. Teeth very large, flat, tessellated. . . . . . . AETOBATIDÆ, 16.

     ee. Teeth very small, flat or tubercular; size enormous, largest of the rays.
                          MANTIDÆ, 17.

FAMILY XIII. PRISTIDIDÆ. (THE SAW-FISHES.)

Rays with elongate body, stout, thick tail, and a long saw-like snout, below which is the inferior mouth with small blunt teeth. Dorsals and caudal well developed. One genus, with 5 or 6 species, in warm seas.

19. PRISTIS Latham. (πρίστης, one who saws; the ancient name.)

   24. P. pectinatus Latham. SAW-FISH. Saw with 25 to 28 pairs of spines. L. 10 feet. West Indies; occasional N. (Lat., comb-toothed.)

FAMILY XIV. RAJIDÆ. (THE SKATES.)

Rays with the disk broad, rhombic, more or less rough; the males usually with about two rows of strong spines on each pectoral ; tail rather stout, with a fold of skin on each side, and two dorsal fins above : caudal fin small or obsolete ; no serrated spine ; no electric organs. Egg in a large leathery case, four-angled, and having two tubular horns at each end. Genera 4, species 40, mostly of the Northern seas.

a. Caudal fin rudimentary; pectorals not confluent, leaving a translucent area at the snout; ventrals deeply notched. . . . . . . RAJA, 20.

20. RAJA (Artedi) Linnæus. (*Raia* or *Raja*, the Latin name.)

a. Middle line of back and tail behind shoulders, unarmed in adult, with a row of spines in young; outline of disk before spiracles obtuse, without acute angle at tip of snout.

  b. Rows of teeth about $\frac{5.5}{6.6}$.

  25. **R. erinacea** Mitchill. COMMON SKATE. TOBACCO-BOX. Spines largest on front of pectorals; smaller ones on head, back, and shoulder girdle. Light brown, with round dark spots. L. 1¼ feet. Smallest and commonest of our skates, from Va. northward. (Lat., hedge-hog.)

  bb. Rows of teeth about $\frac{9.9}{8.8}$.

  26. **R. ocellata** Mitchill. BIG SKATE. Similar to preceding, but much larger, and with additional rows of spines along the back and on sides of tail. Light brown, with dark spots; usually a large white ocellus with a dark centre on P. behind. L. 3 feet. Mass. N.

aa. Middle line of back and tail with a row of spines at all ages; outline of disk before spiracles forming a more or less marked angle at tip of snout.

  c. Angle at tip of snout short, obtuse; teeth $\frac{1.9}{1.0}$; body and tail with strong spines with broad stellate bases.

  27. **R. radiata** Donovan. A median dorsal row of large spines or bucklers; others about head. L. 1½ to 2 feet. N. Atl.; rather rare, S. to Cape Cod. (*Eu.*)

  cc. Angle at tip of snout acute, moderately long; teeth $\frac{9.9}{5.0}$; no coarse spines or bucklers.

  28. **R. eglanteria** Lacépède. Prickles small and sharp; a large spine on each shoulder. Brown, with darker bars and blotches. L. 2 feet. Cape Cod southward; not common. (*Eglantine*, brier-rose.)

  ccc. Angle at tip of snout much produced, blunt; teeth $\frac{3.9}{0.0}$.

  29. **R. lævis** Mitchill. BARN-DOOR SKATE. Spines of body very few and small, on head and back; a row of larger ones on median line of tail; female rougher, as is usual among rays; snout very long, somewhat spatulate. Color brownish, with paler spots mostly ringed with darker. L. 4 feet. Va. N.; not rare. (Lat. smooth.)

FAMILY XV. **TORPEDINIDÆ.** (THE ELECTRIC RAYS.)

  Trunk broad and smooth, the tail short and thick, with rayed caudal and usually two rayed dorsals, the first over or behind ventrals; a large electric organ made up of hexagonal tubes, between head and pectorals. Genera 6, species 15, found in most warm seas; noted for their power of giving electric shocks.

a. Dorsal fins two; ventrals separate; spiracle placed nearly an eye's diameter behind eye. TORPEDO, 21.

21. TORPEDO Duméril.

30. **T. occidentalis** Storer. TORPEDO. CRAMP-FISH. NUMB-FISH. Black, with obscure darker blotches; spiracles with entire edges. L. 3 to 5 feet. Cape Cod S.; not common.

FAMILY XVI. **DASYATIDÆ.** (THE STING-RAYS.)

Disk broad, the pectorals confluent anteriorly, forming tip of snout ; tail, usually whip-like, sometimes short and stout, with or without fins, but never with two dorsals. Tail usually armed with a sharp, retrorsely serrate spine above, near the base (this often duplicated and sometimes wanting): ventral fins entire. Skin smooth or variously rough, the adult roughest. Mouth small, with small teeth. Sexes similar. Genera 10, species 50, in most warm seas. The large spine or "sting" on the tail in most species may inflict a dangerous wound.

a. Tail slender, whip-like, without caudal fin, longer than the disk; "sting" on tail strong. . . . . DASYATIS, 22.
aa. Tail very slender and short, shorter than the very broad disk: sting minute or wanting. . . . . PTEROPLATEA, 23.

22. DASYATIS Rafinesque. (*Trygon* Adanson.)
(δασύς, shaggy or rough; βατίς, skate.)

a. Tail with a fold on its lower margin only, the upper edge rounded.

31. **D. centrurus** (Mitchill). COMMON STING-RAY. CLAM-CRACKER. STINGAREE. Snout not prominent; disk a little wider than long; tail usually not quite twice length of disk. Adult with some stellate tubercles on back and tail. Color olive-brown. ' L. 12 feet. Cape Cod S., common. (κέντρον, spine; οὐρά, tail.)

aa. Tail with a fold of skin on its upper as well as lower margin.

32. **D. say** (Le Sueur). SOUTHERN STING-RAY. WHIP-PAREE. Snout not prominent; disk a little wider than long; tail nearly twice length of disk. Body and tail without large spines. N. Y., S. (To Thomas Say, a distinguished zoölogist.)

23. PTEROPLATEA, Müller & Henle.

33. **P. maclura** (Le Sueur). BUTTERFLY RAY. Disk nearly twice as broad as long, three times as long as tail; sting on tail usually obsolete. Olive-brown, finely marbled and speckled; tail with four dark blotches: front edge of disk with pale half-circular spots. Va. S. (To William Maclure.)

FAMILY XVII. **AETOBATIDÆ.** (THE EAGLE RAYS.)

Pectoral fins interrupted, reappearing on tip of snout as one or two detached appendages or cephalic fins: skull somewhat elevated, so that eyes and spiracles are lateral; teeth large, flat, hexangular, the middle series largest. Otherwise essentially as in *Dasyatidæ*. Genera 3, species 20, in the warm seas.

a. Snout entire.
　　b. Teeth very broad, in one series . . . . STOASODON, 24.
　　bb. Teeth in several series . . . . AETOBATIS, 25.
aa. Snout emarginate; teeth in several series. . . . . RHINOPTERA, 26.

24. STOASODON Cantor. (*Aetobatis* Müller & Henle.)
(στοά, arcade; ὀδούς, tooth.)

34. **S. narinari** (Euphrasen). BISHOP RAY. Disk twice as broad as long. Tail very long, three or four times disk. Brown with many round yellowish spots. Warm seas, N. to Va. (*Narinari*, the Brazilian name.)

25. AETOBATIS Blainville (1816). (*Myliobatis* Duméril, 1817.)
(ἀετός, eagle; βατίς, ray.)

35. **A. freminvillii** (Le Sueur). EAGLE RAY. Skin smooth; color reddish brown. Cape Cod S. Scarce. (For Christian Paulin de Freminville, author of some papers on Plectognaths.)

26. RHINOPTERA Kuhl.

36. **R. bonasus** (Mitchill). COW-NOSED RAY. Cephalic fin emarginate, and placed below level of pectorals, so that the snout appears four-lobed when viewed from the front. Skin nearly smooth. Cape Cod S. "He enters the bay and ranges very extensively the flats where the soft clam lives. These shell-fish he is supposed to devour, for a shoal of cow-noses root up the salt-water flats as completely as a drove of hogs would do." (*Mitchill.*) (*R. quadriloba* Le Sueur.) (Lat., a buffalo.)

FAMILY XVIII. MANTIDÆ. (THE SEA DEVILS.)

Rays of immense size, similar to the *Aetobatidæ*, but with the cephalic fins forming long ear-like appendages, and with the teeth very small. Skin rough. Genera 2, species 7; among the largest of all fishes, found in warm seas.

a. Teeth in lower jaw only; mouth terminal . . . . MANTA, 27.

27. MANTA Bancroft.

(*Manta*, blanket, "a name used at the pearl fisheries of Panama, for an enormous fish much dreaded by the divers, whom it is said to devour, after enveloping them in its vast wings.")

37. **M. birostris** (Walbaum). SEA DEVIL. MANTA. Disk not quite twice as broad as long; tail as long as disk. Brown; disk 12 feet long; its breadth about 20. Tropical seas, N. to Delaware Bay. (Lat. *bis*, two; *rostrum*, snout.)

Subclass HOLOCEPHALI.

This group, defined on page 13, is equivalent to the

Order VI. HOLOCEPHALI.

Skeleton cartilaginous; gill cavity with four clefts within, but externally with a single opening, which is covered by a fold of skin within which is a rudimentary opercle. No spiracles. Jaws without separate teeth, but armed with bony plates. Notochord persistent, the vertebræ consisting of rings around a notochordal sheath. No air-bladder; intestine with a spiral valve; skin smooth, with a highly developed mucous system. Dorsals each with a strong spine. One family. (ὅλος, solid; κεφαλή, head.)

Family XIX. CHIMÆRIDÆ.

Forehead of males with a movable cartilaginous hook, turned forward and armed with prickles at tip. Oviparous, the egg-cases elliptical, with silky filaments. Two genera, 5 or 6 species, in cold waters. Fishes of most singular appearance, unlike anything else.

a. Snout soft, not ending in a cutaneous flap; tail not bent upward.

CHIMÆRA, 28.

28. CHIMÆRA Linnæus.

(Χίμαιρα, Chimæra, a fabulous monster, with the head of a lion, body of a goat, and tail of a serpent.)

38. C. affinis Capello. Color plumbeous. Cold or deep water, S. to Cape Cod. (Eu.) (Lat., related, — to C. monstrosa.)

SUBCLASS **TELEOSTOMI.** (THE TRUE FISHES.)

Skeleton usually bony, sometimes cartilaginous. Skull with sutures; membrane bones (opercle, preopercle, etc.) present; gill openings a single slit on each side; gills with their outer edges free, their bases attached to bony arches, normally four pairs of these, the fifth pair being modified into tooth-bearing pharyngeals; median and paired fins developed, the latter with distinct rays. Ova small; no claspers. Heart developed, divided into an auricle, ventricle, and arterial bulb. Lungs imperfectly developed, or modified to form a swim-bladder, or entirely absent.

We here include under one head the Ganoids and the Teleosts. The former type is chiefly composed of extinct forms. While many of its representatives are extremely dissimilar to the bony fishes, there is a gradual series of transitions, and between the *Halecomorphi* of the Ganoids and the *Isospondyli* of the true Teleosts, the resemblance is much greater than that between the *Halecomorphi* and many other Ganoids The Ganoids are, in fact, the most generalized of the true fishes, those nearest the stock from which the Teleosts on the one hand, and the *Dipnoi* and *Batrachia* on the other, have sprung. The real value or rank of some of the current orders or suborders is still doubtful. (τέλεος, perfect; στόμα, mouth.) Omitting orders not represented in our waters, we have the following analysis of

Orders of True Fishes.

a. Arterial bulb muscular, with numerous valves; optic nerves forming a solid chiasma; ventrals abdominal; air-bladder with a duct; tail strongly heterocercal throughout life; some fins usually with fulcra. (Series GANOIDEI.)

    b. Skeleton cartilaginous: ventrals with an entire series of basilar segments. (*Chondrostei.*)

        c. Maxillary and interopercle obsolete; skin naked: air-bladder cellular. SELACHOSTOMI, VII.

        cc. Maxillary and interopercle present; skin with bony shields; air-bladder simple. . . . . GLANIOSTOMI, VIII.

    bb. Skeleton bony; ventrals with basilar segments rudimentary; air-bladder cellular. (*Holostei.*)

        d. Vertebræ opisthocœlian (concavo-convex); maxillary transversely divided in several pieces; scales rhombic, enamelled plates. GINGLYMODI, IX.

dd. Vertebræ amphicœlian (double concave); maxillary not transversely divided; scales cycloid. . . . . HALECOMORPHI, X.

aa. Arterial bulb thin, with a pair of opposite valves; optic nerves crossing, not forming a solid chiasma. (Series TELEOSTEI.)

  c. Air-bladder (if present) connected by an air-duct with the intestinal canal, this persistent throughout life; ventral fins (if present) abdominal, without spines, their basilar segments rudimental. (Soft-rayed fishes.) (*Physostomi.*)

    f. Shoulder girdle attached to the skull by means of a post-temporal bone (suprascapula); form not eel-like.

      g. Præcoracoid arch, present.

        h. Maxillary bone imperfect, forming the base of a long barbel ; no subopercle nor symplectic bone : four anterior vertebræ much modified, co-ossified, and with an ossicula auditus; supraoccipitals and parietals co-ossified; no scales. . . NEMATOGNATHI, XI.

        hh. Maxillary bone perfect, not entering into a barbel (rarely entirely wanting); subopercle and symplectic bone present.

          i. Anterior vertebræ modified, co-ossified, and with the ossicula auditus. . . . . EVENTOGNATHI, XII.

          ii. Anterior vertebræ similar to the others, separate, and without ossicula auditus. . . . . ISOSPONDYLI, XIII.

      gg. Præcoracoid arch obsolete; anterior vertebræ not modified; parietal bones separated by supraoccipital; head scaly.

                             HAPLOMI, XIV.

    ff. Shoulder girdle not attached to the skull; no præcoracoid arch; parietal bones in contact; maxillary wanting or united with the palatines; form eel-like. . . . . ApodES, XV.

  ee. Air-bladder without duct (in the adult); ventral fins without basal segments, usually anterior in position; spines usually present in the fins; pectoral fins not on the plane of the abdomen; parietal bones usually separated by the supraoccipital. (Spiny-rayed fishes chiefly.) (*Physoclysti.*)

    j. Shoulder girdle connected to the skull by a post-temporal.

      k. Lower pharyngeals co-ossified; no spines; ventrals abdominal; lateral line on side of abdomen. . . . . . SYNENTOGNATHI, XVI.

      kk. Lower pharyngeals separate (or united, and the dorsal fin with spines.)

        l. Gills tufted; pharyngeal bones and most of the branchihyals wanting; skin with bony plates. . . . . LOPHOBRANCHII, XVII.

        ll. Gills pectinate (as usual in fishes).

          m. Superior branchihyals and pharyngeals reduced in number ; ventrals sub-abdominal. . . . . HEMIBRANCHII, XVIII.

          mm. Superior branchihyals and pharyngeals in normal development.

            n. Ventral fins abdominal. . . . . . . PERCESOCES, XIX.

            nn. Ventral fins thoracic or jugular.

              o. Pectoral fins not pediculate, the gill openings in front of them.

                p. Bones of the jaws distinct.

                  q. Cranium normal. . . . . . . ACANTHOPTERI, XX.

                  qq. Cranium twisted, so that both eyes are on the same side of head; no fin spines. . . HETEROSOMATA, XXI.

pp. Bones of jaws co-ossified, the maxillary with the premaxillary, the dentary with the articular.
PLECTOGNATHI, XXII.

oo. Pectoral fins pediculate, the basal bones reduced in number
and elongate, the gills in their axils. PEDICULATI, XXIII.

More than two hundred families are now recognized among the
true fishes. The characters on which family divisions are based
are usually internal, and often difficult for the beginner to ascer-
tain. The boundaries and definitions of many families are also
still uncertain. Instead, therefore, of giving a natural analysis
under each order of the families included within it, I have thought
it best to give instead an Artificial Key by which the student can
recognize any of the families of True Fishes included in this work.
For analytical keys showing, in some degree, the natural characters, the student is referred to Jordan and Gilbert's Synopsis of the
Fishes of North America. A repetition of these analytical tables
would consume considerable space, and would not be of much aid
to any but advanced students.

Artificial Key to the Families of True Fishes included in the Present Work.

SERIES I. VENTRAL FINS PRESENT, ABDOMINAL.

A. Dorsal fins two, the anterior rayed, the posterior adipose.
  B. Body naked; head with 4 to 8 barbels; dorsal and pectoral each with
    a strong spine. . . . . SILURIDÆ, 24.
  BB. Body scaly; no barbels; no spines.
    C. Maxillary wanting, or grown fast to premaxillary; head scaly.
      SYNODONTIDÆ, 32.
    CC. Maxillaries distinct; head naked.
      D. Scales ctenoid; margin of upper jaw formed by premaxillaries
        alone. . . . . PERCOPSIDÆ, 35.
      DD. Scales cycloid; margin of upper jaw formed in part by maxil-
        laries.
        E. Stomach a blind sac, with few pyloric cæca. (Smelt, etc.)
          ARGENTINIDÆ, 33.
        EE. Stomach siphonal, with many pyloric cæca. SALMONIDÆ, 34.
AA. Dorsal fin single, with free spines before it; body naked, or with bony
  plates; ventral rays, I, 1, ... GASTEROSTEIDÆ, 45.
AAA. Dorsal fins two, the anterior of simple rays or spines, the posterior
  chiefly of soft rays; ventrals, I, 5.
  F. Teeth very strong, unequal; a lateral line present. SPHYRÆNIDÆ, 48.
  FF. Teeth small, subequal; no lateral line.
    G. Dorsal spines slender, 4 to 8; anal spine 1. . . ATHERINIDÆ, 47.
    GG. Dorsal spines stout, 4; anal spines, 2 or 3. . . . MUGILIDÆ, 46.
AAAA. Dorsal fin single, of soft rays only (sometimes preceded by fulcra or
  followed by finlets).
  H. Tail evidently heterocercal. (Ganoid fishes.)
    I. Caudal forked, the lower lobe well developed.

J. Body naked; snout spatulate; mouth wide, without barbels; caudal with fulcra. . . . . Polyodontidæ, 20.

JJ. Body with 5 series of bony shields; head with bony shields; mouth inferior, toothless, preceded by 4 barbels; fins with fulcra. Acipenseridæ, 21.

II. Caudal rounded or lanceolate; head with a bony casque.

    X. Scales ganoid (rhombic, enamelled plates); no gular plate; fins with fulcra; dorsal fin short. . . . . Lepisosteidæ, 22.

    XX. Scales cycloid; a bony gular plate; no fulcra; dorsal long.

                                                   Amiidæ, 23.

III. Tail not evidently heterocercal (except in the very young).

  Y. Scales cycloid.

    K. Side of belly with a conspicuous ridge or lateral line; pectoral fins inserted high, on or above the axis of the body; lower lobe of caudal longest; lower pharyngeals united. . Exocœtidæ, 42.

    KK. Edge of belly without conspicuous ridge or lateral line; pectoral fins inserted usually below axis of body; lower pharyngeals separate.

      M. Vent before ventrals; eyes rudimentary. . Amblyopsidæ, 36.

      MM. Vent behind ventrals; eyes normal.

        N. Head more or less scaly.

          O. Upper jaw not protractile, its margin formed by maxillaries posteriorly.

            P. Teeth cardiform, unequal. . . . . . . Esocidæ, 39.

            PP. Teeth villiform, equal. . . . . . . Umbridæ, 38.

          OO. Upper jaw very protractile, its edge formed by premaxillaries alone. . . . . Cyprinodontidæ, 37.

        NN. Head without scales.

          Q. Gill membranes united with the isthmus; lower pharyngeals falciform; mouth toothless; anterior vertebræ coalesced.

            R. Pharyngeal teeth larger, in one or more rows, the main row with less than 8 teeth; dorsal (in native species) with less than 10 rays. . . . . . Cyprinidæ, 26.

            RR. Pharyngeal teeth very numerous, in one row; dorsal rays ten or more. . . . . Catostomidæ, 25.

          QQ. Gill membranes free from the isthmus; lower pharyngeals flattish; anterior vertebræ not modified.

            S. Lateral line present.

              T. Lower jaw with a gular plate; fins with scaly sheaths. Elopidæ, 29.

              TT. Lower jaw without gular plate.

                U. Tongue with canine teeth; mouth terminal, oblique. Hiodontidæ, 27.

                UU. Tongue with blunt teeth; mouth inferior, horizontal. Albulidæ, 28.

            SS. Lateral line wanting.

              V. Mouth very wide, the maxillary reaching much beyond eye; snout short. . . . . Stolephoridæ, 31.

              VV. Mouth moderate, the maxillary scarcely extending beyond eye . . . . Clupeidæ, 30.

YY. Scales none; caudal with a long filament; snout long, tubular, with the small mouth at the end. . . . . Fistulariidæ, 44.

Series II. Ventral Fins present, thoracic or jugular.

A. Eyes unsymmetrical, both on same side of head. . Pleuronectidæ, 89.

AA. Eyes symmetrical.

  B. Gill openings in front of pectorals.

    C. Body more or less scaly, or armed with bony plates.

      D. Ventral fins united into one; no lateral line; gill membranes joined to isthmus. . . . . Gobiidæ, 82.

      DD. Ventral fins separate.

        E. Top of head with a large sucking-disk, modified from the spinous dorsal. . . . . Echeneididæ, 50.

        EE. Top of head without sucking-disk.

          F. Ventral rays, I, 5.

            G. Suborbital with a bony stay which extends across the cheeks to or towards the preopercle; cheeks sometimes entirely bony.

              H. Pectoral fin with 2 or 3 lower rays detached and separate. Triglidæ, 76.

              HH. Pectoral fin entire; slit behind fourth gill small, or wanting.

                I. Dorsal spines, 9 to 17; anal spines three; eyes lateral. Scorpænidæ, 72.

                II. Dorsal spines, four; eyes superior. Uranoscopidæ, 81.

            GG. Suborbital stay wanting; cheeks not mailed.

              K. Dorsal spines all or nearly all unconnected by membrane.

                L. Body elongate, subterete. . . . . Elacatidæ, 51.

                LL. Body oblong or ovate, compressed.

                  M. Caudal peduncle very slender, the fin widely forked. Carangidæ, 56.

                  MM. Caudal peduncle stout, the fin little forked. Stromateidæ, 58.

             KK. Dorsal spines, if present, mostly connected by membrane.

            N. Dorsal and anal each with 4 or more finlets; scales minute. Scombridæ, 55.

           NN. Dorsal and anal without finlets, or with but one each.

              O. Throat with two long barbels . . . . Mullidæ, 67.

              OO. Throat without barbels.

                P. Anal preceded by two free spines (these often obsolete with age).

                  Q. Scales very small, cycloid. . . Carangidæ, 56.

                  QQ. Scales moderate, ctenoid. . . Pomatomidæ, 57

              PP. Anal without free spines.

                R. Tail with a fleshy keel on each side. Carangidæ, 56.

                RR. Tail not keeled.

                  S. Dorsal fin very long, without distinct spines; caudal deeply forked. . Coryphænidæ, 59.

                  SS. Dorsal fin with distinct spines; gill membranes free from isthmus.

                    T. Vomer with teeth.

                      U. Anal spines none; eyes on top of head. Uranoscopidæ, 81.

                      UU. Anal spines, one or two.

V. Anal rays more than 20. . . . . POMATOMIDÆ, 57.

VV. Anal rays less than 20. . . . . PERCIDÆ, 63.

UUU. Anal spines three.

   W. Pseudobranchiæ small, fleshy, covered by skin.

CENTRARCHIDÆ, 62.

   WW. Pseudobranchiæ large, exposed.

      X. Maxillary slipping under preorbital for its whole length.

SPARIDÆ, 66.

      XX. Maxillary not slipping under preorbital for its whole length.

SERRANIDÆ, 64.

UUUU. Anal spines, 4 to 10. . . . . CENTRARCHIDÆ, 62.

TT. Vomer without teeth.

      Y. Teeth setiform (tooth-brush like) ; soft parts of vertical fins densely scaly; body elevated ; dorsal deeply notched.

EPHIPPIDÆ, 71.

      YY. Teeth not setiform.

         Z. Lateral line obsolete ; dorsal spines about 4.

ELASSOMATIDÆ, 61.

         ZZ. Lateral line present.

a. Anal spines 1 or 2 ; a large slit behind fourth gill.

  b. Lateral line extending on caudal fin; snout scaly. . . SCIÆNIDÆ, 68.

  bb. Lateral line not extending on caudal fin; snout scaleless.

PERCIDÆ, 63.

aa. Anal spines 3.

  c. Slit behind fourth gill none; lower pharyngeals completely united : jaws with canines in front. . . . . LABRIDÆ, 70.

  cc. Slit behind fourth gill large.

    d. Anal with more than 15 soft rays; preopercle serrate.

STROMATEIDÆ, 58.

    dd. Anal with less than 15 soft rays.

      e. Maxillary slipping beneath the broad preorbital for its whole length; dorsal spines more than 10. . . . . SPARIDÆ, 66.

      ee. Maxillary not slipping beneath the narrow preorbital.

         y Dorsal spines 12; premaxillaries moderately protractile; pseudobranchiæ large. . . . . LOBOTIDÆ, 65.

         yy Dorsal spines 9 or 10; premaxillaries extremely protractile; pseudobranchiæ concealed. . . . . . . GERRIDÆ, 69.

FF. Ventral fins with or without spine; the number of rays not I, 5.

  x. Upper jaw prolonged in a sword. . . . . . . ISTIOPHORIDÆ, 53.

  xx. Upper jaw not sword-like.

    f. Dorsal fin low, of spines only. . . . . BLENNIIDÆ, 83.

    ff. Dorsal fin of spines anteriorly, of soft rays posteriorly.

      g. Ventral rays 7; vent anterior; dorsal spines 3 or 4.

APHREDODERIDÆ, 60.

      gg. Ventral rays I, 1; dorsal spines free; vent normal; body mailed.

GASTEROSTEIDÆ, 45.

      ggg. Ventral rays I, 4; body scaly; pectoral fin divided to base in two unequal parts. . . . . . CEPHALACANTHIDÆ, 75.

    fff. Dorsal of soft rays anteriorly, with low spines posteriorly.

LYCODIDÆ, 86.

    ffff. Dorsal fin of soft rays only.

h. Dorsal fin very short; body mailed. . . . . . AGONID.E, 74.
hh. Dorsal fin very long; body with small scales.
    i. Dorsal and anal joined to the caudal.
       j. Gill membranes free from the isthmus; ventrals very slender,
           barbel-like. . . . . OPHIDIID.E. 87.
       jj. Gill membranes united to the isthmus. . . LYCODID.E, 86.
    ii. Dorsal and anal free from caudal; tail isocercal. GADID.F, 88.
CC. Body scaleless, smooth or more or less prickly or warty.
    k. Breast with a sucking-disk.
       l. Gill membranes free from isthmus; no spinous dorsal.
                             GOBIESOCID.E, 80.
       ll. Gill membranes attached to the isthmus.
          m. Skin smooth. . . . . LIPARIDID.E, 77.
          mm. Skin warty. . . . . CYCLOPTERID.E, 78.
    kk. Breast without sucking-disk.
       n. Ventrals completely united. . . . . . . GOBIID.E, 82.
       nn. Ventrals separate.
          o. Ventral rays I, 5.
             p. Dorsal and anal with finlets. . . . SCOMBRID.E, 55.
             pp. Dorsal and anal without finlets; two free anal spines.
                             CARANGID.E, 56.
          oo. Ventral rays less than I, 5.
             q. Upper jaw prolonged into a sword. ISTIOPHORID.E, 53.
             qq. Upper jaw not prolonged into a sword.
                r. Suborbital with a bony stay. . . . COTTID.E, 73.
                rr. Suborbital without bony stay.
                   s. Dorsal spines two or three; teeth strong.
                            BATRACHID.E, 79.
                 ss. Dorsal spines 4 to 6; teeth small.
                         GASTEROSTEID.E, 45.
                 sss. Dorsal spines numerous; teeth comb-like.
                            BLENNIID.E, 83.
BB. Gill openings small, behind the pectoral fins, which are pediculate.
    t. Gill openings in or behind lower axil of pectorals; mouth large,
    terminal.
       u. Pseudobranchiæ present; head broad, depressed; mouth very
          large, with large unequal teeth. . . . . . LOPHIID.E, 96.
       uu. Pseudobranchiæ none; head compressed; teeth small.
                            ANTENNARIID.E, 95.
    tt. Gill openings in or behind upper axil of pectorals; mouth small,
    below a projecting snout. . . . . MALTHID.E, 94.

SERIES III. VENTRAL FINS ENTIRELY WANTING.

A. Gill membranes joined to the isthmus, so that the gill openings of the two
sides are not connected.
  B. Dorsal fin single, of spines only (these sometimes slender, like soft rays).
    C. Molar teeth present. . . . . ANARRHICHADID.E, 85.
    CC. Molar teeth none.
      D. Mouth vertical; body naked. . . . CRYPTACANTHODID.E, 84.
      DD. Mouth not vertical; body scaly. . . . . . BLENNIID.E, 83.

BB. Dorsal fins two, the anterior spinous; teeth incisor-like.
BALISTIDÆ, 90.
BBB. Dorsal fin single, of soft rays only.
  E. Snout tubular, bearing the short toothless jaws at the end; body mailed. . . . . SYNGNATHIDÆ, 43.
  EE. Snout not tubular.
    F. Body elongate, eel-shaped; maxillaries and premaxillaries coalescent with vomer and palatines.
      G. Lower jaw projecting ; skin covered with linear imbedded scales arranged at right angles with each other.
ANGUILLIDÆ, 40.
      GG. Lower jaw not projecting; skin scaleless.. . ECHELIDÆ, 41.
    FF. Body not eel-shaped.
      H. Breast with a sucking-disk.
        I. Skin smooth. . . . . LIPARIDIDÆ, 77.
        II. Skin warty. . . . . CYCLOPTERIDÆ, 78.
      HH. Breast without sucking disk.
        J. Teeth in each jaw confluent into one.
          K. Body compressed, the skin rough. . . . . MOLIDÆ, 93.
          KK. Body not compressed, armed with spines.
DIODONTIDÆ, 92.
        JJ. Teeth in each jaw confluent into two. TETRAODONTIDÆ, 91.
AA. Gill membranes free from the isthmus.
  L. Vent at the throat; vertical fins separate. . . AMBLYOPSIDÆ, 36.
  LL. Vent normal.
    M. Caudal fin wanting; body naked.. . . . . TRICHIURIDÆ, 54.
    MM. Caudal fin present.
      N. Upper jaw produced in a sword.. . . . . . . XIPHIIDÆ, 52.
      NN. Upper jaw without sword.
        O. Body ovate, much compressed. . . . . STROMATEIDÆ, 58.
        OO. Body oblong or elongate; gill membranes not united.
          P. Jaws toothless, the lower projecting . AMMODYTIDÆ, 49.
          PP. Jaws with teeth, the lower not projecting. OPHIDIIDÆ, 87.

SERIES GANOIDEI. (THE GANOID FISHES.)

The name *Ganoidei* was first used by Agassiz for those fishes which are armed with bony plates, instead of regular cycloid or ctenoid scales.   Later, Johannes Müller, one of the greatest of systematic zoölogists, restricted the group to those fishes which show more or less distinct reptilian or batrachian affinities, and especially affinities with the mailed fishes of the Devonian and Carboniferous ages.   The group is a heterogeneous one, and one practically scarcely susceptible of definition.   Some of the Ganoids are closely allied to the Teleosts; some approach the *Dipnoi*, and some again resemble the *Holocephali*.   The existence of the solid optic chiasma, the presence of several valves in the arterial bulb, and of a more or less developed spiral valve in the rectum, distinguish the living Ganoids from all Teleosts, but none of these characters can be verified in the extinct forms. It seems to us better not to regard the Ganoids as a separate class or subclass, but to unite them with the Teleosts. (γάνος, splendor, from the enamelled scales.)

ORDER VII. SELACHOSTOMI.

This order contains but one family. (σέλαχος. shark or other cartilaginous fish; στόμα, mouth.)

FAMILY XX. **POLYODONTIDÆ.** (THE PADDLE-FISHES.)

Body fusiform, the skin mostly smooth; snout prolonged in a flat, spatulate blade, which overhangs the broad, terminal mouth; the "spatula" with a reticulated framework; teeth very numerous, minute, disappearing with age; opercle rudimentary, its skin produced in a long flap; gills 4½; no pseudobranchiæ; gill rakers very long, in two rows, separated by membrane; gill membranes connected, free from isthmus; one branchiostegal; spiracles present. C. fin with fulcra; D. posterior; tail heterocercal, the lower lobe nearly as long as the upper; sides of tail with rhombic plates; air-bladder large, cellular; stomach cæcal, the pyloric cæca forming a branching, leaf-like organ. Singular fishes, feeding on mud and minute organisms which they stir up on the bottom with the long oar-like snout. Two species, *Psephurus gladius* of rivers of China, and the following.

a. Gill rakers very fine and numerous; caudal fulcra many, small.
POLYODON, 29.

29. POLYODON (Lacépède) Bloch & Schneider. (πολύς, many; ὀδών, tooth.)

30. **P. spathula** (Walbaum). PADDLE-FISH. SPOON-BILL. DUCK-BILLED CAT. Olivaceous; opercular flap in adult reaching V.; head with flap and spatula more than half length. D. 55, A. 57, V. 45. L. 6 feet. Miss. valley; common in larger streams. (*P. folium* Lac.) (Lat. spatula.)

ORDER VIII. GLANIOSTOMI.

This order contains only the family of Sturgeons. (γλάνις, catfish; στόμα, mouth.)

FAMILY XXI. **ACIPENSERIDÆ.** (THE STURGEONS.)

Body elongate, fusiform, with five rows of bony keeled shields, the skin between these rows with small or minute plates; snout produced; mouth inferior, protractile, toothless; four barbels in a cross-row before mouth; gills 4; an accessory opercular gill; no

3

branchiostegals ; head covered by bony plates joined by sutures ;
gill membranes joined to isthmus ; vertical fins with fulcra ; dorsal
and anal posterior ; tail heterocercal ; air-bladder large, simple ;
stomach not cæcal, with pyloric appendages ; rectum with spiral
valve. Seas and rivers of northern regions ; feeding on small
animals and plants sucked in through the tube-like mouth. Genera
2, species about 20.

The sturgeons change considerably with age. The snout becomes shorter and blunter, the shields smoother, and some of the
shields often fall off or are absorbed in old age.

a. Spiracles obsolete ; snout broad, shovel-shaped, depressed above ; rows of
bony shields coalescent behind the dorsal, so that the depressed tail is
completely mailed ; gill rakers small, fan-shaped, ending in 3 or 4 points.
                                                    SCAPHIRHYNCHUS, 30.

aa. Spiracles present ; snout sub-conic ; rows of bony shields nowhere confluent, the tail not depressed nor mailed ; gill rakers lanceolate.
                                                    ACIPENSER, 31.

30. SCAPHIRHYNCHUS Heckel. (*Scaphirhynchops* Gill.)
(σκάφη, spade ; ῥύγχος, snout.)

40. **S. platyrhynchus** (Rafinesque). SHOVEL-NOSED STUR-
GEON. WHITE STURGEON. Body elongate, tapering into the
slender depressed tail, which extends in the young beyond C. as a
slender filament ; shields sharply keeled ; dorsal shields 15 to 18 ;
lateral, 41 to 46 ; ventral, 11 to 13. L. 5 feet. Miss. Valley, etc.,
common. (πλατύς, flat ; ῥύγχος, snout.)

31. ACIPENSER (Artedi) Linnæus. (Lat., sturgeon.)

a. Plates between vent and A. large, in one or two rows.
  b. Space between dorsal and lateral shields with stellate plates of moderate size in 5 to 10 series: last dorsal shield of moderate size, more
  than half length of one before it.

41. **A. sturio** L. COMMON STURGEON. First dorsal fulcrum
somewhat enlarged, its surface rough ; dorsal shields 9 to 11 :
lateral shields 26 to 31 ; ventral, 9 or 10 ; 2 rows of 2 shields each,
with one median shield between vent and anal. D. 40, A. 26.
L. 8 to 12 feet. N. Atlantic, ascending rivers ; commonest N., S.
to S. C. (*A. oxyrhynchus* Mitchill, the American form ; said to
have usually fewer lateral shields.) (*Eu.*) (Lat., sturgeon.)

  bb. Space between dorsal and lateral shields with minute plates in very
  many series.
  c. Last dorsal shield of moderate size, more than half length of next
  the last ; dorsal shields 15 or 16.

42. **A. rubicundus** Le Sueur. LAKE STURGEON. ROCK STUR-
GEON. First dorsal fulcrum slightly enlarged ; dorsal shields 15 ;
lateral 38, ventral 10 ; 3 shields in a single row between anal fin

and vent. D. 42, A. 27. Changes greatly with age, the young with sharp snout and very rough shields, and the spines strongly hooked ; the adult with blunt snout and small smooth shields, most of them finally lost. L. 6 feet. Miss. Valley, Great Lakes, and N., abundant, ascending rivers in spring, but not entering the sea. (Lat., ruddy.)

cc. Last dorsal shield very small, less than half length of next the last ; dorsal shields 10 to 12.

43. A. brevirostrum Le Sueur. Snout short, bluntish, much shorter than rest of head. Dorsal shields 11 ; lateral, 30 ; ventral, 9 ; one shield between anal and vent. D. 43, A. 24. N. Y. to Fla., scarce. (Lat. *brevis*, short; *rostrum*, snout.)

Order IX. GINGLYMODI.

This order, defined on page 25, contains but one family among recent fishes, although it has many allies among extinct forms; (γίγγλυμος, hinge ; ὀδούς, tooth.)

Family XXII. LEPISOSTEIDÆ. (The Gar-fishes.)

Body subcylindical, covered with rhombic enamelled "ganoid" scales, imbricated in oblique series which run downward and backward. Jaws both elongate, the upper always projecting ; premaxillary forming most of upper jaw, the maxillary transversely divided into several pieces; lower jaw formed much as in reptiles ; both jaws with an outer series of small teeth followed by one or two series of larger teeth of peculiar structure ; close-set, rasplike teeth on jaws, vomer, and palatines ; tongue toothless, broad, emarginate; external bones of head very hard, rugose. Eyes moderate; nostrils near end of snout: pseudobranchiæ present, besides an opercular gill; B. 3: no spiracles; air-bladder cellular, joined by a glottis to the œsophagus, resembling the lungs of reptiles, and used in respiration. Fins with fulcra: D. short, nearly opposite A.; tail heterocereal, produced as a filament in young: vertebræ with ball and socket joint, as in reptiles; pyloric cœca many. One genus now living, with 3 or 4 species. Singular fishes, inhabiting the lakes and larger rivers of Eastern North America. The species are extremely variable in coloration, length of snout, proportions, etc., a fact which has given rise to a multitude of useless specific names.

32. LEPISOSTEUS Lacépède.

(λεπίς, scale; ὀστέον, bone ; more correctly written *Lepidosteus*, but the above is the original word.)

a. Beak long and slender, the snout more than twice length of rest of head.

44. L. osseus (L.). Common Gar-Pike. Long-nosed Gar. Bill-fish. Olivaceous; vertical fins and posterior parts with round black spots, distinct in young; very young with black lateral band. Length of snout 15 to 20 times its least width; large teeth of upper jaw in one row in the adult. Head 3 in length. D. 8, A. 9. V. 6, P. 10. Lat. l. 62. L. 5 feet. Great Lakes to Carolina and Mexico; abundant. (Lat., bony.)

a. Beak shorter and broader, the snout not much longer than rest of head.
  b. Large teeth of upper jaw in one row on each side in adult: (an additional row on the palatines sometimes present in young.)

45. **L. platystomus** Rafinesque. SHORT-NOSED GAR-PIKE. Snout usually 1 to 1½ times rest of head, its length 5 to 6 times its least width. Head 3½ in length, otherwise almost exactly as in *L. osseus*, the color rather darker, the size smaller. L. 3 feet. Miss. valley, etc., less common N. (πλατύς, flat; στόμα, mouth.)

  bb. Large teeth of upper jaw in two series, the inner along outer edge of palatines.

46. **L. tristœchus** (Bloch & Schneider). ALLIGATOR GAR. MANJUARI. Snout usually shorter than rest of head, its least width 3½ in its length, otherwise essentially like the others; but reaching an enormous size. L. 10 feet. Ills. to Mexico and Cuba. (τρίς, three; στοῖχος, row.)

ORDER X. HALECOMORPHI. (THE BOW-FINS.)

This group, characterized on page 26, contains a single family among recent fishes. (Lat. *halec*, herring; μορφή, form.)

FAMILY XXIII. AMIIDÆ. (THE BOW-FINS.)

Body oblong, robust, with thick cycloid scales. Head subconical, bluntish, covered above by a very hard bony helmet; lateral margins of upper jaws formed by the maxillaries, which are divided by a lengthwise suture. Mouth horizontal, its cleft extending beyond the small eye; lower jaw broad, a broad bony striated gular plate placed between its rami; premaxillaries not protractile; jaws each with an outer series of conical teeth, behind them in the lower a band of rasp-like teeth; small teeth on vomer, palatines, and pterygoids: anterior nostril with a short barbel; cheek with a bony shield. B. 10 to 12. No pseudobranchiæ, nor opercular gill; two lanceolate striate appendages on each side of isthmus; gill rakers very short, stout. Lateral line present. Dorsal fin long and low, nearly uniform; no fulcra; anal fin short; tail heterocercal. Vertebræ double-convex, as usual among fishes. Airbladder, somewhat as in the *Dipnoi* and *Batrachia*, cellular, bifid in front, connected by a glottis with the pharynx. No closed oviduct; no pyloric cœca. One species known, in the lakes and sluggish waters of North America, — a voracious fish, remarkably tenacious of life, and with soft and pasty flesh.

33. AMIA Linnæus. (*Amiatus* Rafinesque.) (*ἀμία*, ancient name of some fish.)

47. **A. calva** L. Bow-fin. Mud-fish. Dog-fish. "John A. Grindle." Blackish olive, sides with greenish reticulations, lower side of head with dark spots ; ♂ with a black ocellus edged with orange at base of C. above. Head 3¾ ; depth 4. D. 48. A. 11. Lat. l. 67. ♂ 18 inches ; ♀ 24. Swamps and lakes. Vt. to Dakota, Fla., and Texas ; abundant in lowlands. A fish of great interest to zoölogists, from its relation to earlier types. (Lat., bald.)

SERIES TELEOSTEI.

We now take up the series of *Teleostei* proper, or true Bony-fishes, a group comprising the great majority of existing fishes. It is apparently descended from the Ganoid type, the *Nematognathi* being apparently allies or descendants of the *Glaniostomi*, and the *Isospondyli* of the *Halecomorphi*. As a whole, the *Teleostei* differ from the Ganoids in the more perfectly ossified skeleton, the less heterocercal tail, the degradation of the air-bladder and the arterial bulb, and in the simplicity of the optic chiasma.

The Teleostei are divisible into two great groups, with rather ill-defined boundaries, — the *Physostomi*, or soft-rayed fishes, and the *Physoclysti*, or spiny-rayed. The members of the former group have throughout life a slender duct, by which the air-bladder is joined to the alimentary canal. In most cases the fin-rays are soft, the ventrals abdominal, the pectorals placed low, and the scales cycloid. Although the typical *Physostomi* differ in many ways from the more specialized *Physoclysti*, yet as we approach the junction of the two groups the subordinate differences disappear, leaving finally the presence of the air-duct in *Physostomi* as the only differential character. In view of this close relation of the two groups, several writers, following Professor Gill, have removed as separate orders various aberrant forms, leaving the bulk of both groups in one large order, *Teleocephali*, with numerous suborders. We prefer to regard most of these suborders as distinct orders rather than to treat the heterogeneous group of *Teleocephali* as an "order." (*τέλεος*, perfect ; *ὀστέον*, bone.)

ORDER XI. NEMATOGNATHI.

This order contains several families, which agree in having the subopercle wanting, the anterior vertebræ coalesced, and the maxillary reduced to the bony core of a long barbel. None of the order have scales. (*νῆμα*, thread ; *γνάθος*, jaw.)

FAMILY XXIV. SILURIDÆ. (THE CAT-FISHES.)

Body more or less elongate, naked or with bony plates; margin
of upper jaw formed by premaxillaries only, the rudimentary
maxillaries forming the base of a long barbel; teeth in villiform
bands. Dorsal fin usually present, short, above or before ventrals;
usually an adipose fin behind dorsal. First ray of dorsal and pectorals usually developed as a stout spine. Lower pharyngeals
separate. Air-bladder present, large. A vast family of more
than 100 genera and 900 species, mostly of the rivers and swamps
of warm regions, especially of South America and Africa. A few
species are marine. Many of them are excellent as food, and all
are very tenacious of life.

a. Dorsal short, placed before ventrals; adipose fin present; gill membranes
more or less free from isthmus: body naked.
  b. Anterior and posterior nostrils close together, neither with a barbel, the
  posterior with a valve; palatines with teeth; caudal forked. Marine
  species. (Tachysurinæ.)
    c. Lower jaw with 2 barbels; maxillary barbel band-like; dorsal and
    pectoral spines ending in striated filaments. . AILURICHTHYS, 34.
    cc. Lower jaw with 4 barbels; spines not filamentous. TACHYSURUS, 35.
  bb. Anterior and posterior nostrils well separated, the posterior with a
  barbel; barbels 8; teeth in jaws only. (Ictalurinæ.)
    d. Adipose fin with its posterior margin free.
      e. Premaxillary band of teeth, without backward processes.
        f. Supraoccipital bone prolonged backward so that its emarginate
        apex fits closely around the anterior point of the second inter-
        spinal, thus forming a continuous bony bridge extending from
        the head to the dorsal spine. (Silvery species ; C. deeply
        forked.) . . . . ICTALURUS, 36.
        ff. Supraoccipital bone not reaching the second interspinal, the bony
        bridge more or less interrupted.
          g. Eyes normal. . . . . AMEIURUS, 37.
          gg. Eyes concealed by the skin. . . . . GRONIAS, 38.
      ee. Premaxillary band of teeth, with a lateral backward process on
      each side; lower jaw prominent. . . . . . . LEPTOPS, 39.
    dd. Adipose fin keel-like, adnate to the back, more or less joined to
    caudal fin; a (venom) pore in axil of pectoral. . NOTURUS, 40.

34. AILURICHTHYS Baird & Girard. (αἴλουρος, cat; ἰχθύς, fish.)

48. A. marinus (Mitchill). GAFF-TOPSAIL. SEA CAT. Dusky
bluish. Head short and broad. Maxillary barbels reaching end
of P. spine; P. filament reaching vent, D. filament to adipose fin;
upper lobe of C. longer; palatine teeth in a nearly continuous
band. Head 4¼. D. I. 7. A. 23. L. 30 in. N. Y. to Texas,
common S., not entering streams.

35. TACHYSURUS Lacépède. (*Galeichthys* and *Arius* Cuv. & Val.) (ταχύς, swift; οὐρά, tail.)

a. Teeth all pointed: top of head with a bony occipital shield which is not covered by skin; bands of palatine teeth without backward prolongation on the median line; vomerine bands of teeth not confluent; ante-dorsal shield small, crescent-shaped; eyes well above angle of mouth; species with blue lustre in life. (*Ariopsis* Gill.)

49. **T. felis** (L.). Sea Cat-fish. Interorbital area flattish and smooth, without ridges or granulations; fins not low, the spines more than half length of head; vomerine teeth in a small patch; palatine teeth in a larger one, on each side, the four patches separate; fontanelle prolonged backward as a narrow groove; occipital process long, about ⅓ head, convex at tip, with a median keel; gill membranes not meeting at an angle; maxillary barbel nearly as long as head. L. 24. N. Y. to Mexico; common S. (Lat., cat.)

36. ICTALURUS Rafinesque. (ἰχθύς, fish; αἴλουρος, cat.)

a. Anal fin very long; its rays 32 to 35; its base nearly ⅓ of body.

50. **I. furcatus** (Cuv. & Val.). Chuckle-headed Cat. Silvery, nearly plain; eye small, wholly before middle of head; head 4¼; depth 5. Miss. valley, not uncommon. (Lat., forked.)

aa. Anal fin moderate; its rays 24 to 30; its base 3½ to 4 in body.

51. **I. punctatus** (Rafinesque). Channel Cat. White Cat. Silver Cat. Olivaceous, rarely blackish, the sides silvery, almost always with small round dark olive spots; eye large, not wholly in front of middle of head; mouth small; barbels long; spines strong, serrate; head 4; depth 5. L. 3 feet. Montana to Vt., Ga., and Mexico, very abundant in flowing streams. A handsome fish, the best in the family as food. (Lat., spotted.)

37. AMEIURUS Rafinesque. (*a* privative; μείουρος, curtailed, the tail not notched.)

a. Caudal fin forked (species approaching *Ictalurus*).
   b. Anal rays 25 to 35; humeral process very short and blunt; usually covered by skin, about ⅓ length of pectoral spine.

52. **A. nigricans** (Le Sueur). Great Cat-fish. Mississippi Cat. Flannel-mouthed Cat. Slaty bluish, growing darker with age; body stouter than in the Channel Cat, the head broader, lower, and more depressed, the mouth wider, the caudal less forked, the skin thicker, hiding the bones of the head; head depressed above; supraoccipital above almost reaching second interspinal, the bony bridge broken for a short distance only; anal about as long as head; head 1 in length; depth 5; D. I. 5 or 6; A. 25 to

32. Ontario to Florida and Texas, abundant in lakes and large rivers, reaching 100 pounds or more.

· (*A. ponderosus* Bean, from St. Louis, described from a specimen 5 feet long, weighing 150 pounds, is probably a giant example of this species, differing only in having 35 anal rays. I find 25, 27, 28, and 32 in four specimens of *A. nigricans*.) (Lat., blackish.)

bb. Anal rays 20 to 23; humeral process very rough, more than half length of pectoral spine.

53. **A. albidus** (Le Sueur). WHITE CAT. CHANNEL CAT OF THE POTOMAC. Olive-bluish, silvery below; body stout; head broad, becoming with age very broad, the mouth in old specimens wider than in any other species; C. shallow-forked. L. 24. Penn. to N. C., very abundant in Potomac R. Varies much with age. (Lat., whitish.)

aa. Caudal fin entire or very slightly emarginate. (AMEIURUS.)
d. Anal fin long, of 24 to 27 rays (counting rudiments), its base more than ¼ length of body.

54. **A. natalis** (Le Sueur). YELLOW CAT. Yellowish, greenish, or blackish; body stout, the head short and broad, with wide mouth. Great Lakes to Va. and Texas, common in sluggish streams. L. 15. Excessively variable. (Lat., having large nates, i. e. adipose fin.)

dd. Anal fin moderate, of 18 to 22 rays, its base 4 to 5 in body.
e. Lower jaw projecting.

55. **A. vulgaris** (Thompson). Blackish; head 3½ to 4; A. 20; P. spine 2¼ in head. Great Lakes to Manitoba, essentially as in *A. nebulosus*, except for the form of the mouth; very likely a variety. (Lat., common.)

ee. Lower jaw not projecting.
f. Pectoral spines long, 2 to 2¼ in head; anal rays more than 20.

56. **A. nebulosus** [1] (Le Sueur). COMMON BULLHEAD. HORNED POUT. Dark yellowish brown, varying from yellowish to black sometimes (var. **marmoratus** Holbrook), sharply mottled with dark green and whitish; A. rays usually 21 or 22; its base 4 in body; pectoral spines long. L. 18. New England to Wis., Va.,

---

[1] The Horned Pout are "dull and blundering fellows," fond of the mud, and growing best in weedy ponds and rivers without current. They stay near the bottom, moving slowly about with their barbels widely spread, watching for anything eatable. They will take any kind of bait, from an angle-worm to a piece of a tin tomatocan, without coquetry, and they seldom fail to swallow the hook. They are very tenacious of life, "opening and shutting their mouths for half an hour after their heads have been cut off." They spawn in spring, and the old fishes lead the young in great schools near the shore, seemingly caring for them as the hen for her chickens. "A bloodthirsty and bullying set of rangers, with ever a lance in rest, and ready to do battle with their nearest neighbor." (*Thoreau.*)

and Texas, common, the best known of the smaller Cat-fishes. Introduced into the rivers of Cal. (Lat., clouded.)

ff. Pectoral spines short, 2⅓ to 3 in head (longest in the young); A. 17 to 19.

57. A. melas (Rafinesque). Adult very plump; young more slender. Color usually blackish. A. short and deep, its rays usually 17 to 19, its base nearly 5 in length, its pale rays forming a sharp contrast with the dusky membranes. N. Y. to Kansas, generally common; very close to *A. nebulosus.* (μέλας, black.)

38. GRONIAS Cope. (γρώνη, cavern.)

58. G. nigrilabris Cope. Upper parts, jaws, and fins black; eyes nearly hidden by thick skin; barbels and spines rather short. A. 18. Cave stream, tributary to Conestoga R., E. Penn. A recent descendant of *A. melas* or *nebulosus,* rendered blind by subterranean life. (Lat. *niger,* black; *labrum,* lip.)

39. LEPTOPS Rafinesque. (λεπτός, thin; ὄψ, face.)

59. L. olivaris (Rafinesque). MUD CAT. FLAT-HEAD CAT. RUSSIAN CAT. BASHAW. GOUJON. Yellowish, much mottled with brown. Body slender, the head broad and much depressed, the lower jaw projecting; barbels short; dorsal spine very weak; pectoral spines strong; anal short. A. 12 to 15. C. scarcely emarginate. A very large species, reaching 75 pounds, abundant in sluggish streams, Ohio to Ga. and S. W. A good food fish, of unprepossessing appearance.

40. NOTURUS Rafinesque. STONE CATS.[1] (νῶτος, back; οὐρά, tail.)

a. Premaxillary band of teeth with lateral backward processes, as in *Leptops.* (*Noturus.*)

60. N. flavus (Rafinesque). Yellowish brown, nearly uniform; body elongate; head broad and flat; barbels short; adipose fin deeply notched; a keel on back before it; D. spines short; P. spine retrorse-serrate in front, roughish behind; A. 16. L. 12. Ontario to Va., Neb., and Tenn., not rare in large streams. (Lat., yellow.)

aa. Premaxillary band of teeth without backward processes. (*Schilbeodes* Bleeker.)

b. Pectoral spine serrate on its posterior edge, roughish in front; adipose fin notched.

[1] These little fishes abound in small brooks among logs and weeds. The wounds produced by the sting of their sharp pectoral spines are excessively painful. In the axil is usually a pore, probably the opening of a duct from a poison gland. This matter deserves investigation.

c. Pectoral spines moderate, the inner serræ weak, not half diameter of spine, the outer stronger, retrorse, body elongate; coloration nearly uniform, the fins darker edged.

d. Pectoral spine short and weak, about 3 in head in adult.

61. **N. exilis** Nelson. Head small, rather narrow, depressed, 4 in length; depth 6; pectoral spine retrorse-serrate without, with 6 small teeth within; humeral process obscure; jaws subequal. A. 14 to 17. L. 4. Wis. to Kansas. (*N. elassochir* Swain & Kalb.) (Lat., slim.)

dd. Pectoral spine longer, about 2 (1¾ to 2¼) in head.

62. **N. insignis** (Richardson). Head rather broad, flat and thin, the upper jaw projecting; head 4¼; depth 6. A. 14 to 16. L. 10. Pa. to S. C., common E. (Lat., remarkable).

cc. Pectoral spine very strong, curved, more than half head, its posterior serræ recurved, their length about equal to diameter of spine, the anterior serræ small.

e. Color much variegated; adipose fin deeply notched, but not separated from C.

63. **N. miurus** Jordan. Grayish; top of head, tip of dorsal, middle of adipose fin, and caudal black, the body with four black cross-blotches; head not specially depressed eye 4½ in head; humeral process moderate; pectoral spine 1¼ to 1⅝ in head; head 3⅔. A. 13 to 15. L. 5. E. N. C. to Minn. and La., abundant. (μεί-ουρος, curtailed.)

ee. Color nearly plain brownish, everywhere above covered with fine small dots; adipose fin almost or quite free from caudal.

64. **N. eleutherus** Jordan. Head broad, flat, depressed, the form very much as in *Leptops olivaris :* humeral process obscure; eye 5½ in head; pectoral spine 1¾ to 2 in head; head 3⅛; A. 13. L. 4. White R., Ind., and French Broad R.; 3 specimens known. (ἐλεύθερος, free.)

bb. Pectoral spine entire, grooved behind; adipose fin continuous with the caudal.

65. **N. gyrinus** (Mitchill). Head short, broad and deep; pectoral spine 2 in head; jaws subequal, yellowish brown, not blotched, but with a narrow black lateral streak, sometimes with two above it. A. 15 or 16. L. 5. Hudson R. to Minn. and La., common N. (γυρῖνος, tadpole.)

ORDER XII. **EVENTOGNATHI.** (THE PLECTOSPONDY-LOUS FISHES.)

This group, defined on page 26, contains the great majority of the fresh-water fishes of the world. Its essential character is in the modification of the anterior vertebræ, as in the *Nematognathi,* without the characters of the rudimentary subopercle and maxillary, and the absence of scales, which distinguish the Cat-fishes. The chief families are the *Cyprinidæ* and the *Characinidæ;* the latter, abundant in South America, have an adipose fin and usually teeth in the jaws. (εὖ, well; ἐντός, within; γνάθος, jaw.)

FAMILY XXV. CATOSTOMIDÆ. (THE SUCKERS.)

Body oblong, covered with cycloid scales; head naked; jaws toothless and without barbels, the maxillary forming a large part of the edge of the upper jaw; mouth usually protractile, the lips generally thick and fleshy; lower pharyngeal bones falciform, with many comb-like teeth in one row; branchiostegals 3; gill membranes united to isthmus; dorsal fin rather long, of 11 to 50 rays,[1] without spine; anal short; caudal forked; ventrals abdominal, of about 10 rays; pectorals low; no adipose fin. Alimentary canal long, without cœca. Air-bladder large, divided into two or three parts by transverse constrictions. Genera 11, species about 60, inhabiting the rivers of North America; two species in Asia. The Suckers feed on plants and small animals; the flesh is rather tasteless and full of small bones. They ascend the rivers to spawn in spring, at which time the males have usually the A. and C., and often other parts of the body, covered with tubercles.

a. Dorsal fin elongate, its rays 25 to 50 in number; air-bladder in two parts.
  b. Fontanelle present; body oblong-ovate. (*Ictiobinæ.*)
    c. Dorsal rays 25 to 35; scales large (34 to 41) . . . . ICTIOBUS, 41.
  bb. Fontanelle obliterated by the union of the parietal bones; body elongate. (*Cycleptinæ.*)
    d. Mouth small, inferior, with thick papillose lips; scales small (56).
                                              CYCLEPTUS, 42.
aa. Dorsal fin short, its rays 16 to 18. (*Catostominæ.*)
  e. Air-bladder in two parts; lower pharyngeals slender, with small teeth.
    f. Lips thick, papillose; lateral line complete and continuous; scales small (55 to 115); fontanelle present; mouth small, inferior.
                                      CATOSTOMUS, 43.
    ff. Lips thin, plicate; scales large (40 to 50).
      g. Lateral line wholly wanting, at all ages. . . . ERIMYZON, 44.
      gg. Lateral line imperfect in young, nearly complete in the adult.
                                   MINYTREMA, 45.
  ee. Air-bladder in three parts; fontanelle present; scales large (about 45); lateral line complete.
    h. Mouth normal, the upper jaw protractile; the lips more or less plicate.
      i. Lower pharyngeal bones moderate, the teeth compressed, gradually increasing in size downward. . . . . MOXOSTOMA, 46.
      ii. Lower pharyngeal bones very strong, with the lower teeth much enlarged, subcylindrical and truncate; the upper teeth small and compressed. . . . . PLACOPHARYNX, 47.

[1] In this family, the rudimentary rays before dorsal and anal are not counted.

hh. Mouth singular, the upper lip not protractile, greatly enlarged; the lower split into two separate lobes; pharyngeal bones, etc. as in *Moxostoma.* . . . . LAGOCHILA, 48.

41. ICTIOBUS Rafinesque. BUFFALO-FISHES.

(This genus contains an uncertain number of species, very few of which have been yet well defined. They are large, coarse suckers, especially characteristic of the streams of the Mississippi valley. The group much needs careful study, such as could only be given by a collector resident near some large market). ($ἰχθύς$, fish; $βοῦς$, buffalo.)

a. Mouth large, terminal, protractile forwards; lips thin; lower pharyngeals and teeth weak. (*Sclerognathus* Cuv. & Val.)

66. **I. cyprinella** (Cuv. & Val.). COMMON BUFFALO-FISH. RED-MOUTHED BUFFALO. Body robust, the outline somewhat elliptical; head very large and thick; opercle coarsely striate, nearly half length of head; lips scarcely plicate; color dull brownish olive, not silvery; fins dusky. Head $3\frac{1}{2}$; depth 3. D. 28, A. 9; scales 7–37 to 41–6. L. 3 feet. Miss. valley, etc., common; reaches 20 to 40 pounds weight. (Lat., a small carp.)

aa. Mouth smaller, more or less inferior, protractile downwards, and with thicker lips.

b. Lower pharyngeal bones strong, the teeth comparatively coarse and large, increasing in size downwards; dusky species, not silvery. (*Ictiobus.*)

67. **I. urus** (Agassiz). RAZOR-BACKED BUFFALO. MONGREL BUFFALO. Body not much elevated, the back not keeled, the axis of the body not much farther from back than from line of belly; head thicker and blunter than in *I. bubalus;* eye smaller than in *I. bubalus;* mouth much larger and more oblique, approaching that of *I. cyprinella,* but with lips thicker and plicate, the folds broken up into papillae ; longest dorsal rays scarcely half of base of fin, opercle coarsely striate. Color very dark; fins dark. Head $3\frac{1}{2}$ to 4; depth 3. D. 30; scales 8–41–7. L. $2\frac{1}{2}$ feet. Miss. valley, less common than the others; certainly different from *I. bubalus,* but not always distinguishable by me from *I. cyprinella,* and possibly not really different. (Lat., a wild bull.)

68. **I. bubalus** (Rafinesque). SUCKER-MOUTHED BUFFALO. SMALL-MOUTHED BUFFALO. Body considerably elevated, the back compressed; axis of body much nearer line of belly than back ; head not very blunt, the mouth small and inferior ; eye 4 to 5 in head, rather large ; longest dorsal rays much more than half base of fin in adult; coloration dusky, the fins scarcely black. Head 4; depth $2\frac{2}{3}$; D. 29; scales 8–39–6. L. $2\frac{1}{2}$ feet. Miss. valley, etc., common. (*Bubalichthys bubalus* Agassiz.) (Lat., buffalo.)

bb. Lower pharyngeal bones narrow, with the teeth thin and weak; species of pale coloration, more or less silvery. (*Carp Suckers.*) (*Carpiodes* Rafinesque.)

c. Body subfusiform, the depth about 3 in length, lips thin, silvery white in life, the halves of the lower lip meeting at a wide angle.

69. I. carpio (Rafinesque). Back compressed, little arched; snout not blunt, projecting little beyond the mouth, its length a little more than that of eye; nostrils not close to tip of snout: opercle strongly striate; longest dorsal rays $\frac{2}{3}$ to $\frac{3}{5}$ length of base of fin, the anterior rays sometimes thickened, never filamentous; eye small, $4\frac{1}{2}$ to 5 in head. Head short, 4 in length; depth 3. D. 25 to 27; scales 7–37–5. Color dull silvery, sometimes brassy, some of the scales above often brownish at base. Ohio valley to Texas; probably a valid species, but of doubtful name and synonymy. (Lat., carp.)

cc. Body ovate-oblong, the back elevated, the depth about $2\frac{1}{2}$ in the length.
d. Opercle strongly striate.
e. Lips thin, silver-white in life, the halves of lower lip meeting at a wide angle, as in *I. carpio.*

70. I. difformis (Cope). Similar to *I. velifer*, but with very blunt snout, the maxillary reaching front of pupil; nostril very near tip of snout and above or before upper lip; eyes large ($3\frac{1}{2}$ to 4 in head); dorsal very high. Head 4; depth $2\frac{3}{4}$. Ohio valley.

71. I. thompsoni (Agassiz). Resembles *I. velifer*, but with the head small and pointed, the snout considerably projecting; eye small, 5 to $5\frac{1}{2}$ in head. Back arched. Head $4\frac{1}{4}$; depth $2\frac{1}{4}$. Great Lakes, abundant (specimens examined from Toledo). (For Rev. Zadock Thompson.)

ee. Lips full, thick, flesh-colored in life, the halves of lower lip meeting at an acute angle.

72. I. velifer (Rafinesque). QUILL-BACK. SKIM-BACK. CARP SUCKER. RIVER CARP. Snout sub-conic, projecting; anterior nostril distant from snout more than half an eye's diameter and considerably behind front of upper lip; maxillary reaching about to front of orbit; eye moderate or small, 4 to 5 in head; anterior rays of dorsal always elevated and filamentous, infrequently as long as base of fin. Head $3\frac{3}{4}$ to $4\frac{1}{3}$; depth $2\frac{1}{4}$ to 3. D. 26; scales 7–37–5. Coloration usually pale. Miss. valley, etc., very abundant; variable. (Lat., bearing sails.)

dd. Opercle nearly smooth.

73. I. cyprinus (Le Sueur). CARP SUCKER. Body rather deep, the eye quite small, the dorsal fin high, otherwise essentially as in *I. velifer*. Pa. to Va., chiefly about Chesapeake Bay. (Lat., carp.)

42. CYCLEPTUS Rafinesque. (κύκλος, round; λεπτός, slender; according to Rafinesque, small round mouth.)

74. **C. elongatus** (Le Sueur). BLACK HORSE. GOURD-SEED SUCKER. MISSOURI SUCKER. Head small, short and slender, rounded above; opercles small; eye small; fins large. Color blackish: ♂ in spring covered with small tubercles. Head 7; depth 4½: D. 30: scales 9–56–7. L. 2½ feet. Miss. valley, rather common in larger streams.

43. CATOSTOMUS Le Sueur. FINE-SCALED SUCKERS. (κάτω, inferior ; στόμα, mouth.)

a. Scales very small, much reduced and crowded anteriorly, about 100 in the lateral line. (*Catostomus.*)

75. **C. catostomus** (Forster). NORTHERN SUCKER. Upper lip thin, with 2 to 4 rows of papillæ; snout long, overhanging the large mouth. Males in spring profusely tuberculate and with a broad rosy lateral band. Great Lakes to Alaska, very abundant N.

aa. Scales larger, but small and crowded forwards, about 65 in the lateral line. (*Decactylus* Rafinesque.)

76. **C. teres** (Mitchill). COMMON SUCKER. WHITE SUCKER. Upper lip thin, with 2 or 3 rows of papillæ ; snout shorter than in the preceding, the mouth smaller. Color olivaceous, dusky above; sides rosy in spring. Head 4½; depth 4¼. D. 12. Scales 10–64 to 70–9. L. 18. Canada to Montana and Fla.; commonest of the Suckers, and extremely variable. (Lat., terete.)

aaa. Scales large, scarcely crowded anteriorly, 48 to 55 in the lateral line. (*Hypentelium* Rafinesque.)

77. **C. nigricans** Le Sueur. HOG SUCKER. STONE ROLLER. STONE LUGGER. STONE TOTER. HAMMER-HEAD. CRAWL-A-BOTTOM. HOG MOLLY. HOG MULLET. Head flattened above, concave between eyes; the frontal bone thick, broad and short; eyes small, placed high ; upper lip thick, with 8 to 10 rows of papillæ ; lower fins large. Color brassy olive, the back with dark cross-blotches, disappearing with age; lower fins red. Head 4; depth 4¾. D. 11. Lat. l. 48 to 55. L. 2 feet. Lakes and clear streams, W. N. Y. to Ala. and Kans. (Lat, blackish.)

44. ERIMYZON Jordan. (ἐρι, an intensive particle ; μύζω, to suck.)

78. **E. sucetta** (Lacépède). CHUB SUCKER. SWEET SUCKER. CREEK-FISH. Scales crowded, deeper than long; mandible oblique. Color dusky, brassy below : young with black bands or bars and pale streaks. Head 4 ; depth 2¾ in adult; spring males with 6 tubercles on snout. D. 11 to 13. Scales 43–15 in the northern form, var. **oblongus** Mitchill (the true *sucetta*, southern, with scales

36-15). Mass. to Dakota and S., very common. (Var. *sucetta*, Va. to Fla. and Texas.) (Fr. *sucet*, sucker.)

45. MINYTREMA Jordan. (μινύς, lessened ; τρῆμα, aperture : from the imperfect lateral line.)

79. **M. melanops** Rafinesque. STRIPED SUCKER. Body subterete, little compressed ; mouth small, inferior ; eye small ; scales little crowded forwards. Color dusky, coppery below, a dusky blotch behind dorsal ; each scale with a dark spot at its base, most distinct in adult, these forming longitudinal stripes ; ♂ tuberculate in spring ; lateral line wanting in young, imperfect at 8 inches, nearly complete in adults. Head 4⅓ ; depth 3 to 4¼. D. 12 to 14. Scales 46-13. L. 15. Great Lakes to S. C. and Texas. (μέλας, black ; ὤψ, look.)

46. MOXOSTOMA Rafinesque. RED HORSE. (μύζω, to suck ; στόμα, mouth.)

a. Lips distinctly plicate.
  b. Dorsal large, with 15 to 18 developed rays, its free edge not concave.

80. **M. anisurum** (Rafinesque). WHITE NOSE SUCKER. Body robust, compressed ; mouth large, inferior, the upper lip thin, the lower strongly Λ-shaped ; D. high and large, the first ray about as long as fin ; upper lobe of C. narrow, longer than lower. Color pale ; C. smoky gray ; lower fins red. Head 4 ; depth 3⅓. D. 15 to 18. L. 18. N. C. to Ohio R., Great Lakes, and N. (*Catost. carpio* C. & V., not of Raf. ; *Mox. valenciennesi* Jordan ; *Ptychostomus relatus* and *collapsus* Cope. (ἄνισος, unequal ; οὐρά, tail.)

  bb. Dorsal fin moderate, of 12 to 14 rays ; lower lip full, scarcely Λ-shaped, nearly truncate behind.
    c. Dorsal fin with its free margin nearly straight.
      d. Head large, 4 to 4⅓ in length.

81. **M. macrolepidotum** (Le Sueur). COMMON RED HORSE. WHITE SUCKER. "MULLET." Head broad, flattish above : mouth large, with thick lips ; depth of cheek usually more than half distance from snout to preopercle ; eye large ; edge of D. nearly straight, its first ray shorter than head ; C. lobes subequal. Olivaceous, with bright reflections ; sides silvery ; lower fins always orange-red, C. sometimes so. Head 4 to nearly 5 ; depth 3¼. D. usually 13 ; A. 7. Scales as in other species 5-45-4. L. 2 feet. Chesapeake Bay to Dakota and Ala., very abundant : the western form (var. **duquesnei** Le Sueur) with head and mouth rather larger than in the eastern form, which approaches *M. aureolum*. (μακρός, large ; λεπιδωτός, scaled.)

    dd. Head short and small, 4⅓ to 5¼ in length.

82. **M. aureolum** (Le Sueur). LAKE RED HORSE. Head shorter and smaller : mouth rather small, with thick lips ; snout

bluntish ; eye moderate ; C. lobes subequal; D. rather low, its longest ray less than base of fin. Coloration of preceding, the tail as well as lower fins always red. D. 13 ; depth 3½. Great Lakes, etc. Sometimes confounded with the next, from which it is well distinguished, but it may intergrade with the preceding. (Lat., gilded.)

cc. Dorsal falcate, the free margin deeply incised.

83. **M. crassilabre** [1] (Cope). Form of a *Coregonus*, with deep, compressed body, small head, and sharply conic snout, which overhangs the very small mouth; eye small, 5 in head. D. high, the anterior rays 1⅓ to 1½ times base of fin; free margin of fin concave, so that the fin is decidedly falcate. C. lobes very unequal, the upper always longest ; A. large, falcate, reaching beyond front of C. D. and C. bright red. Head 5 to 5¼ ; depth 3¼ to 3½; lat. l. 45. Ohio R. to N. C. (*Ptychostomus crassilabris, conus,* and *breviceps* Cope; *M. anisura* Jor. & Gilb., not of Raf.) (Lat. *crassus,* thick ; *labrum,* lip.)

bbb. Dorsal fin quite small, of 10 to 12 rays; lower lip thick, truncate behind.

84. **M. cervinum** (Cope). JUMP-ROCKS. JUMPING MULLET. Head very short, rather pointed ; mouth rather large, the lips strongly plicate ; eye small ; fins all small ; free edge of dorsal straight, its longest ray less than head. Color greenish brown, a pale blotch on each scale, these forming continuous streaks; back with brownish blotches ; fins brownish, scarcely red. Head 5 ; depth 4. D. 11. Scales 6–44 to 49–5. L. 10 inches. Va. to Ga., not rare. (Lat., tawny, like a deer.)

47. PLACOPHARYNX Cope. (πλάξ, a broad surface ; φάρυγξ, pharynx.)

85. **P. carinatus** Cope. A large, coarse sucker, externally similar to the species of Moxostoma, from which genus it differs only in the remarkable development of the lower pharyngeals and their teeth ; the bones are very strong, and 6 to 10 of the lower teeth are enlarged, little compressed, with a broad rounded or flattened grinding surface ; the mouth is larger and more oblique than in *M. macrolepidotum* and the lips are thicker. Head broad and flattish above, its upper surface somewhat uneven ; longest rays of dorsal longer than base of fin, 1⅓ in head ; free edge of D. concave ; upper lobe of C. narrower than lower, and more or less longer. Color dark olive-green, the sides brassy ; no silvery lustre ; C. and lower fins orange-red. Head 4 ; depth 3¼. D. 12. Scales 6–45–5. L. 30. Ohio to Ga. and Ark., abundant in larger streams. (Lat., keeled.)

[1] This descrip...s from notes of Dr. C. H. Gilbert, taken from Ohio R. specimens.

48. LAGOCHILA Jordan & Brayton.

(*Quassilabia* Jord. & Brayt.; *Lagochila* being set aside, on account of its similarity to *Lagocheilus*.) (λαγώς, hare; χεῖλος, lip.)

86. **L. lacera** Jordan & Brayton. HARE-LIP SUCKER. RABBIT-MOUTH SUCKER. PEA-LIP SUCKER. CUT-LIPS. SPLIT-MOUTH. Upper lip plicate, much prolonged ; lower reduced to two separate elongate, papillose lobes, the split between them reaching the dentary bones, which have a horny sheath; lower lip separated from upper by a deep fissure at angle ; skin of cheeks sheathing this fissure ; body rather slender, much as in *M. cervinum ;* opercle small; head very small, conical; dorsal low. Color pale, lower fins slightly reddish. Head 5; depth 4⅔. D. 12. Scales 5–45–5. L. 18. Wabash R. (*Evermann*), Scioto R., Clinch R., Chickamauga R., and White R., Ark.; most common in the Ozark Mountains; a most singular fish. (Lat., torn.)

FAMILY XXVI. CYPRINIDÆ. (THE MINNOWS.)

Head naked, body usually scaly; margin of upper jaw formed by premaxillaries only; mouth toothless; barbels 2 to 4 (absent in most of our genera and not large in any); lower pharyngeal bones well developed, falciform, nearly parallel with the gill arches, each provided with one to three series of teeth in small number, rarely more than seven on each side ; belly usually rounded, rarely compressed, never serrated; gill openings moderate, the membranes joined to the isthmus; no adipose fin; dorsal fin (in American species) short, with less than ten rays ; air-bladder usually large, commonly divided into an anterior and a posterior lobe, rarely wanting; stomach without appendages, appearing as a simple enlargement of the intestines.

Fishes of moderate or small size, inhabiting the fresh waters of the Old World and of North America. Genera about 200, species nearly 1,000; excessively abundant where found, both in individuals and in species, and from their great uniformity in size, form, and coloration constituting one of the most difficult groups in zoölogy in which to distinguish species. Ours are mostly of smaller size than those of the Old World, several of the larger European types being represented in America by Catostomoid forms. Our largest eastern species, *Semotilus bullaris*, rarely attains a weight of three or four pounds, and a length of nearly eighteen inches. The smallest species of *Notropis* scarcely reach a length of two inches.

The spring or breeding dress of the male fishes is often peculiar. The top of the head, and often the fins, snout, or other portions of the body, are covered with small tubercles, outgrowths from the

4

epidermis. The fins and other parts are often charged with pigment, the usual color being red, but sometimes satin-white, yellowish, or black.

NOTE. — Young *Cyprinidæ* are usually more slender than adults of the
same species, and the eye is always much larger; they also frequently show
a black lateral stripe and caudal spot, which the adults may not possess. In
the following descriptions, the rudimentary rays of dorsal and anal are not
counted. The fins and scales are often, especially in specimens living in small
brooks, covered with round black specks, parasitic plants. These should not
be mistaken for true color-markings.

No progress can be made in the study of these fish without careful attention
to the teeth, as the genera are largely based on dental characters. The
pharyngeal bones in the smaller species can be removed by inserting a pin (or,
better, a small hook) through the gill opening, under the shoulder girdle.
The teeth should be carefully cleaned with a tooth-brush, or, better, a jet of
water, and when dry may be examined by any small lens. In most cases a
principal row of four or five larger teeth will be found, in front of which is
a set of one or two smaller ones. The two sides are usually, but not always,
symmetrical. Thus, "teeth 2, 4-5, 1," indicate two rows of teeth on each
side, on the one side four in the principal row and two in the lesser, on the
other side five in the main row and one in the other. " Teeth 4-4 " indicates a single row of four on each pharyngeal bone, and so on.

In the *Leuciscine* genera, these teeth, or the principal ones, are "rapta-
torial," that is, hooked inward at the tips. A *grinding* or *masticatory* surface
is an excavated space or groove, usually at the base of the hook. Sometimes
the grinding surface is very narrow and confined to one or two teeth. Sometimes a bevelled or flattened edge looks so much like a grinding surface as to
mislead a superficial observer. In some cases, the edge of the tooth is crenate
or serrate.

Besides the native species here mentioned, representatives of two other
genera have been introduced from Europe, and have become inhabitants of
some eastern streams. These are *Cyprinus* Linnæus, and *Carassius* Nilsson.
The first is distinguished by the very long dorsal, which, like the anal, is preceded by a strong spine, serrated behind. About the mouth are four long
barbels, and the teeth are molar, 1, 3-3, 1. This genus is represented by
the Carp (*Cyprinus carpio* L.). The carp is normally covered with large
scales. In domestication, however, variations have arisen, prominent ones
being the "Leather Carp," naked, and the "Mirror Carp," with a few series
of very large scales.

Carassius Nilsson differs from *Cyprinus*, chiefly in the absence of barbels,
and in having the teeth compressed, 4-4. The Gold-fish (*Carassius auratus*
L.) is originally olivaceous, but only the orange-red variety is valued for
aquaria. Both *Carassius* and *Cyprinus* are native in China.

a. Air-bladder surrounded by many convolutions of the very long alimentary
   canal, which is 6 to 9 times the length of the body. (*Campostominæ*.)
   b. Teeth 4-4, or 1, 4-4, 0, with oblique grinding surface and slight hook;
      peritoneum black (as usual in herbivorous fishes). CAMPOSTOMA, 49.
aa. Air-bladder wholly above (dorsal) of the alimentary canal.
   c. Alimentary canal elongate, more than twice length of body; teeth one-
      rowed, the grinding surface well developed, the hook usually slight
      or wanting; peritoneum usually black. Species chiefly herbivorous.
      (*Chondrostominæ*.)

d. Teeth 5–5 or 4–5; dorsal inserted behind ventrals; scales very small; anal short.

　e. Pseudobranchiæ none; lower jaw thin, with sharp, hard edge; upper jaw protractile, with fleshy covering; lateral line complete; body elongate, subterete. . . . . OXYGENEUM, 50.

　ee. Pseudobranchiæ present; lips thin, normal; upper jaw protractile; lateral line incomplete. . . . . CHROSOMUS, 51.

dd. Teeth 4–4; pseudobranchiæ present; dorsal over ventrals; scales rather large.

　f. First (rudimentary) ray of D. slender, firmly attached to the first developed ray; jaws sharp-edged, the lower with a slight projection in front; scales before D. large (less than 15). HYBOGNATHUS, 52.

　ff. [1] First (rudimentary) ray of D. well developed, bluntish, separated from the first developed ray, to which it is joined by membrane (this character never conspicuous except in adult males; often obscure in young); scales before D. small (more than 20.) PIMEPHALES, 53.

cc. Alimentary canal short, less than twice length of body; teeth hooked, the grinding surface, if present, narrow and rudimentary; peritoneum usually pale ; species mostly carnivorous.

　g. Dentary bones parallel, united for their whole length (the lower jaw reduced to a tongue-like projection, which has a fleshy lobe on each side. (*Exoglossinæ*.)

　　h. Premaxillaries not protractile; upper lip thickened; scales moderate ; teeth 1, 4–4, 1, without grinding surface. EXOGLOSSUM, 54.

　gg. Dentary bones broadly arched, as usual among fishes, and united only at the symphysis. (*Leuciscinæ.*)

　　i. Abdomen behind V. not compressed to an edge, the scales passing over it; anal basis generally short (the rays 7 to 12).

　　　j. Teeth in the main row, 4–4.

　　　　k. Maxillary without traces of barbel.

　　　　　x. [Premaxillaries protractile.]

　　　　　　l. Lower lip thin or obsolete (except in one or two species), not developed as a fleshy lobe on each side.

　　　　　　　m. Mandible, interopercle, and suborbital not evidently cavernous.

　　　　　　　　n. [1] First (rudimentary) ray of D. enlarged and bluntish, separated from the first developed ray by membrane (as in *Pimephales*), this most evident in ♂ ; scales before D. small, about 28; teeth 4–4; [black blotch on front of dorsal and one at base of caudal always present.] . . CLIOLA, 55.

　　　　　　　　nn. First (rudimentary) ray of D. small, closely joined to the first developed ray; teeth 2, 1 or 0, 4–4, 2, 1 or 0; scales rather large; scales before D. large or small (12 to 30). NOTROPIS, 56.

　　　　　　　mm. Mandible, interopercle, and suborbital with conspicuous externally visible cavernous areas (like silvery cross-bars); teeth 1, 4–4, 0; scales large; D. above V. ERICYMBA, 57.

　　　　　ll. Lower lip developed as a fleshy lobe on each side ; teeth 4–4, without grinding surface; D. before V.; isthmus very broad. PHENACOBIUS, 58.

[1] This character is more or less obscure in females and young examples.

xx. [Premaxillaries not protractile; scales very small; barbel present, but minute] . . . . . RHINICHTHYS, 59.
kk. Maxillary with a small barbel at its extremity (rarely obsolete).
  n. Premaxillaries not protractile; teeth 2, 4–4, 2; scales small; dorsal behind ventrals . . . . . . RHINICHTHYS, 59.
  nn. Premaxillaries protractile.
    o. Teeth 4–4, or 1, 4–4, 1, or 1, 4–4, 0; scales not very small.
                                           HYBOPSIS, 60.
    oo. Teeth, 2, 4–4, 2 or 1.
      p. Head transversely convex above; teeth without grinding surface . . . . COUESIUS, 61.
      pp. Head flattened above; teeth with grinding surface; scales large . . . . PLATYGOBIO, 62.
jj. Teeth in the main row 5–5 or 4–5.
  q. Maxillary with a minute barbel placed before its tip; premaxillaries protractile; teeth, 2, 4–5, 2, without grinding surface; caudal fin symmetrical . . . . SEMOTILUS, 63.
  qq. Maxillary without barbel; premaxillaries protractile; anal basis short.
    r. Teeth two-rowed, 2, 4–5, 2, or 2, 5–5, 2, strongly hooked; scales moderate or small . . . . . . PHOXINUS, 64.
    [As above, the head broad and bluntish; the barbel so minute as to be indistinguishable, in the young of SEMOTILUS, 63.]
    rr. Teeth one-rowed, 5–5, with serrate edges; mouth very small, terminal; D. inserted over V. . . . OPSOPŒODUS, 65.
ii. Abdomen behind V., compressed to an edge, the scales not crossing it: anal basis elongate (the rays 12 to 18); teeth 5–5, with grinding surface and serrate edges; gill rakers rather long; no barbels; D. inserted behind V. . . . . . . NOTEMIGONUS, 66.

49. CAMPOSTOMA Agassiz.   (καμπή, curve; στόμα, mouth.)

87. **C. anomalum** (Rafinesque). STONE LUGGER. STONE ROLLER. Brownish, with a brassy lustre above, the scales mottled; a black vertical bar behind opercle; iris orange; D. and A. each with a dusky cross-bar about half-way up, rest of the fin in spring ♂ orange; ♂ in spring with many rounded tubercles on head and body; young mottled brownish, the fins plain; scales crowded forward; intestinal canal six to nine times the total length of the body, its numerous convolutions passing above and around the air-bladder, an arrangement found in *Campostoma* alone among all the vertebrates. D. 8; A. 7. Scales 7–53–8. Teeth 4–4. L. 4 to 8. W. N. Y. to Texas, and Tenn. in small streams, everywhere abundant; one of the most curious of American fishes. Very variable.

50. OXYGENEUM Forbes.   (ὀξύς, sharp; γένος, chin.)

88. **O. pulverulentum** Forbes.   Form of *Moxostoma:* head small, conical; mouth large, terminal; gill rakers slender; eye 4 in head; 31 scales before dorsal; breast scaly. Color pale, the

back and sides dusted with dark specks. Head 4⅛; depth 5.
D. high, 8. A. 7 Lat. l. 63. L. 2½. Illinois R. (Lat., dusted).

51. CHROSOMUS Rafinesque. (χρώς, color; σῶμα, body.)

89. **C. erythrogaster** Rafinesque. RED-BELLIED MINNOW.
Brownish olive, with black spots on the back, a blackish band from
above eye, straight to the tail, sometimes breaking up in spots behind; another below, broader, running through eye, decurved
along the lateral line, ending in a black spot at base of C.; belly
and space between bands bright silvery, brilliant scarlet red in
spring males, as are the bases of the vertical fins; females obscurely marked. D. 8; A. 9. Scales 16–85–10. L. 2½. Penn. to
Dakota and Tenn., abundant in small clear streams; one of the
most beautiful of our fishes; in high coloration the fins are bright
yellow. It is the most desirable of all our minnows for aquarium
purposes, being hardy, graceful, and gaily colored. (ἐρυθρός, red;
γαστήρ, belly)

52. HYBOGNATHUS Agassiz. (ὑβός, gibbous; γνάθος, jaw.)

a. Teeth comparatively long, and scarcely hooked; silvery species. (*Hybognathus.*)
  b. Suborbitals broad, the anterior, about twice as long as deep.
    c. Mouth narrow, its cleft not reaching nearly to eye; lower jaw shorter
    than upper, obtuse at tip.

90. **H. nuchalis** Agassiz. Body rather slender; head rather
short, the profile evenly curved; eye moderate, 4 in head; lateral
line decurved; 13 large scales in front of D.; intestine 7 to 10
times length of body. Silvery green, sides bright silvery, with an
underlying plumbeous shade; fins all pale. Head 4½ to 5; depth 4⅛.
D. 8, A. 7. Scales 5–38–4. L. 4 to 9. N. J. to S. C., Dakota, and
Texas, common near large rivers. Variable; notable varieties are
placita Girard, Arkansas and Missouri rivers, the eye smaller,
5 in head, the snout depressed and blunt, with very small mouth;
var. *regia* Girard, Potomac River, larger (7 inches long), with
deeper body and larger eye, 3¾ in head. (Lat., pertaining to the
nape.)

  cc. Mouth wide, its cleft reaching about to eye; jaws subequal, the lower
  acutish at tip.

91. **H. argyritis** Girard. Silvery. Upper Missouri and Red
R. of North. (Lat., silvery.)

aa. Teeth comparatively short, distinctly hooked; suborbitals very narrow;
  plumbeous species. (*Dionda* Girard.)

92. **H. nubila** (Forbes). Maxillary 3½ in head; snout short, not
very blunt; eye 3 in head; 12 scales before D. Head 4½; depth
4½. Scales 5–37–3. Olivaceous with plumbeous or dusky lateral

band; no caudal spot; fins mostly red. L. 2¼. N. Ill. to Ozark region. (Lat., dusky.)

53. PIMEPHALES Rafinesque. (πιμελής, fat; κεφαλή, head.)

a. Lateral line wanting or more or less imperfect. (*Pimephales.*)

93. **P. promelas** Rafinesque. Body more or less short and deep; head short, blunt, almost globular in adult ♂; V. reaching beyond front of A.; scales before D. about 27. Olivaceous, a black bar across middle of D. (faint in young); a dark shade along caudal peduncle; adult ♂ dusky, the head jet-black, with large tubercles on snout. Head 4; depth 4. D. I. 7. A. 7. Scales 7–47–6. L. 2¼. L. Champlain to Dakota and Texas, abundant in sluggish brooks. Very variable; S. W. specimens (var. *confertus* Girard) have the lateral line almost complete. (πρό, before; μέλας, black.)

aa. Lateral line complete. (*Hyborhynchus* Agassiz.)

94. **P. notatus** (Rafinesque). Body rather elongate; head rather long, the snout abruptly decurved; mouth horizontal, small; V. not to vent; scales before D. small, crowded, about 23. Color olivaceous, little silvery, sides bluish; a dusky shade toward base of D.; a black blotch on front of D., wanting in young; head wholly black in spring males, the snout with 14 large tubercles. Head 4½; depth 5. D. I. 8. A. 7. Scales 6–45–4. L. 4. Quebec to Del., Miss., and Kansas, very abundant, variable. (Lat., marked.)

54. EXOGLOSSUM Rafinesque. (ἔξω, outside; γλῶσσα, tongue.)

95. **E. maxillingua** (Le Sueur). CUT-LIPS. STONE-TOTER. Body rather stout; eye small; head large, with tumid cheeks; lower jaw included. Color dusky, a blackish bar behind head; a dusky shade at base C.; fins plain. Head 4; depth 4½. D. 8. A. 7. Scales 8–53–5. L. 6. Hudson R. to Va., abundant. A curious fish, remarkably distinguished from all other *Cyprinidæ* by its 3-lobed lower jaw. (Lat. *maxilla*, jaw; *lingua*, tongue.)

55. CLIOLA Girard. (A coined name.)

96. **C. vigilax** (Baird & Girard). BULL-HEAD MINNOW. Body rather stout, compressed, with deep tail; head heavy, blunt; snout short, decurved; mouth terminal, slightly oblique; eye 3½ in head; teeth strongly hooked; scales in front of D. small, crowded. Pale olivaceous, with a plumbeous lateral band, always ending in a black spot at base of C.; a conspicuous black spot on middle of front of D. Head 4⅓; depth 4. D. I. 8. A. 7. Scales 8–42–6; 28 scales before dorsal. L. 3. Ind. to Miss. and Texas, very abundant. Resembles *Pimephales notatus*, but distinguished by the short intestine, larger mouth, paler coloration, with more definite markings.

(*Hybopsis tuditanus* Cope; *Alburnops taurocephalus* Hay.)   (Lat., watchful.)

56. NOTROPIS Rafinesque.   (AMERICAN MINNOWS.)
(MINNILUS Rafinesque, etc., etc.)

(As now understood, this genus contains upwards of a hundred species of small Cyprinoids, all of them confined to the waters of E. N. A.   They are feeble fishes, of rather low organization, none of them of any value as food to man, but of great importance as food for the larger predatory fishes.   The species are highly variable, readily affected by surrounding conditions, while the permanently distinctive characters are few.   The identification of species in this group is therefore very difficult, and in the case of young specimens often impossible.   The following analysis must be used with caution, as all characters are subject to occasional or individual variations.)   (νῶτος, back; τρόπις, keel; but the back is not keeled.   Rafinesque's types had been shrivelled by drying.)

a. Teeth 4, 4, or 1, 4–4, 0, or 1, 4–4, 1 (sometimes 1, 4–4, 2 in *N. hudsonius*).
   b. Scales not closely imbricated, not notably deeper than long; D. inserted nearly over V.; A. short, its rays 7 or 8.
      c. Teeth 4–4, the grinding surface more or less developed.
         d. Lateral line usually incomplete; scales before D. large, 13 in number.   (*Hemitremia* Cope.)
            e. Snout very obtuse; lower jaw not projecting.

97. **N. bifrenatus** (Cope).   Body slender, the tail contracted ; upper lip on level of lower part of pupil; jaws subequal, eye large, 3 in head; lateral line very short.   Straw-color, with jet-black lateral band, bordered with orange on snout.   Head 4¼; depth 4¼.   D. 8.   A. 7.   Scales 5–36–3.   L. 2.   Mass. to Md.   (Lat., two-bridled.)

98. **N. anogenus** Forbes.   Very similar to *N. heterodon*, but with lateral line usually complete; the mouth very small and very oblique, the lower jaw included, the upper lip above level of pupil; snout short, blunt.   Dusky, a very distinct lateral band and a black spot at base of C.; a black speck on each pore of lateral line.   Head 4¼; depth 4⅔.   A. 7.   Lat. l. 34 to 37.   L. 1¼.   W. N. Y.   (Ithaca, *Meek*) to Ill.   (*a*, without; γένυς, chin.)

      ee. Snout pointed; lower jaw projecting.

99. **N. heterodon** Cope.   Body rather stout; eye 3 in head; lateral line usually developed about half-way, sometimes nearly perfect.   Olive, sides with dusky plumbeous band, fainter than in preceding.   Head 4 ; depth 4.   A. 8.   Scales 5–36–3.   L. 2¼.   Teeth crenate.   W. N. Y. to Kans.; common.   (Other specimens from Ind. and Ill. have lateral line complete, and teeth 2, 4–4, 2.

Whether a variety or a distinct species is not certainly known.[1])
(ἕτερος. different ; ὀδούς, tooth.)

> dd. Lateral line complete. (*Miniellus* Jordan.)
>> *f*. Lips thin, not fleshy; scales before D. large, in 13 to 17 rows.
>> *g*. Body rather elongate, the depth less than ¼ the length.

100. **N. procne** (Cope). Slender, with the tail long; snout
blunt ; mouth inferior, small; 13 scales before D.; eye large. Olivaceous, a dark lateral band. Head 4¾; depth 5¼. Scales 5–32–3.
A. 7. L. 2½. W. N. Y. to Md. (πρόκνη, a kind of swallow.)

101. **N. fretensis** (Cope). Slender, compressed ; mouth oblique;
eye 3½ in head ; 17 scales before D.; lateral line decurved. Olive,
a plumbeous lateral shade and dark spot at base C. Head 4;
depth 5. A. 8. Scales 6–35–3. L. 2½. Great Lake region (unknown to me.) (Lat., inhabiting straits, i. e. Detroit R.)

102. **N. spectrunculus** (Cope). Body elongate, head large and
broad ; eye 3 in head ; snout thick ; mouth terminal, oblique ; premaxillaries in front on level of middle of pupil ; 15 scales before D.
Olivaceous, dark above, a plumbeous lateral band and distinct
black caudal spot ; ♂ with fins orange. Head 4 ; depth 5¼. A. 9.
Lat. l. 37. L. 3. Tenn. R. (Lat., a little image.)

103. **N. deliciosus** (Girard). Body stoutish, little compressed ;
head rather broad, the mouth small, inferior, horizontal ; snout
obtuse ; eye large, 3 in head ; 12 to 15 scales before D. Pale
olivaceous, sides usually pale ; sometimes with a dusky stripe, but
no dark C. spot. Head 4 ; depth 5. L. 2½. Great Lakes to Va.
and Texas; an insignificant little fish. Variable, running into
several varieties. Var. *deliciosus*, Mo. and S. W., lat. l. 32 to 35;
var. *stramineus* Cope, Miss. Valley, lat. l. 34 to 38 (5–36–4) ; var.
longiceps Cope, Va., lat. l. 33 to 36 ; a distinct lateral stripe, snout
longer and fins higher; var. *volucella* Cope, Mich., snout longer;
fins longer; P. reaching V.

> *gg*. Body rather stout, the depth more than ¼ the length.

104. **N. topeka** Gilbert. Body compressed, stout; snout blunt;
mouth small, terminal, oblique ; eye 4¼ in head ; 14 scales before
D; lateral line anteriorly decurved. Olivaceous, a dusky lateral
streak, ending in a small caudal spot; males with sides and fins
bright red. Head 4 ; depth 3⅝. A. 7. Scales 5–35–4. L. 2¾.
W. Iowa to Kans.

> *ff*. Lips thick, fleshy.

105. **N. phenacobius** Forbes. Mouth small, inferior; body
short and deep; snout long; eye very large, 3½ in head ; breast
naked; fins low. Head 4 ; depth 3¾. A. 8. Lat. l. 35. L. 2¼.

---

[1] See Gilbert, Proc. U. S. Nat. Mus., 1884, p. 207.

Silvery, sides with some black specks.   Illinois R.   (Probably not a *Notropis*.)

 cc. Teeth 1, 4–4, 0; 1, 4–4, 1; or 1, 4–4, 2; the grinding surface more or less developed.

  h. Head comparatively large, 3¾ to about 4 in length ; teeth 1, 4–4, 1 ; species of small size.   (*Alburnops* Girard.)

   i. Eye moderate, 4 in head in adult.

106. **N. gilberti** Jordan & Meck.   Slender, with long tail; head long, flattish above; snout moderate; mouth rather large, little oblique, the lower jaw included.   Scales before D. 17; D. slightly behind V.   Greenish, sides with dusky streak and dark specks. Head 4 ; depth 5.   A. 9.   Scales 5–35–4.   L. 2¼.   Iowa and Mo. (To Prof. Charles Henry Gilbert.)

   ii. Eye very large, 2⅓ to 3 in head.

107. **N. boops** Gilbert.   Body compressed, the back elevated; tail slender; snout short, not blunt ; mouth terminal, very oblique, lower jaw not included; maxillary to front of eye ; D. over V. ; 12 scales before dorsal.   Head 3¾; depth 4⅓.   A, 7.   Scales 5–36–2. Teeth 1, 4–4, 1, with deep, grinding surface, the inner edge strongly crenate.   L. 3.   Olivaceous sides with dusky streak and dark specks. S. Ind. to Iowa and Ark., common S. W. in cold streams.   (βοῦς, bull; ὤψ, eye.)   *N. scabriceps* Jordan & Gilbert, not of Cope.)

  hh. Head short, bluntish, about 5 in length in adult ; species of large size and silvery coloration.   (*Hudsonius* Girard.)

108. **N. hudsonius** (De Witt Clinton).   SPAWN-EATER. "SMELT."   Body elongate, moderately compressed; head short, with blunt snout; eye very large, 3 to 3½ in head; mouth small, subinferior; lateral line slightly decurved; 12 to 18 scales before dorsal; fins rather small.   Pale olive, young always with a round black spot at base of caudal ; sometimes a dark lateral band; fins un- marked.   Head 4½ to 5; depth 4½ to 5.   D. 8.   A. 8.   Scales 5–39–4. Teeth variable, sometimes 2 in one of lesser rows, sometimes none of them with grinding surface.   L. 10.   Lake Superior to N.Y., and S. in coastwise streams to Ga., abundant and very variable.   N. specimens usually have teeth 2, 4–4, 1.   Southern examples, Va. to Ga. (var. *amarus* Girard), usually have teeth 1, 4–4, 1 or 0. The species seldom ascends small streams.   (From Hudson R.)

 bb. Scales very closely imbricated along sides of body, most of them deeper than long ; body usually compressed.

  j. Pharyngeal teeth usually 4–4, their edges serrate.   (*Moniana* Girard.)

109. **N. lutrensis** (Baird & Girard).   Adult with the body deep, strongly compressed, the back arched; young variously elon- gate or elliptical; head short, blunt; mouth moderate, oblique, the lower jaw included; eye small, about 4 in head ; lateral line strongly

decurved; 13 to 15 scales before D. ♂ steel-blue, profusely
tuberculate, belly and fins blood-red; a violet and a crimson crescent behind shoulder; ♀ plain; fins unspotted. Head 3⅔; depth
2⅝ (adult) to 4 (young). A. 8. Scales 6–35–2. Teeth sometimes
1, 4 – 4, 1. L. 3. S. Ill. to Rio Grande, very abundant S. W.; a
very brilliant and very variable little fish. (Lat. *lutra*, otter; first
known from Otter Creek, Ark.)

> *jj.* Pharyngeal teeth 1, 4 – 4, 1, their edges often crenate ; ours with narrow
> grinding surface; adult males with a large black blotch on upper pos-
> terior rays of D. (*Cyprinella*[1] Girard.)
>> *x.* Anal short, its rays 8 or 9 ; D. inserted just behind V.; ♂ in spring
>> with the fins charged with satin-white pigment.

110. **N. whipplei** (Girard). SILVER-FIN. Body subelliptical,
the adult much compressed; head short, not very blunt; mouth
rather small, oblique, the lower jaw shorter; eye small, 4½ in head;
males with high fins. Bluish silvery; scales dusky edged; a dark
vertebral line; dorsal blotch large in adult, wanting in young;
no creamy band across base of C. Head 4¼; depth 4. A. 8.
Scales 5–38–3. Teeth serrate. L. 4. W. N. Y. to Va. and
Minn., S. to Ark., abundant. (To Capt. A. W. Whipple, U. S. A.)

111. **N. galacturus** (Cope). Similar to the preceding, but
larger, more elongate and less compressed, the scales less closely
imbricated, lateral line less decurved; teeth usually not serrate;
the lower jaw included. Color like preceding but more silvery; C.
dusky, its basal third bright creamy yellow. Head 4⅓; depth 4¼.
A. 8. Scales 6–41–3. L. 6. Ozark region, E. to E. Tenn. and
Savannah R. in mountain streams. (γάλα, milk; οὐρά, tail.)

112. **N. camurus** Jordan & Meek. More robust than the preceding, the back elevated; anterior profile steep, the snout bluntly
decurved; mouth small, oblique; teeth crenate. Color much as in
N. whipplei. Head 4⅛; depth 3¼. A. 9. Scales 6–38–4. L. 4.
Ark. R., N. E. to S. Missouri. (Lat., blunt-faced.)

> *aa.* Teeth 2, 4 – 4, 2 ; lateral line complete.
>> *y.* Base of anal short, its rays 7 to 9.
>>> *k.* Scales on sides much deeper than long, especially in the adult, and so
>>> closely imbricated that the exposed edges are very narrow; body
>>> deep ; D. fin inserted above V. (*Luxilus* Raf.)

113. **N. megalops** (Rafinesque). COMMON SHINER; RED-FIN.
DACE. Body short, compressed in the adult, in the young elongate; head heavy, interorbital area rounded; snout bluntish;
mouth moderate, little oblique; lower jaw included; eye moderate,
4 to 5 in head; lateral line decurved; about 20 (15 to 25) scales
before D. Adult steel-blue, with gilt lines in life, sides silvery;

---

[1] Numerous species of this group, some of them very delicately colored, abound in
the rivers of the South.

fins pale; a dark shade behind shoulder; spring males tuberculate, with the belly and lower fins bright rosy. Head 4⅓; depth 3 to 5. A. 9. Scales 6–41–3. L. 8. In all brooks from Maine to Rocky Mts. except those of the Carolinas and Texas; excessively abundant and variable. (*Luxilus cornutus* (Mitchill).) (μεγάλος, big; ὤψ, eye.)

    kk. Scales on sides less closely imbricated,[1] scarcely deeper than long; body not elevated; small fishes often brilliant in the nuptial season. (*Hydrophlox* [2] Jordan & Brayton.)
        m. Teeth with narrow grinding surface; D. inserted more or less behind V.
        n. Lower jaw distinctly projecting beyond upper.

114. **N. coccogenis** (Cope). Body elongate, compressed; head pointed; mouth large, very oblique, the maxillary past front of eye; eye very large, 3½ in head ; 20 scales before D. Olivaceous, silvery below ; males with a scarlet vertical brand on preopercle; a red axillary spot; snout and belly rosy; a dark scapular band; outer half of D. and C. dusky. Head 4; depth 4⅓. A. 9. Scales 7–42–3. L. 5. Mountain streams, Ky. to Ga. (κόκκος, cherry-red; γένειον, cheek.)

        nn. Lower jaw little if at all projecting.

115. **N. zonatus** (Agassiz). Body rather elongate; head long, not acute; jaws equal; maxillary 3 in head, not to eye; snout shortish; eye very large, 3 in head; lateral line decurved; 16 scales before D. ♀ and young olivaceous, with plumbeous lateral band and no caudal spot. ♂ in spring with black lateral band, sides and lower parts flame-red. Head 4⅓; depth 4¾. A. 9. Scales 6–39–4. L. 5. Ozark region.

116. **N. lacertosus** (Cope). Body stout, with large head; mouth wide, the lower jaw projecting; eye large, 3¼ in head, equal to snout or interorbital; maxillary not to eye. Silvery. D. dusky; no red. Head 4; scales 5 above lat. l. L. 4½. Holston R. (Lat., lizard-like.)

117. **N. rubricroceus** (Cope). Red Fall-fish. Head rather pointed; mouth oblique, rather large, the jaws equal; eye large, 3½ in head; lateral line decurved; 19 scales before D. ♂ blue, with black lateral band, the whole body more or less suffused with blood-red; ♀ pale. Head 4; depth 4½. A. 9. Scales 7–38–8. L. 2½. Mountain torrents, on both sides of Great Smoky range. (Lat., saffron-red.)

118. **N. chalybæus** (Cope). Body slender, the back elevated; snout pointed; mouth oblique; lower jaw projecting; lateral line

---

[1] This character is not of much value, as in some of the species the scales become quite closely imbricated in adult specimens.

[2] Numerous species of this type, gaily colored little fishes, are found in the southern streams.

decurved; eye 3 in head; 18 scales before D. Brown, a jet-black lateral band; ♂ orange below. Head 3⅓; depth about 4¼. A. 8. Scales 6–35–3. L. 2. Delaware R. (Lat., steel-colored.)

> *mm.* Teeth without grinding surface; D. fin nearly opposite V.
>> *o.* Base of C. with a black spot ; snout and base of D. red in spring.

119. N. leuciodus (Cope). Slender, the snout rounded; mouth oblique, the lower jaw not projecting ; lateral line nearly straight; 12 scales before D. Silvery, a purplish lateral band. Head 4½. A. 8. Scales 5–39–3. L. 3. Holston R. (λευκός, white.)

>> *oo.* Base of C. without black spot in adult; ♂ without red.
>>> *p.* Eye rather large, 3 to 3⅔ in head.

120. N. jejunus (Forbes). Slender; snout blunt; mouth rather large, oblique; 16 scales before D. Pale, a silvery lateral band over plumbeous. Head 4; depth 4⅔. A. 7. Scales 5–37–3. L. 3. Penn. to Kans. (Lat., hungry.)

121. N. scabriceps (Cope). Stout, head heavy, flattish above, with blunt snout; mouth little oblique; eye 3 in head; lateral line decurved; fins small. Olive, with a silvery plumbeous lateral band. Head 4; depth 4¼. A. 8. Scales 6–38–3. Kanawha R. (Lat. *scaber*, rough ; *ceps*, head.)

>>> *pp.* Eye very large, about 2⅔ in head.

122. N. ariommus (Cope). Body stout, compressed ; head large; snout rather blunt; mouth moderate, oblique, the jaws equal; eye much longer than snout, larger than in any other of our *Cyprinidæ ;* 15 scales before D.; lateral line much decurved. Olivaceous, sides silvery. Head 3¾ ; depth 4¼. A. 9. Scales 6–39–2. L. 5. Ind. to N. Ala. (ἄρι, an intensive prefix ; ὄμμα, eye ; *i. e.* big-eyed.)

> *yy.* Base of anal comparatively elongate, its rays 10 to 12 ; D. inserted behind V.
>> *q.* Scales comparatively small, closely imbricated along sides; scales before D. small, 20 to 30; teeth with narrow grinding surface; nuptial colors brilliant. (*Lythrurus* Jordan.)
>>> *r.* A very distinct roundish black spot at base of first rays of D.

123. N. ardens (Cope). RED-FIN. Body more or less elongate, strongly compressed ; head rather pointed; mouth large, rather oblique, the chin somewhat projecting; eye moderate; D. high; about 30 scales before it; lateral line much decurved. ♀ very pale, the dorsal spot usually distinct; ♂ steel-blue, belly and lower fins brick-red in spring. Head 4¼ ; depth 4¼. D. 8. A. 11 or 12. Scales 9–50–3. L. 3½. Minn. to Va. and Tenn., abundant. Very variable, but the varieties (*lythrurus, atripes, cyanocephalus*) are hardly worthy of separate names. (*Minnilus diplemius* Jordan & Gilbert; not *diplemius* Raf.) (Lat., burning.)

rr. No distinct black spot at base of D. in front.

s. Body rather deep and compressed, the depth $2\frac{2}{3}$ to $4\frac{1}{2}$ in length.

124. **N. umbratilis** (Girard). ♂ with the body very deep; ♀ comparatively elongate; snout short and blunt; mouth terminal, wide, oblique; lower jaw included; eye about 4 in head; 30 scales before D. Olivaceous, thickly dusted with black specks; fins in ♂ all jet-black; paler or dusky in ♀; body and fins flushed with red in spring. Head 4; depth (male) $2\frac{2}{3}$. A. 10 or 11. Lat. l. 40. L. 3. Ill. to Kansas and S., locally common. Very variable. (*N. macrolepidotus* Forbes (Ill.) seems to be the same, but with 19 scales before D.) (Lat., shady.)

ss. Body elongate, the depth even in males about $\frac{1}{6}$ the length.

125. **N. lirus** Jordan. Very slender; eye 3 in head. Color pale, silvery; sides with a band of metallic blue; series of black dots on bases of D. and A.; males with red fins. Head $4\frac{1}{2}$; depth $5\frac{1}{4}$. A. 10. Scales 8–45–4. L. $2\frac{1}{2}$. Tenn. and Ala. (λειρός, pale.)

qq. Scales comparatively large, not closely imbricated; scales before D. large, in about 15 rows; teeth without grinding surface; D. inserted behind V.; mouth oblique, the lower jaw scarcely shorter; elongate, silvery species, the males usually with snout and base of dorsal rosy. (*Notropis.*[1])

t. Fins moderate, the ventrals extending beyond middle of dorsal.

126. **N. photogenis** (Cope). Slender, compressed; mouth oblique, the jaws subequal; maxillary not quite to orbit; lateral line decurved; eye large, $3\frac{1}{3}$ in head, as long as snout. Greenish, sides silvery. Head $4\frac{1}{4}$; depth $5\frac{1}{2}$. A. 10. Scales 6–40–3. L. 3. Penn. to W. Va. and S. (φώς, light; γένυς, cheek.)

127. **N. telescopus** (Cope). Similar to preceding, the D. farther forward, not much behind V., midway between snout and C.; eye very large, $2\frac{2}{3}$ in head; mouth oblique, the jaws subequal; scales above dark-edged. Head $4\frac{1}{4}$; depth 5. A. 10. Scales 5–38–3. L. $3\frac{1}{2}$. Tenn. R. (τηλεσκόπος, far-seeing.)

128. **N. dilectus** (Girard). Body moderately elongate, the back scarcely elevated, the tail slender; head longer than in related species, rather pointed; mouth rather large, oblique, the jaws subequal; eye moderate, 4 in head. Olivaceous, sides silvery; vertebral line faint. Head $4\frac{1}{4}$; depth $4\frac{2}{3}$. A. 10. L. $2\frac{3}{4}$. Ohio to Neb. and Ark. Common. Much smaller than *N. atherinoides*, with longer head. (*Alburnellus rubrifrons* and *percobromus* Cope.) (Lat., delightful.)

129. **N. atherinoides** (Rafinesque). Body comparatively elongate, compressed, the back not elevated; head short, blunt; mouth moderate, very oblique, maxillary reaching front of eye; eye large,

[1] The species of this group are extremely closely related, and in some cases scarcely distinguishable.

$3\frac{1}{4}$ in head, about equal to snout ; fins low; dorsal well behind ventrals : lateral line decurved. Greenish, pale above; sides silvery ;
a dark vertebral line. Head $4\frac{2}{3}$; depth $5\frac{1}{2}$. A. 11. Scales 5–38–3.
L. 5. Great Lakes to Tenn., abundant in lakes and rapids in
rivers. Variable. (*Alburnus rubellus* Agassiz.) (*Minnilus dine-
mus* Raf., with shorter snout and smaller eye, is probably the same,
as also *Alburnellus jaculus* Cope; the latter, from Michigan and
S., is slenderer, depth 6 in length.) (Like *Atherina.*)

130. **N. arge** (Cope). Eye very large, longer than snout, 3 in
head ; lateral line nearly straight , head large, the snout not very
blunt ; mouth large, the chin projecting. Pale, the silver band on
sides bounded above by a blackish line ; a dark vertebral streak.
Head $4\frac{1}{3}$; depth 6. A. 11. Scales 5–39–3. L. $3\frac{1}{2}$. Wabash Valley (Evermann) and S. Mich. ; slenderer than *N. atherinoides*,
with much larger eye, but very likely a variety. (ἀργής, shining
white.)

    tt. Fins all small, the short V. not reaching vent, and barely to middle
      of D.

131. **N. micropteryx** (Cope). Very slender, compressed; head
rather pointed ; mouth large, oblique, the jaws subequal; eye moderate, $3\frac{1}{4}$ in head; lateral line decurved; D. inserted well behind
V. Pale olive ; sides bright silvery, base of C. dusky. Head $4\frac{1}{4}$;
depth $5\frac{1}{2}$. A. 10. Scales 6–39–2. L. $2\frac{3}{4}$. Ozark region, E. to
E. N. C., in mountain streams. (μικρός, small; πτέρυξ, fin.)

    57. ERICYMBA Cope. (ἔρι, an intensive particle;
               κύμβη, cavity.)

132. **E. buccata** (Cope). Body rather elongate, little compressed, head long, with broad, prominent snout; mouth small,
subinferior, the lower jaw shorter. Suborbitals broad, silvery,
crossed by conspicuous translucent or silvery mucous channels, as
are also the interopercle and lower jaw; 15 scales before D. ; lateral line straightish; eye large, 4 in head. Olivaceous, sides silvery;
sexes alike. Head 4; depth 5. D. 8. A. 8. Scales 5–33–3.
L. 4. Mich. to Kans. and W. Fla., abundant in small, clear brooks,
remarkably distinguished by the structure of the bones of the head.
(Lat., big-jawed.)

    58. PHENACOBIUS Cope. (φέναξ, deceptive; βίος, life.)

a. Scales rather large, 40 to 52.
  b. Breast scaly.

133. **P. teretulus** (Cope). Body slender, subterete ; snout
thick, decurved ; mouth small; eye large, high up, $3\frac{1}{2}$ in head.
Yellowish, darker above, a plumbeous lateral band. Head $4\frac{2}{3}$.
depth $4\frac{2}{3}$. D. 8. A. 7. Scales 6–43–5. L. $3\frac{1}{2}$. W. Va. (Lat.,
terete.)

CYPRINIDÆ. — XXVI. 63

bb. Breast naked.

134. P. mirabilis (Girard). Body rather slender, the caudal peduncle short; snout blunt, prominent; eye 4 in head. Pale greenish, a silvery lateral band and a conspicuous black spot at base of C. Head 4½; depth 4⅓. A. 7. Scales 6–50–5; lateral line varying from 43 (var. *scopifer* Cope) to 52. L. 4. Ill. R. to N. Texas, abundant; sexes similar. (Lat., wonderful.)

aa. Scales small, about 60 in lateral line; breast naked.

135. P. uranops Cope. Body very slender; tail long; head long, flattish above; snout broad, blunt; mouth inferior, larger than in other species; eye large, 3½ in head, placed high and behind middle of head; 24 scales before D. Head 4¾; depth 6. A. 7. Scales 7–60–6. L. 4. Tenn. R. (οὐρανός, sky; ὤψ, eye.)

59. RHINICHTHYS Agassiz. (ῥίν, snout: ἰχθύς, fish.)

a. Snout long and prominent, projecting notably beyond the mouth, about twice length of eye in adult.

136. R. cataractæ (Cuv. & Val.). LONG-NOSED DACE. Body elongate, subterete; eye nearly median, 5 in head; barbel evident; P. enlarged in males. Dusky olive, irregularly mottled; no distinct lateral band; a dusky spot on opercle; male with lips, cheeks, and lower fins crimson in spring. Head 4½; depth 5⅓. D. 8. A. 7. Scales 14–65–8. L. 6. Mass. to Va. and Montana, in clear mountain streams. Larger than the next and with longer snout. (*Leuciscus nasutus* Ayres.) (Lat., of the cataract; first taken at Niagara.)

aa. Snout moderate, projecting little beyond the small mouth; its length 1⅓ times eye.

137. R. atronasus (Mitchill). BLACK-NOSED DACE. Body moderately elongate; head rather large; eye small, 4½ in head; fins small; barbel minute, sometimes obsolete. Blackish, the scales mottled above; a black or brown lateral band, bordered above and below by pale; spring ♂ with this band and lower fins crimson, the color changing to orange in summer. Head 4; depth 4½. D. 7. A. 7. Scales 4–63–8. L. 3. Maine to Iowa and Ala., very abundant in all clear brooks. Variable. (Lat. *ater*, black; *nasus*, nose.)

60. HYBOPSIS Agassiz. (*Nocomis* Girard; *Ceratichthys* Baird.) (ὑβός. gibbous; ὄψις, face.)

a. Species of moderate or small size, the mouth inferior, horizontal. Color silvery; preorbital broad; sexes more or less alike. (*Hybopsis.*)

  b. Eye moderate or small, 3¼ to 5 in head; barbel very long (rarely duplicated); lower lip rather thick; D. usually more or less behind V.; small, slender species.

    c. Teeth 4–4.

      d. Dorsal fin without black blotch; scales large.

e. Lower lobe of C. chiefly black; upper lobe pale; color pale, unspotted.

138. **H. gelidus** (Girard). Very slender; snout long, thick, blunt, overhanging the rather large mouth; barbel as long as eye; eye 4½ in head; fins all high; P. as long as head; C. deeply forked. Head 4; depth 5½. A. 7. Lat. l. 44. L. 2. Missouri River, abundant in the river channels but not ascending brooks; a singular little fish. (Lat., frigid.)

ee. Lower lobe of C. pale, like the upper; body dusted with dark specks.

139. **H. hyostomus** (Gilbert). Body and head very slender; snout long, acute, projecting beyond mouth for half its length; mouth short, wide, inferior; eye 3¼ in head; barbel long; P. large, other fins small; 13 scales before D. Head 4; depth 5½. A. 8. Lat. l. 37. L. 2½. Silvery, dusted with dark specks. Ind. to Iowa. (Similar species are *II. æstivalis* Girard, Ark. to Mexico, still more slender, with longer snout and much smaller eye, 4 in head, and *II. tetranemus* Gilbert, Kansas, nearly like *II. æstivalis,* but with two barbels on each side.) (ὑς, hog; στόμα, mouth.)

dd. Dorsal fin with a large black blotch on its last rays; scales small.

140. **H. monachus** (Cope). Body slender; head long and slender; eye 4½ in head; 24 scales before D. Olivaceous, sides silvery; a black spot at base of C.; no lateral band; scales not speckled. Head 4; depth 5¼. A. 8. Scales 8–56–4. L. 4. Tenn. R. (Lat., solitary.)

bb. Eyes very large, 2? to 3 in head; barbels conspicuous; D. inserted more or less before V.; body not conspicuously speckled.

f. Teeth 4–4; sides with dark blotches.

141. **H. dissimilis** (Kirtland). Body very long and slender; head long, the snout blunt at tip, projecting beyond the small mouth; eye 2¾ in head; P. long; 22 scales before D. Olivaceous, with dusky lateral band, along which are several large round dusky spots, the most distinct at base of C. Head 4½; depth 5¼. D. 8. A. 7. Scales 6–47–5. L. 5. Lake Erie to Ky. and Iowa. (Lat., unlike.)

ff. Teeth, 1, 4–4, 1 or 0.

g. Sides with a dark lateral band overlaid by silvery.

142. **H. amblops** (Rafinesque). Body slender, the head large, flattish above; eye longer than snout, 3 in head; mouth small; snout bluntly decurved; 16 scales before D. Greenish; sides with a blackish or plumbeous band extending around snout, overlaid by silvery. Head 4; depth 5. Scales 5–38–4. L. 4. Ohio Valley to Ala., common. Smaller than the next, and somewhat different in color. (ἀμβλύς, blunt; ὤψ, face.)

gg. Sides bright silvery, without dusky shade.

143. **H. storerianus** (Kirtland). Body rather elongate; back elevated; tail long. Head short, broad between eyes; eye equal to snout, three in head; preorbital broad, conspicuous, silvery; snout abruptly decurved, its tip fleshy; lateral line decurved: fins high. Light olive, sides brightly silvery; fins all pale. Head 4½; depth 4. D. 8. A. 8. Scales 5–42–4. L. 4 to 8. Ohio to Neb. and Tenn., abundant in larger streams. (*Ceratichthys lucens* Jordan.) (To David Humphreys Storer, author of Fishes of Mass.)

aa. Species of large size, little silvery, the mouth nearly terminal; D. slightly behind V. (*Nocomis* Girard.)

144. **H. kentuckiensis** (Rafinesque). HORNY HEAD. RIVER CHUB. JERKER. Robust; head large, broad above, the snout long, bluntish; mouth large, little oblique, the lower jaw shorter; eye small; suborbitals narrow; barbel evident; scales not crowded forwards, 18 before D. Bluish olive, with coppery shades; a dark bar behind opercle; fins pale orange, unspotted; young with a black spot at base C. Adult males in spring with a much swollen crest and large tubercles; a round crimson spot on each side of head. Head 4; depth 4½. D. 8. A 7. Scales 6–41–4. Teeth 4–4 or 1, 4–4, 1. L. 10. Penn. to Dakota and Ala., very abundant in the rivers, rarely in small brooks; variable. (*Ceratichthys biguttatus* Kirtland.)

61. COUESIUS Jordan. (To Dr. Elliott Coues.)

a. Scales small, about 68 in the lateral line.

145. **C. plumbeus** (Agassiz). Body rather elongate; head small; snout bluntish; mouth rather small, terminal; eye 4 in head; D. above V. Dusky, a plumbeous lateral band, fins plain. Head 5; depth 5. D. 8. A. 7. Scales 11–68–7. L. 6. Teeth usually 2, 4–4, 2. N. N. Y. (*Mather*) to L. Superior, chiefly in or near cold lakes. (Lat., leaden.)

aa. Scales larger, about 60 in the lateral line.

146. **C. dissimilis** (Girard). Body more robust, with lateral line more decurved. Mouth oblique, subterminal, resembling that of *Semotilus*. Dusky. Head 4½; depth 4½. Lat. l. 60. L. 6. Minn. to Montana.

62. PLATYGOBIO Gill. (πλατύς, broad; Lat. *gobio*, gudgeon.)

147. **P. gracilis** (Richardson). FLAT-HEADED CHUB. Body elongate; head short, small, very broad and depressed above, the interorbital area 2 in head; mouth large, oblique; eye small, 6 in head; fins large; 23 scales before D. Very pale, sides silvery,

5

young with dusky lateral shade. Head 4¼: depth 4¾. D. 8 A. 8.
Scales 6–50–5. L. 12. Missouri Basin, abundant in river channels, N. to Saskatchewan, S. to Cairo, Ill. (Lat., slender.)

63. SEMOTILUS Rafinesque. (σῆμα, banner; the remainder, according to Rafinesque, means " spotted.")

a. Scales scarcely crowded anteriorly, about 8–45–5; no black spot at base of dorsal in front.

148. **S. bullaris** (Rafinesque). FALL-FISH. CHUB. ROACH.
D. inserted midway between nostril and base of C.; barbel very
small; eye 4½ in head: 22 scales before D. Bluish above, sides
silvery; fins plain. Head 4; depth 4. D. 8. A. 8. L. 18. Quebec
to Va., abundant E., the largest of the *Cyprinidæ* E. of the Rocky
Mts. On the Pacific slope are species (*Ptychocheilus, Mylopharo-
don*, etc.) 5 to 6 feet in length. "The chub is a soft fish: it tastes
like brown-paper salted." (*Thoreau.*) (Lat., *bulla*, bubble.)

aa. Scales small, crowded anteriorly, about 10–54–7; lat. l. 52 to 65; a roundish black spot at base of D. in front.

149. **S. atromaculatus** (Mitchill). HORNED DACE. CREEK
CHUB. D. inserted midway between pupil and base C.; body
robust; head large and broad; barbel minute, not evident in the
young; mouth large, lower jaw included; eye small; 30 scales before D. Dusky, little silvery, a dark bar at shoulder; young with
dark lateral band; ♂ more or less red and with coarse tubercles
in spring. Head 3¾; depth 4. D. 7. A. 8. L. 12, or less. W.
Mass. to Dakota. Va. and La., very abundant, especially in small
clear brooks. Variable. (*Semotilus corporalis* of authors, not of
Mitchill.) (Lat. *ater*, black; *maculatus*, spotted.)

64. PHOXINUS Agassiz. DACE.

(As here understood, a very large genus, one of the largest in
Ichthyology, comprising a great number of species, mostly of
Europe, Asia, and Western North America, distinguished from
Notropis, in general by the better developed dentition; the teeth
2, 4 – 5, 2, or 2, 5 – 5, 2, and by the larger size of the body; the
scales being in general smaller than in *Notropis*. We here unite
Squalius (lat. l. complete) with *Phoxinus* (lat. l. incomplete).
When we consider European species only, the two genera appear
to be widely separated, but the intergradation is almost perfect
when American species are taken into account. (Old name from
φοξός, tapering.)

a. Lateral line complete (*Squalius* Bonaparte).
  b. Teeth without grinding surface; caudal peduncle rather slender; anal
    basis short.

c. Mouth very wide, the lower jaw much projecting, the maxillary reaching to below pupil; body elongate, compressed: D. well behind V.: scales quite small: size small. (*Clinostomus* Girard.)

    d. Scales very small, 63 to 70 in the lateral line.

150. **P. elongatus** (Kirtland). Body elongate, compressed: head long, pointed; mouth larger than in any other of our *Cyprinidæ.* Eye 4 in head; lateral line decurved. Dusky bluish, mottled with paler; a broad black lateral band, the front half of which is bright crimson in spring males. Head 4; depth 5. A. 9. Scales 10–70–5. L. 4. Penn. to Minn., chiefly northward, in clear brooks. (*Clinostomus proriger* Cope.)

    dd. Scales larger, 48 to 55 in the lateral line.

       e. Mouth very large, the gape half head, the maxillaries reaching to opposite middle of orbit.

151. **P. estor** (Jordan & Brayton). Body elliptical, compressed; head very large; eye 4 in head; lateral line decurved; 23 scales before D. Dark olive, mottled with darker; sides silvery; no broad black lateral band; males largely crimson. Head $3\frac{2}{3}$; depth $4\frac{1}{4}$. A. 8. Scales 8–50–5. L. 4. Cumberland and Tenn. Rivers (Lat., devourer.)

       ee. Mouth smaller, the maxillaries not reaching to opposite middle of eye.

       f. Body deep, the depth in adult $3\frac{2}{3}$ in length.

152. **P. vandoisulus** (Cuv. & Val.). Head large; eye $3\frac{1}{2}$ in head; bluish, some scales irregularly blackish; no black lateral band; spring males rose-red, especially anteriorly. Head $3\frac{2}{3}$. A. 8. lat. l. 53. Va. to Ga., common. (Fr. *vandoise,* dace.)

       ff. Body rather slender, the depth in adult $4\frac{1}{4}$.

153. **P. funduloides** (Girard). Head and mouth smaller than in any of the preceding species. Eye 3 in head. Dusky, a dark lateral band with a pale streak above it; males red below in spring. Head $4\frac{1}{4}$. A. 8. Scales 9–48–4. Penn. to N. C. (Lat., like Fundulus.)

    cc. Mouth moderate, terminal, oblique, the chin usually not projecting; premaxillary below level of pupil, the maxillary not reaching pupil. (*Tigoma* Girard.)

       g. Anal short, with about 8 rays.

154. **P. margaritus** (Cope). Body robust, little compressed: head blunt, thick, rounded; mouth small, the maxillary not to eye; eye rather large; lateral line decurved; dorsal behind ventrals. Dusky, sides plumbeous silvery, crimson in spring males. Head 4; depth 4. A. 8. Scales 11–58–8. L. 3. Susquehanna R. A pretty fish, similar to the typical species of *Phoxinus* in all respects, but the lateral line is complete. (Lat., pearly.)

aa. Lateral line incomplete. (*Phoxinus.*)
  g. Scales very small, 75 to 90 in the lateral line.

155. **P. neogæus** Cope. Body robust, little compressed; head
very large, broad, with blunt snout; mouth moderate, oblique;
the chin projecting, the maxillary beyond front of orbit. Eye 3½
in head; dorsal well behind ventrals. Blackish, sides plumbeous
with a dusky lateral band ; lower parts crimson in spring males.
Head 3¾ ; depth 4⅓. A. 8. Scales 18–80–11. L. 3. Miss. Valley,
rare ; the few specimens known, from Mich., Wis., and Ark. (*νέος*,
new ; *γέα*, world : this being a near relative of the " Minnow " of
Europe, *P. phoxinus* L.).

  gg. Scales moderate, 40 to 45 in the lateral line.
    h. Body not very slender, the depth about 4 in length.

156. **P. flammeus** Jordan & Gilbert. Head rather short, the
snout bluntish ; mouth small, oblique ; the jaws equal, the maxillary to front of eye ; lateral line with pores on 14 scales ; color of
preceding ; the males largely scarlet ; dark spot at base of C.
Head 4. A. 8. Scales 7–43–5. L. 2½. Tenn. R. (Lat., flaming.)

    hh. Body slender, the depth 5¼ in length.

157. **P. milnerianus** Cope. Mouth larger, the maxillary about
to pupil. Color of preceding ; a dark spot at base of C. Head 4 ;
eye 3½ in head. A. 8. L. 2½. Upper Missouri R. (To James
W. Milner, of the U. S. Fish Comm.)

65. OPSOPŒODUS Hay. (*ὀψοποιέω*, to feed daintily; *ὀδούς*,
tooth.)

158. **O. emiliæ** Hay. Body elongate, compressed ; head short,
the snout blunt and rounded ; mouth very small, terminal, oblique, smaller than in any of our *Cyprinidæ* ; jaws equal ; eye
very large, 3 in head; D. behind V.; P. very small ; breast naked ;
16 scales before D. ; lateral line usually incomplete. Yellowish,
sides silvery ; a dark lateral stripe ; D. with a black blotch on its
last rays. Head 4⅓ ; depth 4¾. D. 9. A. 8. Scales 5–40–3. L.
2¼. S. Ind. to Ark. and Miss. ; not common. (*Trycherodon megalops* Forbes.) (To Mrs. Emily Hay.)

66. NOTEMIGONUS Rafinesque.
(*νῶτος*, back ; *ἡμι*, half ; *γῶνος*, angle.)

159. **N. chrysoleucus** (Mitchill). GOLDEN SHINER. BREAM.
Body more or less elongate, much compressed ; head short, low,
compressed : mouth small, oblique, the maxillary not to eye; eye
moderate, or large, 3 to 4 in head ; lateral line much decurved.
Greenish above, sides silvery with golden reflections; fins yellowish. Sexes similar. Head 4½ ; depth about 3. D. 8. A. 13

(12 to 14). Scales 10–51–3. L. 12. Maine to Dakota and La., everywhere abundant in sluggish or weedy waters.
S. E. (N. C. to Ala.) occurs var. bosci (Cuv. & Val.) with A. longer, about 16 ; the scales larger, 8–43–2, and the lower fins scarlet in males. The two forms intergrade and both are very variable. (*Cyprinus americanus* L., 1766, not of 1758) (χρυσός, gold ; λευκός, white.)

ORDER XIII. **ISOSPONDYLI.** (THE SALMON, HERRING, ETC.)

This order contains a great variety of soft-rayed fishes, which agree in lacking the modified vertebræ and the falciform pharyngeals of the preceding order, and in having a more complex structure of the shoulder-girdle than the *Haplomi*. There are 20 or 25 families, most of them marine ; some in the deep seas. (ἴσος, equal ; σπόνδυλος, vertebra).

FAMILY XXVII. **HIODONTIDÆ.** (THE MOON-EYES.)

Body oblong, much compressed, covered with large, silvery cycloid scales ; head naked ; mouth terminal, oblique ; margin of upper jaw formed by intermaxillaries mesially and by maxillaries laterally ; maxillaries entire ; no barbels ; no adipose fin ; lateral line distinct ; abdomen compressed, not serrated ; moderate sized teeth on jaws, vomer, sphenoid, hyoid, pterygoid, and palatine bones ; tongue with sharp canines ; gill rakers few, short, thick ; eye very large ; gill openings wide ; one pyloric appendage ; airbladder simple ; no oviducts. One genus, with three species, inhabiting our Western Streams and the Great Lakes, handsome fishes, of little value as food.

67. HIODON Le Sueur. (ὑοειδής, hyoid (bone) ; ὀδών, tooth.)

a. Belly strongly carinate, both before and behind V. ; D. very small, of nine developed rays.

160. **H. alosoides** (Rafinesque). Body deep, closely compressed ; snout blunter than in other species ; eye moderate, $3\frac{1}{2}$ in head ; P. short, nearly as long as head, about reaching V. ; longest dorsal ray about half longer than base of fin ; sides with golden lustre. Head $4\frac{1}{2}$; depth $3\frac{1}{2}$. D. 9. A. 32. Scales 6–56–9, L. 12. Ohio Valley to Saskatchewan R., common N. (Lat., *alosa*, shad ; εἶδος, form.)

aa. Belly scarcely carinate before V. ; dorsal rays (developed) 12.
b. Belly carinate between V. and A.

161. **H. tergisus** Le Sueur. MOON-EYE. SILVER BASS. TOOTHED HERRING. Snout rounded, shorter than the large eye, which is $3\frac{1}{2}$ in head. Olivaceous, sides brilliantly silvery. Head

4½: depth 3. D. 12. A. 28. Scales 5-56-7. L. 15. Great Lakes and Mississippi Valley, abundant; one of our most beautiful fresh-water fishes. (Lat., polished.)

bb. Belly nowhere carinate.

162. **H. selenops** Jordan & Bean. Body elongate, less compressed; eye very large, 2½ in head. Head 4½; depth 4. D. 12 A. 27. Cumberland R. to Ala. (σελήνη, moon ; ὤψ, eye.)

FAMILY XXVIII. **ALBULIDÆ.** (THE LADY-FISHES.)

Body elongate, little compressed, covered with small, silvery scales : head naked; snout conic, pig-like, overhanging the small, inferior mouth; maxillary short, with supplemental bone; preorbital very broad ; sides of upper jaw formed by maxillaries; eye large, with an adipose eyelid; gill rakers tubercle-like; preopercle with membranous edge; villiform teeth on jaws, vomer, and palatines ; coarse blunt teeth on tongue and roof of mouth ; lateral line present ; belly flattish, not carinate; D. moderate, inserted before V.; A. very small; no adipose fin; C. forked. One species, in most warm seas.

68. **ALBULA** (Gronow) Bloch & Schneider. (Lat., white.)

163. **A. vulpes** (L.). LADY-FISH. BONE-FISH. MACABI. A band of elongate scales along middle of back; brilliantly silvery. Head 3¾; depth 4. D. 15. A. 8. Scales 9-71-7. L. 30. Warm seas, N. to Cape Cod. (Lat., fox.)

FAMILY XXIX. **ELOPIDÆ.** (THE TARPUMS.)

Body elongate, more or less compressed, covered with silvery, cycloid scales; mouth large, terminal, the lower jaw prominent; maxillary long, of three pieces, forming side of upper jaw; an elongate bony plate between branches of lower jaw (as in *Amia*); bones of mouth almost all with villiform teeth : eye large, with an adipose eyelid ; gill rakers long and slender; belly not compressed, covered with ordinary scales ; D. over or behind V.: C. forked; no adipose fin. Genera 2, species 4 or 5, in warm seas.

a. Body elongate, with small scales ; A. smaller than D.; pseudobranchiæ present. (*Elopinæ.*) . . . . ELOPS, 69.
aa. Body oblong, compressed, with very large scales; no pseudobranchiæ; last ray of D. much produced. (*Megalopinæ.*). . . MEGALOPS, 70.

69. **ELOPS** Linnæus. (ἔλοψ, name of some sea-fish.)

164. **E. saurus** L. TENPOUNDER. Silvery, darker above; gular plate about three times as long as broad; eye large ; tail very long; C. deeply forked. Head 4¼; depth 5½. D. 20. A. 13. Scales 12-120-13. L. 36. Warm seas, N. to Cape Cod; remarkable for

the development of membranous sheaths at bases of fins and elsewhere. (σαῦρος, name of some sea-fish.)

70. MEGALOPS Lacépède. (μέγαλοψ, large eye.)

165. **M. atlanticus** Cuv. & Val. TARPUM. TARPON. GRANDE
ÉCAILLE. SILVER-FISH. SABALO. SAVANILLA. Brilliantly
silvery. Mouth large, its cleft oblique, extending beyond the very
large eye; lower jaw very prominent: D. inserted behind V.; dorsal filament as long as head. Head 4; depth 4. D. 12. A. 20.
Lat. l. 42. L. 6 feet. West Indies and Gulf Coast, occasional N.
to Cape Cod; remarkable for its enormous scales, sometimes three
inches across.

FAMILY XXX. **CLUPEIDÆ.** (THE HERRINGS.)

Body oblong, covered with cycloid scales; head naked; side of
upper jaw formed by maxillaries; maxillaries composed of 2 or 3
pieces; teeth feeble or wanting; dorsal moderate; anal often very
long; caudal forked; no lateral line; no gular plate; branchiostegals 6 to 15; the tips of the larger ones abruptly truncate; pseudobranchiæ present; gill rakers long and slender; gill openings
wide.

Genera about 17; species 120; found in most seas, many species entering fresh water to spawn, a few remaining permanently.
Many are highly valued as food fishes. It is probable that the
Clupeidæ are more numerous in individuals than any other family
of fishes.

a. Maxillary large, of about three pieces; mouth terminal, the jaw scarcely
shorter; carnivorous fishes, with simple not muscular stomach.
  b. Belly rounded, with ordinary scales; the body subterete, supplemental
  bones of maxillary very narrow. (*Dussumieriinæ.*)
    c. V. small, behind D.; teeth small, persistent. . . . ETRUMEUS, 71.
  bb. Belly compressed to an edge and more or less serrated; body compressed; bones of maxillary broad. (*Clupeinæ.*)
    d. Scales with their posterior edges entire and rounded.
      e. Last ray of D. not produced; scales loosely attached; vertebræ 47
      to 56. . . . . CLUPEA, 72.
      ee. Last ray of D. produced in a long filament; scales rather firm;
      vertebræ 48. . . . . OPISTHONEMA, 73.
    dd. Scales with their posterior margins vertical, and pectinate or fluted;
    head very large; D. small, posterior. . . . . BREVOORTIA, 74.
aa. Maxillary short and narrow, with a single supplemental bone; mouth
small, inferior, the lower jaw much shorter; mud-eating fishes, with the
stomach muscular, like the gizzard of a fowl. (*Dorosominæ.*)
  f. Last ray of D. produced in a long filament. . . . . DOROSOMA, 75.

71. ETRUMEUS Bleeker. (From the Japanese name.)

166. **E. sadina** (Mitchill). ROUND HERRING. Mouth small,
reaching front of orbit; eye large; fins all very small. Bluish,

sides silvery. Head 4 ; depth 6. D. 18 ; A. 13. L. 5. N. Y. to
Fla., scarce. (*Alosa tercs* Dekay.) (Corruption of sardine.)

72. CLUPEA (Artedi) Linnæus. (Lat., herring.)

a. Vomer with an ovate patch of minute teeth; serratures on belly very
weak.

167. **C. harengus** L. COMMON HERRING. Body elongate,
the scales deciduous; cheeks longer than high; upper jaw scarcely
emarginate; gill rakers X + 40; D. inserted before V.; lower fins
small ; peritoneum dusky. Blue, silvery on sides. Head 4½;
depth 4⅓. D. 18. A. 17. Lat. l. 57. Scutes, 28 + 13. L. 12.
N. Atlantic, everywhere, S. to Cape Cod; spawns in the sea. (*Eu.*)
(Low Latin, herring.)

aa. Vomer without teeth ; ventral serratures very strong; upper jaw emarginate.
  b. Cheeks notably longer than deep, the preopercle produced forward
    below ; body not very deep; depth 3¼ or more in length. (*Pomolobus*
    Rafinesque.)
    c. Teeth in jaws all disappearing with age, a small patch sometimes re-
      maining on tongue.
    d. Peritoneum pale.

168. **C. mediocris** Mitchill. TAILOR HERRING. FALL HER-
RING. MATTOWACCA. Head rather long, the profile straight and
not very steep ; form more elliptical than in the next and less
heavy forwards; opercles less emarginate below ; fins low. Bluish
above ; sides with faint longitudinal streaks. Head 4 ; depth 3⅔.
D. 15. A. 21. Lat. l. 50. Ventral scutes, 20 + 16. Cape Cod to
Fla., chiefly S.

169. **C. pseudoharengus** Wilson. ALEWIFE. GASPEREAU. *Can*
BRANCH HERRING. WALL-EYED HERRING. Body deep, heavy
forward ; head short, nearly as deep as long ; eye large, 3½ in head;
gill rakers long, about X + 35 ; first ray of D. about equal to base
of fin; lower lobe of C. longer; fins rather high. Bluish, sides
silvery, with faint dark streaks along rows of scales; a round dark
spot at shoulder. Head 4⅔; depth 3¼. D. 16. A. 19. Lat. l. 50.
Scutes 21 + 14. L. 15. Newfoundland to S. C., abundant, entering
streams to spawn ; landlocked in lakes of W. N. Y. (*Clupea vernalis* Mitchill, 1815, but according to Dr. Gill, the paper of Alex-
ander Wilson was published before 1814. See McDonald, Nat.
Hist. Aquat. Anim. 580, 594.) (ψεῦδος, false; herring.)

    dd. Peritoneum black.

170. **C. æstivalis** Mitchill. GLUT-HERRING. BLUE-BACK.
BLACK-BELLY. SUMMER HERRING. Very close to the preceding, the body more elongate, the fins lower, and the eyes smaller,
the back darker. First ray of dorsal not equal to base of fin.

Head 5; depth 3½. With *C. pseudoharengus*, but running later, less abundant and much less valuable as a food-fish. (Lat., belonging to summer.)

cc. Teeth on jaws; usually persistent at tip of both jaws; peritoneum pale.

171. **C. chrysochloris** Rafinesque. Skip-jack. Body elliptical; head slender, rather pointed; lower jaw strongly projecting; maxillary reaching posterior part of eye; eye large, 4½ in head; fins moderate; gill rakers not numerous, rather stout, about X + 23; opercles striate. Bright blue, sides with golden reflections. Head 3⅞; depth 3⅘. D. 16. A. 18. Lat. l. 52. Scutes 20 + 13. L. 18. Miss. Valley, etc., abundant and resident in larger streams, introduced into Great Lakes. Also in Gulf of Mexico. A handsome but lean and poor fish in the rivers, becoming excessively fat in salt water. (χρυσός, golden; χλωρός, green.)

bb. Cheeks little if at all deeper than long, the preopercle scarcely prolonged forward below; body deep; depth of body 2¾ to 3; teeth few or none. (*Alosa* Cuvier.)

172. **C. sapidissima** Wilson. Shad. Body rather deep; mouth large, the jaws subequal; gill rakers very long and slender, X + 40 to 60; fins low. D. nearer snout than C. Bluish, sides more or less silvery; usually a dark blotch behind opercle, and often several in a row behind this; peritoneum white. Head 4½; depth about 3. D. 15. A. 21. Lat. l. 60. Scutes 21 + 16. L. 30. Atlantic coast from the Miramachi to the Alabama, ascending rivers to spawn; one of the best of food-fish. Introduced in Ohio R. etc. (Superlative of Lat., *sapidus*, good to eat.)

73. OPISTHONEMA Gill. (ὄπισθε, behind; νῆμα, thread.)

173. **O. oglinum** (Le Sueur). Thread Herring. Body compressed; belly strongly serrate; jaws toothless; dorsal filament about as long as head. Bluish, silvery below; a bluish shoulder spot; dark streaks along scales of back. Head 4; depth 3½ D. 19. A. 24. Lat. l. 50. Scutes 17+11. L. 12. West Indies, N. to Cape Cod.

74. BREVOORTIA Gill. (To James Carson Brevoort, late of Brooklyn, N. Y.)

174. **B. tyrannus** (Latrobe). Menhaden. Mossbunker. Bug-fish. Fat-back. Body compressed, deep, heavy anteriorly; no teeth; gill rakers very long and slender; scales very closely imbricated, irregularly arranged; fins small. Bluish, sides silvery or brassy; fins yellowish; a dark scapular blotch, behind which are usually smaller spots. Head 3½; depth 3. D. 19. A. 19. Lat. l. 60 to 80. Scutes 20+12. L. 20. Cape Cod to Florida; very abun-

dant, spawning in sea ; used for oil and manure. (A parasitic
crustacean. *Oniscus prægustator* Latrobe, is found in the mouth of
this fish. The names of both species refer to this fact; the ancient
Roman Emperors (*tyranni*) having had their tasters (*prægusta-
tores*) to try their food before them, to prevent poisoning.)

75. DOROSOMA Rafinesque. (δορός, lance ; σῶμα, body.)

175. **D. cepedianum** (Le Sueur). GIZZARD SHAD. HICKORY
SHAD. MUD SHAD. WHITE-EYED SHAD. HAIRY-BACK. Body
deep, compressed: the scales thin, deciduous; head small; snout
short, blunt ; mandible enlarged at base : gill rakers very slender,
not very long ; an adipose eyelid; D. about median, its filament
about as long as head ; C. widely forked, its lower lobes longer ;
belly sharply serrate. Bluish, sides silvery; young with a round
dark shoulder spot. Head 4⅕; depth 2⅓ (2 to 3); eye 4½ in head.
D. 12. A. 31. Lat. l. 56. Scutes 17+12. L. 15. Cape Cod to
Mexico; abundant S. entering all rivers ; permanently resident
throughout the Miss. Valley. A handsome, mud-loving fish, nearly
worthless as food. (To Bernard Germain Etienne, Comte de La
Cépède, afterwards "Citoyen Lacépède.")

FAMILY XXXI. **STOLEPHORIDÆ.** (THE ANCHOVIES.)

Body elongate, compressed, with thin, deciduous scales ; mouth
very large; the pointed, pig-like snout, usually extending beyond
it ; maxillary very long and slender, of about 3 pieces, extending
backward far beyond the eye; premaxillaries small; teeth usually
very small; eye large, well forward; gill rakers long and slender.
B. 7 to 14; no lateral line ; belly rounded, or weakly serrate ; no
adipose fin ; C. forked. Small fishes swimming in large schools,
abundant in all warm seas. Genera 9, species about 65, most of
them belonging to *Stolephorus.*

a. Gill membranes scarcely connected; gill openings very wide; no pectoral
filaments; A. moderate, beginning behind D.; lower jaw included; maxillary not extending beyond gill openings; teeth very small or wanting.
STOLEPHORUS, 76.

76. STOLEPHORUS Lacépède. (*Engraulis* Cuvier.) (στολή,
a stole; φορός, bearing. in allusion to the silvery band.)

a. Body compressed, moderately elongate, the depth more than one-fifth the
length: insertion of D. nearer C. than tip of snout.
b. Anal long, its rays about 26, its base 3⅔ in body.

176. **S. mitchilli** (Cuv. & Val.). Snout rather blunt, little projecting: body much compressed; both jaws with teeth ; eye very
large. Pale, a narrow, diffuse, silvery lateral band, little broader
than pupil. Head 3½; depth 4. D. 14. A. 26. Lat. l. 37. L. 2½.

Cape Cod to Texas, common S. (To Prof. Samuel Latham Mitch-ill, of New York, an early ichthyologist.)

bb. Anal fin moderate, its rays about 20, its base 4? in body.

177. **S. browni** (Gmelin). Snout pointed, considerably pro-jecting; belly somewhat serrated; eye 3½ in head; teeth in both jaws. Translucent, silvery band : sharply defined, about as broad as eye. Head 3¾; depth 4¾. D. 15. A. 20. Lat. l. 40. L. 6. Cape Cod to Brazil, exceedingly abundant S. (To Mr. P. Browne, author of Nat. Hist. of Jamaica, in 1756.)

aa. Body elongate, less compressed, the depth less than one-fifth the length : insertion of D. midway between snout and C.

178. **S. argyrophanus** (Cuv. & Val.). Tail long and slender; snout pointed, projecting; belly not serrated; eye 4 in head; teeth in jaws present, feeble ; anal short, its base 5½ in body ; silvery lateral band broad, diffuse, broader than eye. Head 3¾ ; depth 6. D. 14. A. 19. L. 4. Wood's Holl, Mass., and S. (S. *eurystole* Swain & Meek.) (ἄργυρος, silver ; φαίνω, to show.)

FAMILY XXXII. **SYNODONTIDÆ.** (THE LIZARD-FISHES.)

Body elongate, subterete, covered with cycloid scales; head de-pressed; mouth very wide, its margin formed by the slender pre-maxillaries; the maxillaries closely joined to them ; teeth usually strong, cardiform, the large ones often depressible ; no barbels ; sides of head usually scaly; adipose fin usually present; D. short, median; C. forked. Air-bladder small or wanting; skeleton weakly ossified ; no phosphorescent spots. Ovaries with an oviduct. Genera 6 or 8; species about 25, mostly of deep waters in warm regions. (More or less related to this family are several others : *Stomiatidæ, Scopelidæ, Chauliodontidæ*, etc., found in the deep waters off our coasts. Most of these deep-sea forms are provided with phosphorescent spots. A very full account of them has been lately published by Dr. Günther, — Deep-sea Fishes of the Chal-lenger Exped.)

a. Teeth not barbed; maxillary not dilated behind; teeth on palatines in a single band on each side; shore-fishes. . . . . . . SYNODUS, 77.

77. **SYNODUS** (Gronow) Bloch & Schneider.
(συνόδους, ancient name of some fish).

179. **S. fœtens** (L.). LIZARD-FISH. SNAKE-FISH. Dorsal slightly higher than long; snout longer than broad; lower jaw in-cluded; scales of cheeks in 7 rows ; ventrals 2½ in head. Olivaceous, back mottled ; top of head vermiculated ; V. and mouth yellow. Head 4¼ ; depth 6 or 7. D. 11. A. 11. Scales 4–64–6. L. 12. Cape Cod to Fla., on sandy coasts. (Lat., ill-scented.)

FAMILY XXXIII. **ARGENTINIDÆ.** (THE SMELTS.)

The smelts may be looked upon as reduced *Salmonidæ*, the only important difference being in the form of the alimentary canal. The stomach is a blind sac, the œsophagus and the pylorus opening close together, and the pyloric cœca are very few or wanting, Genera 7 or 8, species about 20, chiefly small fishes of the Northern Seas, some of them descending to considerable depths. All are silvery and none have phosphorescent spots.

a. Mouth large; V. before middle of D.
   b. Scales very small, some of them modified in males; teeth feeble; P. large, of 16 to 20 rays, adipose fin with long base. . MALLOTUS, 78.
   bb. Scales moderate, all alike; teeth stronger, those on tongue enlarged ; P. moderate, of about 12 rays; adipose fin short. . . OSMERUS, 79.

78. MALLOTUS Cuvier. (μαλλωτός, villous).

◀80. M. villosus (Müller). CAPELIN. ICE-FISH. Dusky, sides grayish. Old males with scales above lateral line and on side of belly, elongate, closely imbricate, forming villous bands. Head 4¼; depth 6. D. 12. A. 18. Lat. l. 150. L. 12. Arctic, S. to Maine.

79. OSMERUS (Artedi) Linnæus. (ὀσμηρός, odorous.)

a. Vomer with 2 to 4 fang-like teeth; lat. l. about 68.

181. **O. mordax** (Mitchill). SMELT. FROST-FISH. Greenish, sides with a silvery band ; back with dark points; teeth strong, gill rakers shortish, ⅔ eye. Head 4; depth 6¼. D. 10. A. 15. Lat. l. 68. L. 12. Nova Scotia to Va., entering rivers, sometimes land-locked. (Lat., biting.)

FAMILY XXXIV. **SALMONIDÆ.**[1] (THE SALMON.)

Body oblong, covered with cycloid scales; head naked; mouth terminal or subinferior, of varying size ; teeth various ; maxillary with supplemental bone, forming side of upper jaw ; pseudobranchiæ present ; no barbels; D. median ; an adipose fin ; C. forked ; V. median; lateral line present ; belly not compressed ; vertebræ about 60. Stomach siphonal, with 15 to 200 pyloric cœca. Eggs large; no oviduct. Genera 8 ; species about 80 ; peculiar to the northern regions, most of them in fresh waters, the larger species ascending rivers to spawn. In beauty, activity, gaminess, quality as food, and even in size of individuals, different members of this group stand easily with the first among fishes.

a. Jaws toothless or nearly so; scales large; A. rather elongate (10 to 12 rays); maxillary short and broad. . . . . COREGONUS, 80.

---

[1] For a detailed account of the fishes of this family see Jordan, Science Sketches, p. 35.

aa. Jaws with distinct teeth ; scales smaller.

   b. Dorsal very long and high, of about 20 rays ; scales medium ; tongue toothless. . . . . THYMALLUS, 81.

   bb. Dorsal moderate, its rays 9 to 15; tongue with teeth; teeth strong; A. short, of 9 to 11 developed rays.

     c. Vomer flat, its toothed surface plane, the teeth on its shaft in one or two rows, sometimes deciduous ; species black-spotted, with conspicuous scales. . . . . SALMO, 82.

     cc. Vomer boat-shaped, the shaft strongly depressed, without teeth : scales very small, more or less imbedded; species with red or gray spots. . . . . SALVELINUS, 83.

80. COREGONUS (Artedi) Linnæus. WHITE-FISHES.

(This genus contains about forty species, lake-fishes of northern regions, usually spawning in shallow waters or in brooks in late fall or winter. All are excellent food-fishes, and all are very variable.) (The old name, of uncertain origin.)

a. Lower jaw included ; premaxillaries broad, placed more or less vertically, or the lower edge turned inward ; the cleft of the mouth less than one-third the head.

   b. Gill rakers short, thickish, about $X + 16$; preorbital broad, wider than pupil; maxillary short, broad, not reaching to eye; the supplemental bone narrowly elliptical; supraorbital broad ; mouth very small. (*Prosopium* Milner.)

   182. **C. quadrilateralis** Richardson. ROUND-FISH. PILOT-FISH. SHAD-WAITER. MENOMONEE WHITE-FISH. Body subterete, the back broad; maxillary $5\frac{1}{2}$ in head ; head long, the snout compressed and bluntly pointed ; preorbital wider than pupil. Dark bluish, sides paler. Head 5 ; depth $4\frac{3}{4}$. D. 11. A. 10. Scales 9–85–8. N. II. to L. Superior, Alaska, and N. Throughout the Rocky Mountains is found a closely related species, *C. williamsoni* Girard, with shorter snout and longer maxillary. (Lat., 4-sided).

   bb. Gill rakers numerous, long and slender, $X + 20$ to 25; preorbital long and narrow; maxillary rather long, the supplemental bone ovate. (*Coregonus.*)

     c. Tongue toothless; body robust, elevated at the shoulders in the adult; the head very small, especially in old examples.

   183. **C. clupeiformis** (Mitchill). COMMON WHITE-FISH. Snout bluntish, obliquely truncate; preorbital not half pupil; maxillary past front of orbit, 4 in head; eye large; gill rakers $\frac{2}{3}$ eye. Color pale, scarcely silvery. Head 5 to 6; depth $2\frac{1}{2}$ to 4. D. 11. A. 11. Scales 8–74–9. L. 30. Great Lakes and N.; by far the most valuable of the American white fishes. Very variable; feeds on minute organisms. (*Clupea*, herring ; *forma*, shape.)

     cc. Tongue with about 3 series of small teeth; body rather elongate; the back scarcely elevated.

184. **C. labradoricus** Richardson. SAULT WHITE-FISH.
MUSQUAW RIVER WHITE-FISH. "WHITING." Head com-
pressed, rather long; mouth rather small, the jaws equal; maxillary
to front of pupil; eye 4½ in head : supraorbital narrow ; D. high
in front, its last rays short. Bluish, sides little silvery; fins dusky.
Head 4⅔ ; depth 4¼. D. 11.   A. 11.   Scales 9-80-8. L. 20.
White Mts. to Labrador and L. Superior; abundant N.

> aa. Lower jaw projecting; premaxillaries narrow, not vertically placed; pre-
> orbital elongate; gill rakers very long and slender, about X + 30; the
> cleft of the mouth 2½ to 3½ in the head.
>> d. Body elongate, herring-shaped; scales small, uniform, the free edges
>> convex. (*Argyrosomus* Agassiz.)
>>> e. Lower fins pale, or tipped with dusky.
>>> f. Scales brilliantly silvery, without dark specks.

185. **C. hoyi** (Gill). LAKE MOON-EYE. CISCO OF LAKE
MICHIGAN. Head rather long, lower jaw barely included ; maxillary 3 in head to middle of pupil; eye very large, 3½ in head; lower
jaw little projecting; gill rakers nearly as long as eye ; fins low ;
free edge of D. very oblique. Color bluish, sides brilliantly silvery,
as in *Hiodon* and *Albula*. Head 4¼ ; depth 4¾. D. 10.   A. 10.
Scales 7-75-7.   L. 12.   Smallest and prettiest of our white-fishes,
from Skaneateles L., N. Y., to L. Mich., in deep water.   (To Dr.
Philo R. Hoy, of Racine, Wis.)

> ff. Scales more or less punctulate with darker.

186. **C. artedi** Le Sueur. LAKE HERRING. CISCO. MICHI-
GAN HERRING. Maxillary 3½ in head, reaching middle of pupil;
eye 4 to 5 in head. Bluish or greenish, sides silvery ; scales and
fins with dark specks. Head 4½; depth 4¼. D. 10.   A. 12.   Scales
8-75 to 90-7.   L. 15.   Great Lakes to Labrador ; very abundant,
usually in shoal waters, also land-locked in lakes of N. Ind. and
Wis. (var. **sisco** Jordan), where it lives in deep water, spawning near shore in December. (To Peter Artedi, the "father of
ichthyology.")

> ee. Lower fins blue-black.

187. **C. nigripinnis** (Gill). BLUE-FIN. Body more robust than
in *C. artedi;* mouth large; eye 4 in head; teeth present, minute.
Dark bluish, sides silvery, punctulate. Head 4¼; depth 3¾. D. 10.
A. 12.   Scales 9-88-7.   L. 20.   L. Mich., in deep water. (Lat.,
niger, black; *pinna*, fin.)

> dd. Body short, deep, compressed; the curve of back similar to that
> of belly; scales large, larger forwards, closely imbricated ; the
> free margin little convex. (*Allosomus* Jordan.)

188. **C. tullibee** Richardson. TULLIBEE. "MONGREL WHITE-
FISH." Head much as in *C. nigripinnis;* maxillary as long as
eye; jaws equal when closed; eye as long as snout, 4½.   Bluish,

sides white; centre of each scale silvery; outside dotted, the sides thus with faint pale stripes. Head 4; depth 3. D. 11. A. 11. Scales 8-74-7. L. 18. Great Lakes, N., scarce. (An Indian name.)

81. THYMALLUS Cuvier. (θύμαλλος, ancient name of the Grayling.)

189. **T. signifer** Richardson. AMERICAN GRAYLING. Body compressed, rather elongate; head short, subconic; mouth moderate, the maxillary to middle of eye; jaws subequal; teeth on tongue disappearing with age; eye large, 3 in head; a bare space on breast; gill rakers slender, X + 11; D. very high, especially in males, highest in specimens from far North. Purplish gray, with small black spots; ventrals dusky, with pale lines; dorsal highly variegated, with crimson and dusky streaks and greenish and rose-colored spots. Head 4¾; depth 4⅔. D. 20. A. 10. Scales 8-90 to 100-9. L. 18. Arctic America, in clear, cold streams. The Michigan Grayling is var. **ontariensis** Cuv. & Val. (= *T. tricolor*, Cope), with rather longer head and lower dorsal, its height rarely greater than depth of body; in a few streams in N. Mich. and Montana; a remnant perhaps of the glacial fauna. (Lat., bearing a banner.)

82. SALMO (Artedi) Linnæus. (Lat., salmon, originally from *salio*, to leap.)

(Besides the native Salmon, the following species have been introduced into waters within our limits: *Salmo fario* L., the " Brook-trout," or " Brown Trout," of Europe, with the vomerine teeth well developed and the scales rather large, about 120. *Salmo gairdneri* Richardson, the " Rainbow Trout " of California, similar to the last, and with about 130 scales. *Salmo mykiss* Walbaum, the Red-throated or Rocky Mountain Trout, with larger mouth and the scales about 175. Besides this, the great Salmon of the Columbia, the Quinnat or King Salmon, *Oncorhynchus tschawytscha* (Walbaum), has been introduced. This, the most valuable of all *Salmonidæ*, may be known by the presence of 16 anal rays, and by the black spots on back and upper fins. Its scales are about 145.)

a. Marine Salmon, anadromous, with the vomerine teeth little developed, those on the shaft of the bone few and deciduous; scales large (lat. l. 120); C. deeply lunate, truncate in old age; no hyoid teeth; sexual differences strong; breeding males with the lower jaw hooked upwards, the upper emarginate or perforate, to receive its tip (*Salmo*).

190. **S. salar** Linnæus. COMMON SALMON. Mouth moderate, maxillary reaching past eye, 2½ to 3 in head; preopercle with a distinct lower limb. Brownish above, the sides silvery; many black spots on head, body and fins, these sometimes X-shaped; sides with

red patches in males; young (parr; smolt) with dark cross-bars
and red spots. Head 4; depth 4. B. 11. D. 11. A. 9. Scales
23-120-21 ; vert. 60 ; pyl. cæca 65; usual weight 15 pounds, but
often much larger. N. Atlantic, S. to N. Y. and France, ascending all suitable rivers; often (var. **sebago** Girard), land-locked in
lakes. One of the best known and most valued of food-fish. Variable. (An old name, from *salio*, to leap.) (*Eu.*)

83. SALVELINUS (Nilsson) Richardson. CHARRS. (An old name, allied to the German, *Sälbling*.)

(The species of this group are in general smaller, finer, hand-
somer, and more wary than the Salmon, and they inhabit in general
colder waters. Besides the native species, attempts have been
made to introduce the following : *Salvelinus alpinus* L., the European charr, Sälbling, or Ombre Chevalier, a species very close to
S. oquassa and *S. malma* (Walbaum), the "Dolly Varden Trout,"
or " Bull Trout" of the Rocky Mountain slope; very close to *S.
fontinalis*, the back as well as the sides, with red spots).

a. Hyoid bone (base of tongue) with a band of strong teeth (besides the
usual teeth around edge of tongue); head of the vomer with a raised
crest, which projects backward, free from and parallel with the shaft;
this crest with teeth; lake trout, very large, spotted with gray. (*Cristivomer* Gill & Jordan.)

191. **S. namaycush** (Walbaum). GREAT LAKE TROUT.
MACKINAW TROUT. SALMON TROUT (of the Lakes, not of
England, nor of Oregon, nor of the Gulf of St. Lawrence).
LONGE. TOGUE. Head very long; mouth very large, the maxil-
lary reaching much beyond eye, 2 in head ; teeth very strong ;
C. well forked. Dark gray, varying in shade; everywhere with
round pale spots; head above, and D. and C. reticulate with
darker; eye large. Head $4\frac{1}{4}$; depth 4. B. 12. D. 11. A. 11.
Lat. l. 185 to 205. L. 3 feet or more. Great Lake region, and
lakes from New Brunswick to Montana, British Columbia and
Alaska, abundant, variable. A food-fish of high value. In Lake
Superior is found var. **siskawitz** Agassiz, the Siscowet, similar, but
less elongate and inordinately fat. (Indian name.)

aa. Hyoid bone with a very few feeble teeth or with none; vomer with teeth
on its head only and without posterior crest; red-spotted species. (*Salvelinus.*)

b. Hyoid teeth none; head large, 4 to $4\frac{1}{2}$ in length; red spots of body on
sides only.

192. **S. fontinalis** (Mitchill). BROOK TROUT.[1] SPECKLED
TROUT. Head large, the snout bluntish; mouth large, the maxil-

---

[1] "This is the last generation of trout fishers. The children will not be able to find
any. Already there are well trodden paths by every stream in Maine, in New York,
and in Michigan. I know of but one river in North America by the side of which you

lary reaching beyond eye; eye large; C. lunate, forked in young.
Dusky greenish, sides with red spots mostly smaller than pupil;
back mostly unspotted, barred or mottled with dark; D. and C.
mottled or barred; lower fins dusky, with an orange band followed
by a darker one; belly mostly red in males. Very variable. Sea-
run individuals (var. **immaculatus** H. R. Storer) are silver-gray,
nearly plain, and they reach a large size. Specimens from Dublin
Pond, N. H. (var. **agassizii** Garman) are likewise pale, looking like
Lake Trout. Head 4½; depth 4½. D. 10. A. 9. Scales 37–230–30.
Gill rakers 6 + 11. L. 5 to 20. Greatest weight about 11 pounds.
Our finest game fish, abounding in clear cold streams from Maine
to Dakota and N. to Arctic Circle; S. in Mts. to Chattahoochee R.
(Lat., living in fountains.)

    bb. Hyoid teeth present, feeble, often lost; head smaller (about 5 in length);
        mouth small, the maxillary scarcely reaching past middle of eye.
    c. Gill rakers curled at the ends.

    193. S. aureolus Bean. SUNAPEE LAKE TROUT. Maxillary
reaching middle of eye, 2⅔ in head; eye a little longer than snout,
4⅔ in head; P. largest in ♂. Brownish, sides silver gray, with small
orange spots above and below lateral line; C. grayish; belly orange;
A. orange, edged before with white; V. orange, with white band
on outer rays; no mottlings anywhere. Head 4⅕; depth 4⅕. D. 9.
A. 8. Scales 35–210–40. L. 12, or more. Sunapee Lake, N. H.,
very close to *S. oquassa*, but reaching a larger size. (Lat., gilded.)

    cc. Gill rakers straight.

    194. S. oquassa (Girard). BLUE-BACK TROUT. RANGELEY
LAKE TROUT. Body elongate, compressed; head small, flattish
above; eye 3½ in head; P. and V. not elongate; C. deeply lunate;
opercles without striæ. Dark blue, the red spots smaller than
pupil, on sides only; traces of dark bars on sides; lower fins variegated as in other charrs. Head 5; depth 5. D. 10. A. 9. Lat.
l. 230. Gill rakers 6 + 11. L. 12. Smallest and prettiest of our
Salmonidæ, and most like the European *Salvelinus alpinus*, found
only in the Rangeley Lakes in S. W. Maine, and (*S. naresi* Günther), in some lakes in Arctic America. Perhaps a variety of *S.
stagnalis* Fabricius, of Greenland. (From Oquassoc, one of the
Rangeley Lakes.)

will find no paper collar or other evidence of civilization. It is the Nameless River.
Not that trout will cease to be. They will be hatched by machinery and raised in
ponds and fattened on chopped liver, and grow flabby and lose their spots. The
trout of the restaurant will not cease to be. He is no more like the trout of the
wild river than the fat and songless reed-bird is like the bobolink. Gross feeding
and easy pond life enervates and depraves him. The trout that the children will
know only by legend is the gold-sprinkled living arrow of the White-water, able to
zigzag up the cataract, able to loiter in the rapids, whose dainty meat is the glancing
butterfly." (*Myron W. Reed.*)

82   TELEOSTEI : HAPLOMI. — XIV.

FAMILY XXXV. **PERCOPSIDÆ.** (THE TROUT PERCHES.)

Body elongate, covered with moderate-sized, thin, strongly ctenoid scales; head naked; no barbels; opercles well developed; gill openings wide; an adipose fin; mouth small, horizontal; teeth very small, villiform; no teeth on vomer or palate; margin of upper jaw formed by premaxillaries alone, these short and not protractile; gill rakers tubercle-like; cavernous structure of the skull highly developed, as in *Stelliferus*, *Acerina* and *Ericymba*; fins much as in *Salmonidæ*; pellucid; branchiostegals 6; stomach siphonal with about 10 pyloric cæca; ova large; no oviduct. A single species inhabiting cold fresh waters in the northern U. S. Interesting little fishes, with the general characters of *Salmonidæ*, but having the mouth and scales decidedly Perch-like.

84. PERCOPSIS Agassiz. (πέρκη, perch; ὄψις, appearance.)

195. **P. guttatus** Agassiz. TROUT PERCH. Silvery; upper parts with rounded dark spots made up of minute dots; lower jaw included; tail long. Head 3¾; depth 4⅓. D. 11. A. 8. Lat. l. 50. L. 10. Great Lakes and tributaries, rarely S.; Ohio R. (Jordan); Potomac R. (Baird); Delaware R. (Abbott); Kansas (Gill). (Lat., spotted.)

ORDER XIV. **HAPLOMI.** (THE PIKE-LIKE FISHES.)

This order differs from the other soft-rayed fishes, chiefly in the simpler structure of the shoulder girdle, which lacks the præcoracoid arch. There is never an adipose dorsal; the dorsal is posterior in position and the head is depressed and usually more or less scaly. The pseudobranchiæ are wanting or glandular. The group is made up chiefly of fresh-water species. (ἁπλόος, simple; ὦμος, shoulder.)

FAMILY XXXVI. **AMBLYOPSIDÆ** (THE CAVE FISHES.)

Body elongate, with long depressed head; mouth large, the lower jaw projecting; premaxillaries scarcely protractile, forming whole edge of upper jaw; teeth villiform; eyes sometimes rudimentary and concealed under the skin; head naked, with papillary ridges; body with small, cycloid scales, irregularly arranged; no lateral line; D. far back, opposite A.; C. rounded; V. small, or wanting; vent at the throat, as in *Aphredoderus;* gill membranes joined to isthmus; stomach cæcal, with pyloric appendages; some (and probably all) viviparous. Genera 3; species 5.

Fishes of small size living in subterranean streams and ditches of the central and southern U. S., probably remnants of an ancient fauna.

a. Eyes rudimentary, concealed under the skin; body colorless; one pyloric
cæcum.

   b. Ventrals present, small. . . . . AMBLYOPSIS, 85.

   bb. Ventrals entirely wanting. . . . . TYPHLICHTHYS, 86.

aa. Eyes well developed; body colored; no ventrals; two pyloric cæca.

                              CHOLOGASTER, 87.

85. AMBLYOPSIS DeKay. (ἀμβλύς, obtuse; ὄψις, vision.

196. **A. spelæus** DeKay. BLIND FISH OF THE MAMMOTH
CAVE. Head 3; depth 4½; D. and A. equal, well developed; head
and body with papillary ridges; scales small; colorless. D. 10. A.
9. V. 4. P. 11. L. 2 to 5. Subterranean streams of Ky. and
Ind., Mammoth Cave, etc. (Lat., living in caves.)

   " If the *Amblyopses* be not alarmed, they come to the surface to feed, and
swim in full sight, like white aquatic ghosts. They are then easily taken by
the hand or net if perfect silence be preserved, for they are unconscious of the
presence of an enemy except through the sense of hearing. This sense is
however very acute; for at any noise, they turn suddenly and hide beneath
stones at the bottom. They take much of their food near the surface, as the
life of the depths is apparently very sparse. This habit is rendered very easy
by the structure of the fish, for the mouth is directed upwards, and the head
is very flat above, thus allowing the mouth to be at the surface." (*Cope.*)

86. TYPHLICHTHYS Girard. (τυφλός, blind; ἰχθύς, fish.)

197. **T. subterraneus** Girard. General character of *A.
spelæus*, but the head rather blunter and broader forwards; the
mouth smaller. D. 8. A. 8. P. 12. L. 2. Caves and wells in Ky.,
Tenn., Ala.; as common as the preceding, of which it is perhaps a
variation.

87. CHOLOGASTER Agassiz. (χωλός, maimed; γαστήρ, belly.)

198. **C. agassizii** Putnam. Eyes large; uniform light brown;
fins speckled. P. a little more than half way to D. Head 4;
depth 4. D. 9. A. 9. L. 1¼. Subterranean streams in Tenn.
and Ky. A closely related species (*C. cornutus* Ag.), is known
from a rice-ditch in S. C. (For Louis Agassiz.)

199. **C. papilliferus** Forbes. Yellowish brown, dark above;
sides with three dark streaks, the middle streak pale behind head;
C. dark, with cross-rows of white specks ; eye small, 6 in head,
above and well behind maxillary; P. reaching half way to D.; body
with tactile papillary ridges. Head 3½. L. 1 inch. Cave spring,
Union Co., Ill. (Lat., bearing *papilla*.)

FAMILY XXXVII. **CYPRINODONTIDÆ.** (THE KILLI-
FISHES.)

   Body oblong, depressed in front, more or less compressed behind, covered with adherent cycloid scales; no lateral line; head
scaly; mouth small, terminal, extremely protractile; the edge of

upper jaw formed by premaxillaries; teeth various; gill membranes
somewhat connected, free from isthmus; B. 4 to 6; D. single,
inserted posteriorly, rarely preceded by a spine; C. not forked;
stomach siphonal, without pyloric cæca; sexes unlike ; some species ovoviviparous. Genera 30; species 140; in fresh and brackish
waters of all warm regions. Most of them are small in size, and
some species of *Heterandria* are perhaps the smallest of fishes. The
species here mentioned are carnivorous, surface swimmers; many
southern species feed on mud and slime.

  a. Intestinal canal short, but little convoluted; dentary bones firmly united;
       teeth fixed; carnivorous species.
    b. Oviparous species, the anal fin of the male not modified into an intromit-
         tent organ. (*Cyprinodontinæ.*)
      c. Teeth in a single series, incisor-like, notched; dorsal of 10 or 11 rays,
           the first ray small; gill openings restricted above; body stout and
           deep. . . . . CYPRINODON, 88.
      cc. Teeth pointed; ventrals present; air-bladder present.
        e. Teeth in more than one series.
          f. Dorsal rather large, well forward; its rays usually 11 to 18, the
               first above or in front of A. . . . . . . FUNDULUS, 89.
          ff. Dorsal small and posteriorly placed, its rays 7 to 10; the first
               generally behind front of the small anal; size small.
                                                    ZYGONECTES, 90.
        ee. Teeth in one series; D. inserted before A.; D. and A. short, of 9
             to 13 rays. . . . . LUCANIA, 91.
    bb. Ovoviviparous species, the anal fin of the male advanced and modified
         into a sword-shaped intromittent organ. (*Anablepinæ.*)
      g. Eye normal, not divided by crosswise partition; jaws short;
           fins small; D. inserted behind A. . . . GAMBUSIA, 92.

88. CYPRINODON Lacépède. (κυπρῖνος, carp; ὀδών, tooth.)

200. **C. variegatus** Lacépède. Body short, deep, compressed;
humeral scale 4 times size of others; ♂ steel-blue, more or less
copper-red below; C. with black bar at base and tip; ♀ olivaceous, sides silvery, with irregular dark cross-streaks; a dark spot
on D. behind. Head 3; depth in adult about 2. D. 10. A. 10.
Scales 25–12. L. 2 to 4 inches, southern specimens being larger
and more brightly colored. Cape Cod to Texas, in brackish
waters.

89. FUNDULUS Lacépède. (Lat., *fundus*, bottom; they
often bury in the mud.)

  a. Scales comparatively large, about 36 in a lengthwise series, 13 in a cross-
       series.
    b. Branchiostegals 6; ♂ with dark cross-bars and a black dorsal spot; ♀
         with longitudinal black bands.

201. **F. majalis** (Walbaum). KILLIFISH. MAY-FISH. Head
long. with long snout; D. moderate; A. very high in ♂ ; eye moderate; ♂ olivaceous, brassy on sides; with about 12 bars of color

of back; a black spot on D.; lower fins sometimes yellow, and top of head black; ♀ much larger than ♂, paler, a black band on level of eye with two shorter bands below it; one or two black cross-bars at base of C. Head 3⅘; depth 4. D. 12. A. 10. Scales 36–13. L. 6. Cape Cod to Fla.; the largest of the genus, common in shallow bays. S. occurs *F. similis* Baird & Girard, with scales 33, and both sexes resembling ♂ of *F. majalis*. (Lat., pertaining to May.)

    bb. Branchiostegals 5: ♂ with silvery spots and bars; ♀ nearly plain olivaceous; young with black cross-bars.

    202. **F. heteroclitus** (Linnæus). COMMON KILLIFISH. MUMMICHOG. MUD-FISH. Body short, deep, the head short, broad; eye about equal to snout; ♂ dark green, sometimes orange below, sides with scattered yellowish spots, sometimes running into silvery cross-bars; vertical fins dark, with pale spots, usually a black spot on D.; young ♂ with 9 or 10 silvery bars; young ♀ with 9 or 10 black bars; adult ♀ nearly plain. Head 3⅔; depth 3½. D. 11. A. 10. Scales 35–12. L. 2 to 5. Maine to Mexico; everywhere common along shore, in shallow water; S. specimens (var. *grandis* Baird & Girard) larger and brighter. (*F. pisculentus* Mitchill; *F. nigrofasciatus* Le Sueur.) (ἕτερος, different; κλιτύς, slope.)

    aa. Scales moderate, 43 to 50 in longitudinal series.
      c. Dorsal inserted before A.; sides with many dark cross-bands.

    203. **F. diaphanus** (Le Sueur). Body rather slender; head slender, flat above; fins low. Olivaceous, sides silvery, with 15 to 25 narrow dark cross-bands; fins nearly plain. Western specimens, var. **menona** Jordan & Copeland (Ohio, W.) have the bands very distinct, and somewhat irregular; the back always spotted; sometimes silvery cross-bands replace the darker. E. specimens (Cayuga L., N. Y. Bay) have the back plain, the bands faint and regular. Head 3½ to 4; depth 5. D. 13. A. 11. Scales 46–12. L. 4. Great Lakes and tributaries, E. to coast of Mass., S. to N. Ind., W. to Colorado, ascending clear streams to their sources, also in lakes and river mouths.

    cc. D. inserted over front of A.; sides with regular series of orange or brown spots.

    204. **F. catenatus** (Storer). STUD-FISH. Body long, compressed; head broad; color greenish; ♂ with an orange spot on each scale, ♀ with smaller brown spots, these forming continuous stripes. Head 4; depth 4½. D. 14. A. 15. L. 7. Mountain streams, E. Tenn. and Ozark region; very pretty. In Alabama R. is a still brighter species (*F. stellifer* Jordan), with scattered orange spots. (Lat., with chain-like lines.)

    aaa. Scales very small, about 60 in a longitudinal series; sides barred.

205. **F. zebrinus** Jordan & Gilbert. Body slender, the head
long; fins low; greenish, sides silvery white, with 14 to 18 cross-
bars of the color of the back; fins plain. Head $3\frac{2}{3}$; depth $4\frac{2}{3}$.
D. 14. A. 13. Scales 60-21. L. 3. Kansas to Texas. (Lat.,
like a zebra.)

90. **ZYGONECTES** Agassiz.   (ζυγόν, yoke; νήκτης, swimmer ;
they being said to swim in pairs.)

a. Sides with a broad blue-black lateral band; vertical fins dotted.

206. **Z. notatus** (Rafinesque). Top-minnow. Body rather
elongate; head low; snout long; eye 3 in head; fins moderate;
outer teeth enlarged; lateral band darkest in $\male$, serrated in young;
back dotted ; a translucent spot on top of head. Head 4 ; depth
$4\frac{1}{2}$. D. 9. A. 11. Scales 34-11. L. 3. Mich. to Ala. and Tex.,
abundant in quiet waters. (Lat., noted, *i. e.* spotted.)

aa. Sides with 10 to 12 dark vertical bars, but without longitudinal stripes;
D. in $\male$ with a large black ocellus, edged before with white, behind
by yellow.

207. **Z. luciæ** (Baird). Body rather elongate; green-yellow
below. Head about $3\frac{1}{2}$; depth about 4. D. 8. A. 9. Scales un-
described, probably about 35. L. 1. Beesley's Point, N. J., not
lately recognized. (Named for Lucy Baird, daughter of Professor
Baird.)

aaa. Sides with about 10 stripes of orange-brown following the rows of scales,
a spot on each; $\male$ with the lines interrupted and with 9 dark cross-
bars; a black blotch below eye.

208. **Z. dispar** Agassiz. Body deep, compressed; head short,
very broad ; fins low ; D. much smaller than A. Bluish olive,
lateral stripes wavy. Head $3\frac{3}{4}$; depth $3\frac{1}{2}$. D. 7. A. 9. Scales
35-10. L. $2\frac{1}{2}$. Lakes and ponds, Ohio to Iowa. (Lat., dissimilar.)

aaaa. Sides plain olivaceous, without spots or lines.

209. **Z. sciadicus** (Cope). Body short, deep; fins small. Head
$3\frac{1}{2}$; depth about $3\frac{1}{4}$. D. 10. A. 12. Scales 39-13. L. 2. Platte
R. etc.

91. **LUCANIA** Girard.   (A name of euphony without meaning.)

210. **L. parva** (Baird & Girard). Rain-water Fish. Body
rather deep, $\male$ dark olive. D. dusky orange with black and orange
ocellus at base in front ; other fins chiefly orange ; $\female$ larger, the
fins plain. Head $3\frac{1}{2}$; depth about 3. D. 10. A. 10. Scales 26-8.
L. 2. L. I. to Key West, in tide pools, etc. (Lat., small.)

92. GAMBUSIA Poey.

(From the Cuban word *Gambusino*, which signifies *nothing*, with the idea of a joke or farce. Thus people say, " one fishes for Gambusinos," when he catches nothing. *Poey.*)

211. **G. patruelis** (Baird & Girard). Top-Minnow. Body plump; tail rather long ; snout broad ; eye about 3. Olivaceous, usually a dark streak along upper part of side; a blackish area below eye, usually distinct; D. and C. mostly with dark cross-streaks; usually a dusky blotch on sides in females (the dark interior showing through translucent skin); small specimens often uniform yellowish. Head 3⅔; depth 3¼ to 4. D. 7 to 9. Scales 28–7. L. 2½ ♀. ♂ 1. Very abundant in all lowland waters from the Potomac to Ill. and the Rio Grande. The males are scarce and very small, the anal process about as long as head. The young are born at the length of about ⅓ inch, in the spring. The gravid females are recognized without difficulty, the others are easily mistaken for *Zygonectes*, and have been repeatedly described as such. (Lat., cousin.)

Family XXXVIII. UMBRIDÆ. (The Mud-minnows.)

Body formed as in *Fundulus;* head large, flattened above; mouth moderate, the premaxillaries not protractile, the maxillaries forming lateral margin of upper jaw; jaws, vomer, and palatines with villiform teeth; gill openings wide ; gill rakers obsolete; scales cycloid on head and body; no lateral line; C. rounded; P. narrow. Intestinal canal without cæca; air bladder simple. Oviparous, sexes similar. Carnivorous fishes living in mud in the clear waters of sluggish streams and ponds in cool regions, extremely tenacious of life. One genus with 2 species, *Umbra crameri* of Austria and *U. limi.*

" A locality which, with the water perfectly clear, will appear destitute of fish, will perhaps yield a number of mud fish on stirring up the mud at the bottom and drawing a seine through it. Ditches in the prairies of Wisconsin, or mere bog-holes, apparently affording lodgment to nothing beyond tadpoles, may thus be found filled with Mud-minnows." (*Baird.*)

93. UMBRA (Kramer) Müller. (Lat., shade.)

212. **U. limi** (Kirtland). Mud-minnow. Dog-fish. Ventrals slightly before D. ; A. much smaller than D. The typical form (Great Lakes and W.) is dull olive green, with about 14 narrow pale bars, faint in young; black caudal bar faint; lower jaw pale ; the Eastern form, var. **pygmæa** DeKay (Conn. to N. C.), with narrow pale lengthwise streaks instead of bars; dark caudal bar

very evident; lower jaw black. Head 3¾; depth 3¾. B. 6. D. 14.
A. 9. Scales 35-15. L. 4. N. C. to Conn. and Ontario, W. to
Ind. and Minn., in cool weedy streams and swamps. (Lat., of the
mud.)

FAMILY XXXIX. ESOCIDÆ. (THE PIKES.)

Body elongate, somewhat compressed, with rather small, cycloid
scales; lateral line present, more or less imperfect; head long, the
snout much prolonged and depressed; mouth very large, the lower
jaw longest; upper jaw not protractile, most of its edge formed by
the maxillaries; premaxillaries, vomer, and palatines with bands of
more or less movable cardiform teeth; lower jaw with strong, unequal teeth; tongue with small teeth; head naked above, scaly on
sides; gill rakers tubercular; B. 12 to 20; D. opposite A. as in
other *Haplomi;* C. emarginate; P. small; intestinal canal simple,
with cæca; air-bladder present. One genus, with 5 species, one
in the fresh waters of both continents, the rest all American. All
are noted for their voracity, "mere machines for the assimilation
of other organisms." The flesh is white, flaky, and excellent.
The Pike is "a solemn stately ruminant fish, lurking under the
shadow of a lily-pad at noon, with still, circumspect, voracious eye,
motionless as a jewel set in water, or moving slowly along to take
up its position; darting from time to time at such unlucky fish or
frog or insect as comes within its range, and swallowing it at a gulp.
Sometimes a striped snake, bound for greener meadows across the
stream, ends its undulatory progress in the same receptacle."
(*Thoreau.*)

94. ESOX (Artedi) Linnæus.   (An old name of the Pike).

a. Cheeks and opercles entirely scaly.

b. Branchiostegals 12 (11 to 13): scales 105 to 108; D. 11 or 12; A. 11 or
12; snout short, the middle of eye nearer tip of lower jaw than edge of
opercle; species of small size.

213. E. americanus Gmelin.   Head short, 3½; snout 2¼ in
head; eye 2⅔ in snout.   Dark green, sides with about 20 distinct
curved dusky bars; fins plain. Depth 5½. L. 12. Mass. to Fla., in
coastwise streams.

214. E. vermiculatus Le Sueur.   LITTLE PICKEREL.   Head
longer, 3¼; snout 2¼ in head; eye 2¼ in snout. Olive green; sides
with many darker curved streaks, usually distinct and more or less
reticulate; fins mostly plain; depth 5½. L. 12. Miss. Valley, etc.,
very abundant in small streams and bayous. (Lat., having marks
like worm-tracks.)

bb. Branchiostegals 14 to 16; D. 14 (developed rays); A. 13; scales
about 125; snout long, the middle of eye midway between chin and
edge of opercle.

215. E. reticulatus Le Sueur. EASTERN PICKEREL. Head long, $3\frac{1}{2}$; snout $2\frac{1}{4}$ in head, eye $3\frac{1}{2}$ in snout. Greenish, with numerous narrow dark lines and streaks, mostly horizontal and more or less reticulated; fins plain; depth 5. L. 30. Me. to Ala., abundant in coastwise streams, not W. of Alleghanies. (Lat., having a net-work of marks.)

aa. Cheeks entirely scaly; lower half of opercles bare; B. 14; D. 16 or 17; A. 13 or 14; scales about 123.

216. E. lucius L. PIKE; NORTHERN PICKEREL. Head long, $3\frac{1}{4}$; snout $2\frac{3}{4}$ in head; eye 3 in snout; eye placed as in preceding. Grayish, with many round whitish spots; the young with pale bars; D., A. and C. spotted with black; a white horizontal streak bounding naked part of opercle. Depth 5. L. 30 to 50. N. Eur., Asia, and N. Am from L. Champlain to N Ind. and N. W. to Alaska; abundant, N. (Eu.) (Lat., pike.)

aaa. Cheeks as well as opercles scaleless on the lower part; B. 17 to 19; D. 17, A. 15; scales about 150.

217. E. masquinongy (Mitchill). MUSKALLUNGE. MASKINONGY. Head large, $3\frac{2}{3}$; snout $2\frac{1}{4}$ in head; eye 4 to 5 in snout; eye placed as in E. reticulatus. Dark gray, sometimes (var. **immaculatus** Garrard) immaculate, usually with small round blackish spots on a paler ground; fins spotted with black. Depth 6. L. 8 feet. A magnificent fish, one of the largest in fresh waters. Great Lake region and N. W.; occasional in the Ohio valley. " A long, slim, strong, and swift fish, in every way fitted for the life it leads, that of a dauntless marauder." (Hullock.) (The Indian name.)

ORDER XV. **APODES.** (THE EELS.)

Scapular arch free from the cranium; no præcoracoid arch; body much elongate, with many vertebræ; no ventral fins; maxillaries and premaxillaries united with other bones or else wanting; pharyngeal and opercular bones more or less deficient; no fin spines; gill openings narrow; no pseudobranchiæ; scales minute or wanting. A large group, as yet of uncertain boundaries, composed of degenerate Physostomi, its origin and relationship as yet, however, uncertain. Most of the Eels are tropical and marine, and many belong to the deep seas. Numerous genera and species not here included occur in the deep waters off our coast. (a, privative; πούς, foot.)

FAMILY XL. **ANGUILLIDÆ.** (THE TRUE EELS.)

Body compressed, covered with small, imbedded scales, linear in form, placed obliquely, some of them at right angles to others; lateral line present; head long; mouth large, the lower jaw projecting; teeth small, subequal, in bands on jaws and vomer; pterygoid bones slender; tongue free in front; nostrils lateral; lips full; opercles developed; vertical fins confluent; D. beginning well behind head; P. present; gill openings moderate. Sexual organs inconspicuous. One genus with four or more species, crawling in the mud and ooze of brackish and fresh waters of most regions, absent on the Pacific coast of America. They are among the most voracious of fishes. "On their hunting excursions, they overturn alike huge and small stones, beneath which they find species of shrimp and cray-fish, of which they are excessively fond. Their noses are poked into every imaginable hole in their search for food, to the terror of innumerable small fishes." (*W. H. Ballou.*)

The eels often move for a considerable distance on land, in damp grass. High waterfalls, dams, and other obstructions are often passed in this way. It is thought that eels spawn only in the sea, and that the female spawns once and then dies. The females are larger than the males, paler in color, with smaller eyes and higher fins.

95. ANGUILLA Thunberg.

218. **A. anguilla** (L.). EEL. Brown, more or less tinged with yellowish. Head 8½. L. 40. N. Atlantic, from Maine to Brazil, ascending all streams; found throughout Mississippi valley, never in the open sea. The American Eel (var. **rostrata** Le Sueur) has the distance from front of D. to front of A. a little less than head; in the European form this is a little greater, the D. being a little farther back in the former. (*Eu.*) (Lat., eel.)

FAMILY XLI. ECHELIDÆ. (THE CONGER EELS.)

Eels closely related to the *Anguillidæ*, but without scales, and with the ovaries in the female evident, and with comparatively large eggs similar to those of fishes generally. D. commencing not far behind head. Genera 3 or 4; species about 10, all strictly marine. **Leptocephalus morrisi** Gmelin, a translucent, ribbon-shaped creature, with very small head, and no generative organs, is occasionally taken on our coasts. This is thought to be a stage of arrested development of the young of *Echelus*, a larval form which goes on increasing in size without ever reaching the characters of the perfect animal.

a. Jaws with an outer series of close-set teeth; lower jaw not projecting; dorsal beginning behind root of pectorals. . . . . . . ECHELUS, 96.

96. ECHELUS [1] Rafinesque. (ἔγχελυς, eel, softened into *Echelus*.)

[1] In strictness, the name *Leptocephalus* should supersede *Echelus*, but there may be some doubt as to the identification of *L. morrisi*, and for the last hundred years *Leptocephalus* has been used as a general name for these peculiar immature forms.

219. **E. conger** (L.). CONGER EEL. Cleft of mouth reaching beyond middle of the large eye; dark brown above, paler below. D. and A. usually pale, with broad, black margin; P. dusky, pale-edged; pores of lateral line whitish; body sometimes wholly black. L. 6 feet. Open sea, not rare on our coast. (*Eu.*) (γόγγρος, Conger, the ancient name.)

SERIES **PHYSOCLYSTI.**

We now begin the division of fishes in which the air-bladder in the adult loses all connection with the alimentary canal. This character in itself is of slight importance, but it is associated with gradual modifications in other respects, of such character that the typical Physoclyst is quite unlike the average Physostome. Most of the Physoclysts have spines in some of the fins; the ventral fins are normally thoracic, each with a spine and five rays, while the pectorals are inserted high. But there are many exceptions to each of these characters. We commence the series with the forms most closely related to the *Haplomi* and other soft-rayed forms. (φῦσα, bladder; κλειστός, closed.)

ORDER XVI. **SYNENTOGNATHI.** (THE SYNENTOGNA-THOUS FISHES.)

Physoclistous fishes without spines in the fins, with the ventrals abdominal and the lower pharyngeals fully united. This peculiar transitional group contains a single family divided by osteological characters into two strongly marked groups, called families by Dr. Gill. These are the *Belonidæ* and the true *Exocœtidæ* or *Scomberesocidæ*. (σύν, together; ἐντός, within; γνάθος, jaw.)

FAMILY XLII. **EXOCŒTIDÆ.** (THE NEEDLE-FISHES. FLYING-FISHES.)

Body oblong, compressed, with cycloid scales; a ridge, apparently representing the lateral line, running along side of belly; head scaly; premaxillaries not protractile, but with a hinge at base, forming most of margin of upper jaw; teeth various. D. posterior, similar to anal; ventrals inserted posteriorly; P. inserted high; C. usually forked, the lower lobe the longer; gill openings wide; pseudobranchiæ hidden; air-bladder large; intestinal canal simple. Genera about 11; species about 100; in all warm seas, some of them endowed with remarkable power of flight.

a. Jaws with sharp, unequal teeth; both jaws much produced; *no finlets;* maxillaries grown fast to premaxillaries; ovary single. (*Beloninæ.*)
    x. Gill rakers none; no teeth on vomer; D. and A. falcate; C. lunate.

b. Body little compressed, its breadth more than ⅔ its greatest depth.
                                                TYLOSURUS, 97.
  bb. Body much compressed, its breadth not half its greatest depth.
                                                ATHLENNES, 98.
aa. Jaws with minute teeth or none.
  c. Maxillary grown fast to premaxillary; one or both jaws produced in
    a long beak.
    d. D. and A. with finlets; scales small; both jaws produced. (*Scom-
      beresocinæ*). . . . . SCOMBERESOX, 99.
    dd. D. and A. without finlets; upper jaw short, the lower much pro-
      duced. (*Hemiramphinæ.*)
      e. Pectorals moderate, not longer than head without beak; body
        rather stout; sexes similar. . . . . . HEMIRAMPHUS, 100.
      ee. Pectorals very long, twice head without beak; V. short; body
        very slender, almost band-like   . EULEPTORHAMPHUS, 101.
  cc. Maxillary distinct from premaxillary ; both jaws short. (*Exocœ-
    tinæ.*)
    f. Roof of mouth with teeth; body elliptical in cross-section; V.
      long, inserted behind middle of body; D. high, its base about
      equal to anal base; snout and lower jaw short.
                                                PAREXOCŒTUS, 102.
    ff. Roof of mouth nearly toothless; body quadrate in cross-section;
      P. long, about reaching base of C.
      g. Ventrals inserted anteriorly, much nearer tip of snout than
        base of C., small, not used as organs of flight.
                                                HALOCYPSELUS, 103.
      gg. Ventrals inserted posteriorly, nearer base of C. than snout,
        used as organs of flight. . . . . . EXOCŒTUS, 104.

97. TYLOSURUS Cocco. (τύλος, callus, *i. e.* keel; οὐρά, tail.)

a. Caudal peduncle depressed, with a dermal keel.
  b. D. and A. short, each of 14 to 16 rays; last rays of D. low; jaws slender.

220. T. marinus (Bloch & Schneider). GAR-FISH. BILL-FISH. NEEDLE-FISH. SILVER GAR.

Scales and bones green : green, a silvery lateral band ; a dark
bar on opercles; P. pale. Head 2¾; depth 5½ in head. Lat. l. 300.
L. 4 feet. Maine to Texas, abundant ; ascending rivers.

  bb. D. and A. long, each of 21 to 24 rays; last rays of D. sometimes ele-
    vated; caudal keel black.

221. T. acus (Lacépède). HOUND-FISH. AGUJON. Beak

long, twice rest of head. Green, no lateral stripe. Lat. l. 380. L.
5 feet. West Indies and Mediterranean; occasional N. to Cape
Cod. (*Eu.*) (Lat., needle.)

98. ATHLENNES [1] Jordan & Fordice. (ἀβλεννής, without mu-
cosity, an ancient epithet applied to *Belone belone.*)

222. A. hians (Cuv. & Val.). Jaws long and very slender, the

upper arched at base, so that the mouth cannot be closed; tail not

---

[1] This name was inadvertently printed "*Athlennes,*" and may remain so ; "*Ablen-
nes*" was intended.

keeled; eye very large, scales minute; D. elevated behind. Green, sides silvery; young with round dark spots. D. 25. A. 26. Lat. l. 520. L. 40. W. Indies, occasional N. (Lat., gaping.)

99. SCOMBERESOX Lacépède. (Scomber + Esox.)

a. Jaws produced in a slender beak; the snout longer than rest of head.

223. S. saurus (Walbaum). Saury. Skipper. Fins small; C. forked. Olive, sides with distinct silvery band. Head $3\frac{1}{2}$; depth 9. D. 9⁻VI. A. 12⁻VI. Lat. l. 110. L. 18. Open Atlantic, not rare; in large schools, skipping along the surface. (*Eu.*) (An old name, " lizard-fish.")

100. HEMIRAMPHUS Cuvier. (ἡμι-, half; ῥάμφος, beak.)

a. Ventrals inserted midway between eye and base C.; A. about as long as D., both with 14 to 16 rays; last ray of D. not produced.

224. H. unifasciatus Ranzani. Half-beak. Green; lower jaw red; sides with a silvery band. Head $4\frac{1}{2}$; depth 6 to $7\frac{1}{2}$. Lat. l. 54. L. 12. W. Indies, etc.; the typical form with shortish jaw, from Florida Keys, S. Var. **roberti**, Cuv. & Val., more slender, with longer lower jaw, longer than rest of head, ranges N. to Cape Cod. (Lat., one-banded.) (From Va., S., occurs H. *balao* Le Sueur, with V. midway between middle of P. and base of C.)

101. EULEPTORHAMPHUS Gill. (εὐλεπτός, very slender; ῥάμφος, beak.)

225. E. longirostris (Cuvier). Lower jaw much longer than rest of head; no lateral band. Head $6\frac{2}{3}$; depth 10. D. 22. A. 19. L. 18. Open sea, occasional N. to Cape Cod. (Lat., long-snouted.)

102. PAREXOCŒTUS Bleeker. (παρά. near; *Exocœtus.*)

226. P. mesogaster (Bloch). Second ray of P. divided; D. very high. Blue; sides silvery; D. largely black, other fins pale. Head $4\frac{2}{3}$; depth 5. D. 12. A. 13. Lat. l. 38. L. 6. Open sea, N. to R. I. (μέσος, middle; γαστήρ, the position of V.)

103. HALOCYPSELUS Weinland. (ἅλς, sea; κύψελος, swallow.)

227. H. evolans (L.). Second ray of P. divided; A. nearly as long as D.; D. low; P. dark above, pale below; other fins pale; V. white. Head 4; depth $5\frac{1}{2}$. D. 13. A. 13. Lat. l. 42. L. 9. Open sea, N. to Cape Cod. (Lat., flying away.)

104. EXOCŒTUS[1] (Artedi) Linnæus. (Flying-fishes.)

(The flying-fishes live in the open sea, swimming in large schools. They will " fly " a distance of from a few rods to more than an

---

[1] For a detailed account of the American Flying-fishes, see Jordan & Meek, Proc. U. S. Nat. Mus., 1885, p. 44.

eighth of a mile. rarely rising more than 3 or 4 feet.   Their movements in the water are extremely rapid ; the sole source of motive
power is the action of the strong tail while in the water.   No force
is acquired while the fish is in the air.   On rising from the water,
the movements of the tail are continued until the whole body is out
of the water.   While the tail is in motion, the pectorals seem to be
in a state of rapid vibration, but this is apparent only, due to the
resistance of the air to the motions of the animal.   While the tail
is in the water, the ventrals are folded.   When the action of the
tail ceases, the pectorals and ventrals are spread and held at rest.
They are not used as wings, but act rather as parachutes to hold
the body in the air.   When the fish begins to fall, the tail touches
the water, when its motion again begins, and with it the apparent
motion of the pectorals.   It is thus enabled to resume its flight,
which it finishes finally with a splash.   While in the air it resembles a large dragon-fly.   The motion is very swift, at first in a
straight line, but later deflected into a curve.   The motion has no
relation to the direction of the wind.   When a vessel is passing
through a school of these fishes, they spring up before it, moving
in all directions, as grasshoppers in a meadow.[1]

The young of different species often have long fleshy barbels at
tip of the lower jaw.   These are lost with age.   They were formerly
placed in a separate "genus," *Cypselurus* Swainson.   (ἐξώκοιτος,
sleeping outside ; an old name of some fish imagined to sleep on
the beach at night.)

a. Anal long, its base nearly equal to that of D., its first ray opposite first of
   D.; anal rays 11 or 12; dorsal rays 11 or 12.   (*Exocœtus.*)
   b. Second ray of P. simple, as well as the first; 4th and 5th rays longest;
      V. largely black.
      c. Second ray of P. scarcely longer than first.

228. **E. exsiliens** Müller (1776).   V. $2\frac{1}{3}$ in body, reaching C. ;
P. $1\frac{1}{3}$; eye large.   Head 4 ; depth $5\frac{1}{4}$.   Scales 48.   L. 10.   P.
and V. marbled with black and white ; D. with black spot anteri-
orly ; A. white : a dark blotch at base C.   Open sea, occasional N.
(*E. exiliens* Gmelin, 1788.)   (Lat., leaping out.)   (*Eu.*)

   cc. Second ray of P. nearly half longer than first.

229. **E. rondeletii** Cuv. & Val.   V. $3\frac{1}{2}$ in body, reaching last
A. ; P. $1\frac{2}{3}$ in body ; eye moderate.   Head $4\frac{1}{2}$: depth $5\frac{1}{4}$.   Scales 50.
Ventrals chiefly *black;* P. dusky ; no black on D. or A.   Open sea,
frequently N.   (*Eu.*)   (For Guillaume Rondelet, one of the fathers
of ichthyology.)

   bb. Second ray of P. divided; 3d or 4th longest.

---

[1] Observations on the flight of these fishes have been made under very favorable
conditions by Prof. C. H. Gilbert, and the writer.   Several species have been thus
observed, especially the largest of the group, *E. californicus* Cooper.

d. Ventrals chiefly black, inserted midway between eye and base C.

230. **E. vinciguerræ** Jordan & Meek. P. dusky, uniform or with a small white cross stripe ; D. and A. without black. Head 4⅓ ; depth 6¼. Scales 48. L. 12. Atlantic, N. to Grand Banks. (*Eu.*) (To Dr. Decio Vinciguerra, of Rome.)

dd. Ventrals nearly white; inserted midway between opercle and tail.

231. **E. volitans** L. P. dark brown, with an oblique whitish band from axil to middle of fin ; D. and A. without black. Head 4¼ ; depth 6½. D. 12. A. 11. Scales 55. L. 12. Atlantic, N. to Grand Banks, frequent. (*Eu.*) (Lat., flying.)

aa. Anal short, its base half to two-thirds that of dorsal, its first ray behind first of D.; anal rays 9 or 10; dorsal 12 to 14. (*Cypselurus* Swainson.)
   e. Second ray of pectoral divided (first simple); 3d and 4th longest; V. midway between eye and tail; P. without round black spots; young with barbels.
   f. D. and A. plain whitish; V. pale.

232. **E. heterurus** Rafinesque. P. with an oblique white band on lower half. Head 4⅔ ; depth 5⅓. Scales 58. L. 12. Atlantic, the commonest species on our coasts. (*Eu.*) (ἕτερος, unequal ; οὐρά, tail.)

ff. D. and A. blotched or spotted with black; V. chiefly black.

233. **E. furcatus** Le Sueur. P. black, with a white band ; C. with 3 dusky cross-bars. Head 4½ ; depth 5¼. Scales 46. L. 12. Warm seas, N. to Cape Cod. (*Eu.*) (Lat., forked.)

ee. Second ray of P. simple, like the first; V. chiefly black.

234. **E. gibbifrons** Cuv. & Val. Snout more bluntly rounded than in any other species, 4½ in head ; V. midway between eye and C. ; P. dusky, paler at base ; vertical fins plain, rather dusky. Head 4 ; depth 5½. Scales 46. L. 12. N. Atlantic, rare. (Lat., *gibbus*, swollen; *frons*, front.)

ORDER XVII. **LOPHOBRANCHII.** (THE TUFT-GILLED FISHES.)

Gills contracted, tufted, composed of small rounded lobes, attached to the gill-arches ; pharyngeal bones reduced in number; mouth very small, toothless, at the end of a tubular snout; posttemporal grown fast to skull ; anterior vertebræ modified, with expanded apophyses; gill covers reduced to a simple plate ; skin with bony plates arranged in rings; fins small. Two families, the E. Indian *Solenostomatidæ* have spinous dorsal and ventral fins; ours lack both. (λόφος, tuft ; βράγχια, gills.)

FAMILY XLIII. **SYNGNATHIDÆ.** (THE PIPE-FISHES.)

Body elongate, covered with bony rings; gill openings reduced to a small aperture behind upper part of opercle ; no spinous dor-

sal nor ventral fins; caudal small or wanting; anal minute, of 1 or
2 rays; tail long. Male fishes with an egg-pouch, usually placed
on under side of tail and formed of two folds of skin which meet on
the median line. The eggs are received from the female into this
pouch, and retained for some time after hatching, when the pouch
opens and the young, then ⅓ to ½ inch long, escape. Genera about
14; species 150, in all warm seas. (σύν, together; γνάθος, jaw.)

a. Axis of head in a line with axis of body. (Syngnathinæ.)
  b. Humeral bones united below; C. present; P. well developed; D. oppo-
      site vent; shields not spinous. . . . . SIPHOSTOMA, 105.
aa. Axis of head forming an angle with axis of body; the head and neck
      horse-shaped, or like that of a "Knight" at chess. (Hippocampinæ.)
  c. Body compressed; occiput with a narrow, bony crest, surmounted
      by a star-shaped coronet; shields tubercular or spinous; egg-sac
      in male at base of tail, which is prehensile and without fin.
                                                        HIPPOCAMPUS, 106.

105. SIPHOSTOMA Rafinesque. PIPE-FISHES. (σίφων, tube;
                    στόμα, mouth.)

235. **S. fuscum** (Storer). COMMON PIPE-FISH. Top of head
slightly keeled; D. covering 4 body rings and 5 behind vent; rings
18 to 20 + 36 to 40; dorsal rays 36 to 40; snout moderate. Head
9; L. 7. Olivaceous, sides mottled. Newfoundland to Va., common. Numerous other species occur S. (Lat., dusky.)

106. HIPPOCAMPUS Rafinesque. SEA HORSES.

(These small fishes inhabit grassy bays and often the open sea
in warm regions. They are wont to twist the very prehensile tail
around pieces of floating sea-weed or eel-grass. They are thus often
drifted to great distances in the sea. The species are very similar to each other, and not easily distinguished.) (Ancient name
from ἵππος, horse, and κάμπος, a wriggling creature.)

236. **H. hudsonius** DeKay. SEA HORSE. Dusky, unspotted,
but with grayish blotches, edged with blackish; D. with dark band;
snout 1⅓ in head; spines on head weak, with cirri; spines all bluntish. D. 19, on 3½ of the 11 body rings. L. 6. Cape Cod to Fla.,
not common. Several other species occur S.

ORDER XVIII. **HEMIBRANCHII.** (THE HALF-GILLED
                    FISHES.)

Gills normal, but the branchihyals and pharyngeals reduced in
number; V. more or less abdominal. A small group of 5 or 6 families, intermediate between the Lophobranchii and the true Acanthopteri. (ἡμι-, half; βράγχια, gills.)

FAMILY XLIV. **FISTULARIIDÆ.** (THE TRUMPET-
· FISHES.)

Body elongate; naked, with some bony plates; snout produced
in a long dilatable tube, with the short jaws at the end; teeth
minute; no spinous dorsal; C. forked, its middle rays produced in
a long filament; V. small, with 6 rays. Tropical seas; one genus,
3 species.

107. FISTULARIA Linnæus. (Lat., *fistula*, a tube.)

237. **F. tabaccaria** L. TRUMPET-FISH. Brown, with blue
spots. Head 2⅓. D. 14. A. 13. Warm seas, rarely N. to N. Y.
(Lat., pertaining to tobacco-pipe.)

FAMILY XLV. **GASTEROSTEIDÆ.** (THE STICKLEBACKS.)

Body elongate, with slender tail, naked or shielded with bony
plates; head large, compressed, the snout not tubular; mouth
moderate, the chin prominent; teeth sharp, in jaws only; suborbital large. B. 3; opercles unarmed. D. with 2 to 15 free
spines; A. with one spine; V. subabdominal, I, 1. P. short, well
behind gill openings, preceded by an area covered with smooth
skin. Genera 5, species about 20, in fresh and brackish waters of
Northern regions; small fishes, lively, greedy and quarrelsome,
and exceedingly destructive to the spawn of large fishes. Most
of them build nests, which they defend with much spirit.

a. Innominate bones joined, forming a median plate on belly, behind V.
   b. Gill membranes joined, their border free from isthmus; spines small.
     c. Dorsal spines 7 to 11, divergent; pubic bones long, weak, widely di-
      vergent; body slender, mostly naked.   . . . PYGOSTEUS, 108.
     cc. Dorsal spines 5, in right line; pubic bones short, widely divergent;
      body stout, naked. . . . . EUCALIA, 109.
   bb. Gill membranes not free from isthmus; dorsal spines 3 or 4, strong,
     divergent; pubic bones broad, little divergent; form robust; skin
     usually mailed. . . . . GASTEROSTEUS, 110.
aa. Innominate bones not joined, each extending as a strong process under
   skin, outside of V., the area between them flat and not bony; pubic bones
   weak; dorsal spines 4, divergent; gill membranes joined to isthmus;
   tail very slender; skin smooth. . . . . APELTES, 111.

108. PYGOSTEUS Brevoort. (πυγή, rump; ὀστέον, bone.)

238. **P. pungitius** (L). NINE-SPINED STICKLEBACK. Olivaceous, punctulate and irregularly barred with black. Tail keeled; eye
large. Head 4; depth 5 to 6. D. IX – I, 9. A. I, 8. L. 2⅓. New
York to L. Mich., N. to Greenland, in fresh waters and entering
sea. (*Eu.*) (Lat., pungent.)

109. EUCALIA Jordan. (εὖ, good; καλιά, nest.)

239. **E. inconstans** (Kirtland). BROOK STICKLEBACK. ♂ in
spring jet black, reddish-tinged; ♀ olivaceous, mottled and dotted,

7

no dermal plates, the bones and spines all feeble; tail keeled. Head
3½; depth 4. D. IV - I, 10. A. I, 10. L. 2½. N. Y. to Kansas
and Greenland, abundant N. W. in small brooks; S. to Greensburg,
Ind. (*Shannon.*) Var. **cayuga** Jordan (W. N. Y.) has V. spines
longer, longer than innominate bones, and other trifling differences.

110. GASTEROSTEUS (Artedi) Linnæus.

(γαστήρ, belly; ὀστέον, bone.)

a. Sides partly covered with bony plates, the tail naked.
  b. Lateral plates 2 to 7.
    c. Ventral spine without cusp at base; lateral plates 2 or 3.

240. **G. wheatlandi** Putnam. No mucous pores; tail compressed. Blackish. D. II, I, 10 to 12. A. I, 8. Cape Cod, N.
scarce. (To Dr. Richard H. Wheatland, of Salem, Mass.)

  cc. Ventral spine with a strong cusp at base behind; lateral plates
    about 7.

241. **G. gymnurus** Cuvier. Tail keeled. Grayish, dotted.
D. II, I, 12. A. I, 8. L. 2½. Newfoundland to Greenland, etc.
(*G. dimidiatus* Reinhardt.) (*Eu.*) (γυμνός, naked; οὐρά, tail.)

  bb. Lateral plates 15; tail keeled.

242. **G. atkinsii** Bean. Slender; V. long. Head 3⅓; depth 5.
D. II, I, 11. A. I, 8. L. 1½. Maine. (To Charles G. Atkins,
Fish Commissioner of Maine.)

aa. Sides entirely covered with (28 to 33), bony plates; tail keeled; V. spine
  with cusp at base.

243. **G. aculeatus** L. COMMON STICKLEBACK. Olivaceous,
sides silvery; back dotted; opercles striate; rugose plates at base
of spines; spines serrate. Head 3½; depth 4½. D. II - I, 13. A.
I, 9. L. 4. N. Y. to Greenland and Europe, abundant, variable.
(*Eu.*) Perhaps all the preceding are forms or varieties of this.
(Lat., bearing prickles.)

111. APELTES DeKay. (a, privative : πέλτη, shield.

244. **A. quadracus** (Mitchill). Olive, mottled; males nearly
black, the V. red in spring; body plump, with long slender tail;
skin naked. Head 4; depth 4. D. III, I, 11. A. I, 8. L. 2.
N. J. to Labrador; abundant along coast. (Lat., four-spined.)

ORDER XIX. PERCESOCES.

This group comprises *Physoclysti*, which have the general characters of the great group of *Acanthopteri*, but in which the ventral
fins are abdominal, the pelvic bone not being attached to the
shoulder-girdle. Scales cycloid, opercles unarmed. The spinous
dorsal is short and sometimes (*Ophiocephalidæ*) wanting. (Lat.,

Perca, perch ; *Esox*, pike ; intermediate between Pikes and Perches.)

FAMILY XLVI. **MUGILIDÆ**. (THE MULLETS.)

Body oblong, with large cycloid scales ; no lateral line ; mouth small, nearly toothless; upper jaw protractile; gill membranes free from isthmus ; gill rakers long, slender ; pseudobranchiæ large. Dorsals separate, the anterior with four spines ; anal similar to soft dorsal, its spines 2 or 3. Air-bladder large; intestinal canal long; vertebræ 11 + 13 = 24. Genera 5, species 75; in fresh waters and seas of warm regions, feeding on mud.

a. Jaws with tooth-like cilia; stomach muscular, gizzard-like; anal spines 3.
MUGIL, 112.

112. MUGIL (Artedi) Linnæus. (Ancient name from *mulgeo*, to suck.)

a. Adipose eyelid well developed. (*Mugil.*)
  b. Soft D. and A. nearly naked; A. III, 8.

245. **M. cephalus** L. STRIPED MULLET. COMMON MULLET. Silvery, darker above ; dark stripes along the rows of scales ; a dusky blotch on base P. Head 4 ; depth 4. D. IV - I, 8. Scales 40–13. L. 24. Warm seas ; common N. to Cape Cod, ascending streams. (*M. albula* L.) (*Eu.*) (An old name, from κεφαλή, head.)

  bb. Soft D. and A. scaly; A. III, 9.

246. **M. curema** Cuv. & Val. WHITE MULLET. BLUE-BACK MULLET. LIZA. Silvery; scales without dark stripes ; a dark spot at base P.; P. not nearly reaching D. Head 4 ; depth 4. D. IV - I, 8. Scales 38–12. L. 18. Warm seas. N. to Cape Cod, scarce N. (*M. brasiliensis* Günther, not of Agassiz.) (A Brazilian name.)

FAMILY XLVII. **ATHERINIDÆ**. (THE SILVERSIDES.)

Body elongate, compressed, with cycloid scales ; no lateral line ; mouth moderate ; teeth small ; opercles unarmed ; gill membranes free; pseudobranchiæ present; gill rakers slender. Dorsals well separated, the first of 3 to 8 slender spines ; A. similar to soft D., with one spine ; V. I, 5; air-bladder present ; vertebræ numerous. Genera 8 ; species 50, fishes living in schools along coasts of warm regions, a few in rivers. (ἀθερίνη, the old name from ἀθήρ. a dart.)

a. Premaxillaries freely protractile; their posterior end broad; teeth in bands. none on vomer; a silvery band along side.
  b. Jaws both short, the upper scarcely longer than eye.. . MENIDIA, 113.
  bb. Jaws both produced in a short beak; the upper about half longer than eye. . . . . LABIDESTHES, 114.

113. MENIDIA Bonaparte.   (An old name, from μήνη, moon.)

a. Scales entire: soft D. and A. naked.

b. Anal rather long, its rays I, 22, to I, 25.

247. **M. notata** (Mitchill).   COMMON SILVERSIDE.   FRIAR.
Body slender: transparent green ; scales speckled.   Head 5 ;
depth 6.   D. IV – I, 8.   Scales 46–10.   L. 5.   Maine to Va., very
common N.   (Lat., marked.)

bb. Anal rather short, I, 15, to I, 18.

248. **M. beryllina** (Cope).   First D. over vent, nearer base C.
than snout.   Head 4¼ ; depth 6.   D. "V – I, 11."   L. 2½.   Potomac R., only the type known.   (Lat., beryl-color.)

aa. Scales with ragged edges especially on back ; soft D. and A. scaly.

249. **M. laciniata** Swain.   Green ; back with dark points forming streaks along rows of scales.   Head 4¾ ; depth 5½.   D. IV – I, 8.
A. I, 19 to I, 21.   Scales 50–7.   L. 5.   Va. to S. C. ; probably a
var. of *M. vagrans* Goode & Bean, S. C. to Texas, which has A. I,
14 to I, 18.   (Lat., gashed.)

114. LABIDESTHES Cope.   (λαβίς, a pair of forceps ;
ἐσθίω, to eat.)

250. **L. sicculus** Cope.   BROOK SILVERSIDE.   Translucent
green, back dotted ; silver band very distinct ; body very slender ;
scales entire.   Head 4½ ; depth 6.   D. IV – I, 11.   A. I, 23.   Lat. l.
75.   L. 3½.   Mich. to Perdido Bay and Kans., abundant in quiet
waters ; a most graceful little fish.   (Lat., dry, *i. e.* found in half-
dry pools.)

FAMILY XLVIII.   **SPHYRÆNIDÆ.**   (THE BARRACUDAS.)

Body elongate, subterete, with small, cycloid scales.   Head very
long. pointed ; mouth large, with unequal teeth, some of them very
large ; lower jaw projecting, a very strong tooth at tip ; gill rakers
obsolete ; gill openings wide ; lateral line present ; air-bladder
large ; P. short ; V. I, 5.   Dorsals separate, the first with 5 stout
spines ; A. with one spine ; C. forked ; vertebræ 24.   One genera
with 15 species ; voracious pike-like fishes of warm seas, some of
them very large, all excellent as food.

115. SPHYRÆNA (Artedi) Bloch.   (Ancient name, from
σφῦρα, hammer.)

a. Scales small, 130 to 150 in lateral line.

251. **S. borealis** DeKay.   LITTLE BARRACUDA.   Olivaceous,
silvery below ; young with dusky blotches ; P. not nearly reaching
D. ; maxillary not nearly to eye.   Head 3 ; depth 8.   D. V – I, 9.
A. I, 9.   L. 12.   Cape Cod to Va., not rare N.   (Lat., northern.)

aa. Scales moderate, about 110 in lateral line.

252. **S. guachancho** Cuv. & Val. P. about reaching spinous D. Head 3¼; depth 7. D. V - I, 9. A. I, 8. L. 24. West Indies, rarely N. (The Spanish name.) We place next a family of uncertain relationship.

FAMILY XLIX. **AMMODYTIDÆ.** (THE SAND LANCES.)

Body elongate, compressed, with small, cycloid scales; lateral line along side of back; mouth large, toothless, the chin projecting; upper jaw very protractile; gill membranes separate, free; gill rakers long and slender; pseudobranchiæ large; D. long and low, of soft rays only; A. similar, shorter; C. forked; no ventrals; P. low. No air-bladder. Vertebræ 63. Genera 4; species 8. Small fishes swimming in large schools, and burying themselves, by a quick movement, in sand. Coasts of N. regions. The relations of the family are still uncertain. They may be *Anacanthini, Percesoces,* or possibly allies of the Scombroids. In many regards, especially the structure of the gills, they resemble *Sphyrœna.*

a. Body with many transverse oblique folds; a fold of skin along edge of belly; vomer unarmed. . . . . AMMODYTES, 116.

116. **AMMODYTES** (Artedi) Linnæus. (ἄμμος, sand; δύω, dive.)

253. **A. tobianus** L. SAND LANCE. LANT. Olivaceous; a steely lateral stripe; P. reaching front of D. Head 4¾; depth 10. D. 60. A. 28. Lateral folds 125 to 130. L. 6. North Atlantic and Pacific, S. to N. J.; common N. (*Eu.*) The American form (var. *americanus* DeKay) has dorsal beginning a trifle further back. (An old name, unexplained.)

ORDER XX. **ACANTHOPTERI.** (THE SPINY-RAYED FISHES.)

This order contains the great bulk of the spiny-rayed fishes, and includes a far greater variety of forms than any other of the so-called orders. In all, the ventrals, if present, are thoracic, or jugular, normally I, 5, the opercles and pharyngeals are well developed, the gills normal, usually 4 in number, and the premaxillary forming the whole border of the mouth. Usually the anterior rays of D. and A. are simple or spine-like. (ἄκανθα, spine; πτερόν, fin.)

The various suborders of this group have not yet been fully defined or generally adopted. The following ten, of varying value, may be recognized for the fishes discussed in the present work: *Discocephali, Scombriformes, Perciformes, Pharyngognathi, Epelasmia, Cataphracti, Haplodoci, Xenopterygii, Scyphobranchii,* and *Anacanthini.*

DISCOCEPHALI. — Of these various suborders, we notice first the DISCOCEPHALI, a small group characterized by a singular modification of the dorsal fin.

FAMILY L. ECHENEIDIDÆ. (THE REMORAS.)

Body fusiform, elongate, with minute smooth scales; mouth wide, with villiform teeth; lower jaw projecting. Spinous dorsal changed into a sucking disk placed on top of head and composed of a double series of transverse movable cartilaginous plates. Opercles unarmed, P. placed high. V. I, 5. D. and A. long, similar; gillrakers short; no pseudobranchiæ; no air-bladder. Vertebræ more than 24. Genera 3, species 10. Of the open seas, attaching themselves to sharks, sword-fishes, tunnies, and floating objects, and thus carried for great distances in the sea. The relationships of this group are still uncertain. Their resemblance to *Elacate* is such that they apparently should not be placed very far away from the next family.

a. Body slender; vertebræ 14 + 16 = 30; disk of 21 to 25 laminæ; not more
    than ½ body. ECHENEIS, 117.
aa. Body robust; vertebræ 12 + 15 = 27; disk more than ½ body, of 16 to 18
    laminæ.
    b. Pectoral rays normal, soft. . . . . REMORA, 118.
    bb. Pectoral rays stiff, broad, ossified. . . . . RHOMBOCHIRUS, 119.

117. ECHENEIS (Artedi) Linnæus. (ἐχενηΐς, an ancient name meaning one who holds ships back.)

254. **E. naucrates** L. SUCKING-FISH. PEGADOR. Blackish, belly dark; a black lateral stripe; corners of C. pale. Head 5½; disk 3⅔, shorter than D.; width between P. 7½. D. XXI to XXV - 32 to 41. A. 32 to 38. L. 30. Warm seas. N. to Cape Cod; the commonest species, on sharks, etc. (*Eu.*). (ναυκράτης, pilot.)

118. REMORA Gill. (Ancient name, meaning one who holds back.)

a. Dorsal about XVIII - 23. (*Remora.*)

255. **R. remora** (L.). REMORA. SUCKING-FISH. Uniform dusky: head 4; disk 2⅔, longer than D.; width between P. 5¼. A. 25. L. 12. Warm seas, N. to N. Y. (*Eu.*)

aa. Dorsal about XVI - 30. (*Remoropsis* Gill.)

256. **R. brachyptera** (Lowe). SWORD-FISH REMORA. Light brown. Head 4; disk shorter than dorsal; width between P. 6½. A. 26. Warm seas, rarely N.; on sword-fish. (βραχύς, short; πτερόν, fin.)

119. RHOMBOCHIRUS Gill. (ῥόμβος, rhomb; χείρ, hand.)

257. **R. osteochir** (Cuvier). SPEAR-FISH REMORA. Light brown; mouth small; disk very large. Head 5; disk 2¼; width

between P. 5. D. XVIII - 21. A. 20. W. I., rare N. (ὀστέον, bone; χείρ, hand.)

The position of the next family is still uncertain. Common opinion places it between the Remoras and the mackerel-like fishes.

FAMILY LI. **ELACATIDÆ.** (THE COBIAS.)

Body elongate, fusiform, with very small, smooth scales; head long, low; mouth moderate; jaws with bands of small teeth; chin projecting; lateral line present, wavy. Dorsal spines about 9, low, all separate; second D. and A. long; two weak anal spines; V. I, 5. C. forked; no air-bladder; no sucking disk; pyloric cæca branched. One species, in all warm seas.

120. **ELACATE** Cuvier. (ἠλακάτη, spindle.)

258. **E. canada** (L.). COBIA; CRAB-EATER; SERGEANT-FISH. Dusky, sides with a broad black band. Head 4¼; depth 5⅔; D. IX, 33. A. II, 25. L. 5 feet. Warm seas, N. in summer.

We now begin the great series or suborder of SCOMBRIFORMES or mackerel-like fishes, with one of the most aberrant members of the group.

FAMILY LII. **XIPHIIDÆ.** (THE SWORD-FISHES).

Body elongate, naked; bones of upper jaw consolidated into a long stiff " sword "; teeth disappearing with age; D. long, without distinct spines, the rays enveloped in the skin; the fin divided into two in the adult; A. similarly divided; tail slender, keeled; C. widely forked; V. wanting. Gills peculiar, the laminæ of each arch joined in one plate by reticulations; air-bladder simple; pyloric cæca numerous. Vertebræ short, the neural and hæmal spines normal; ribs very few. One species, a very large fish of the open sea, much valued as food.

121. **XIPHIAS** Linnæus. (ξιφίας, ancient name from ξίφος, sword.)

259. **X. gladius** L. SWORD-FISH. Dark bluish. Head 2¼; depth 5½; snout 3. D. 40-4. A. 18-14. L. 15 feet or more. Open sea, N. to Nova Scotia. (*Eu.*) (Lat; sword.)

FAMILY LIII. **ISTIOPHORIDÆ.** (THE SAIL-FISHES.)

Similar to the Sword-fishes, but with rudimentary scales, small persistent teeth, and ventral fins of 1 or 2 rays; air-bladder sacculated; rays of fins distinct, not embedded in skin. Vertebræ " elongate hour-glass-shaped; neural and hæmal spines flag-like; ribs well-developed." Two genera, with 5 species. These are smaller than the *Sword-fishes*, but similar in character and habits.

a. Ventral rays united into one; D. low. . . . . . TETRAPTURUS, 122.
aa. Ventral rays 2 or 3; D. very high. . . . . . . ISTIOPHORUS, 123.

122. TETRAPTURUS Rafinesque.   (τέτρα-, four ; πτερόν, fin ;
ουρά, tail.)

260. **T. albidus** Poey.   SPEAR-FISH.   BILL-FISH.   Blue-black;
head (with sword) 2⅖; depth 7½.   D. III, 39–6.   A. II, 13–6.   L. 8
feet.   W. I., N. to Cape Cod.   (Lat., white.)

123. ISTIOPHORUS Lacépède.   (ἱστίον, sail ; φορέω, to bear.)

261. **I. americanus** Cuv. & Val.   SAIL-FISH.   SPIKE-FISH.
Bluish-black; dorsal very high, its membrane with round black
spots.   Sword, from eye, 2⅖ times rest of head, nearly twice as
broad as deep.   Head 2⅔; depth 6.   D. XLI – 7.   A. 9–7.   L. 6 to
8 feet.   Warm seas, N. to Cape Cod.

FAMILY LIV.   **TRICHIURIDÆ.**   (THE SCABBARD-FISHES.)

Fishes closely related to the *Scombridæ*, but having the vertebræ
very numerous, and the dorsal fin long and low, its spines and soft
rays indistinguishable from each other, and without finlets.   Ventral fins rudimentary or wanting.   Genera 6 ; species about 15, in
the warm seas.

a. No caudal fin; tail tapering to a point; no ventrals; teeth very strong,
unequal, some of them barbed. . . . . TRICHIURUS, 124.

124. TRICHIURUS Linnæus.   (τρίχιον, a little hair ;
ουρά, tail.)

262. **T. lepturus** L.   SCABBARD-FISH.   CUTLASS-FISH.   SIL-
VER EEL.   Silvery, D. dark-edged ; snout long ; lower jaw longer.
Head 7½; depth 16.   D. 135.   A. very low, 100.   Warm sea, N. to
N. Y.   (λεπτός, thin ; ουρά, tail.)

FAMILY LV.   **SCOMBRIDÆ.**   (THE MACKERELS.

Body subfusiform or compressed, with small cycloid scales, those
at the shoulders sometimes enlarged, forming a corselet ; lateral line
present.   Head pointed ; mouth large, not protractile ; teeth sharp,
large or small ; opercles unarmed; gill openings very wide ; pseudobranchiæ large.   Dorsals two, the first of slender spines, the second
usually followed by detached finlets; tail slender, keeled, its fin
widely forked; V. thoracic I, 5.   Vertebræ in increased number, 30
to 70 ; pyloric cæca many.   Coloration metallic, the sexes similar.
Genera about 17 ; species about 70.   Fishes of the high seas, many
of them cosmopolitan, coming to northern shores to spawn, and
often irregular in their visits.   Most of them are valued as food,
but the red, oily flesh of some is very coarse.

c. Finlets present (5 to 10 in number) behind D. and A.; dorsal spines less than 25. (*Scombrinæ.*)

  b. Caudal peduncle with median keel, a small keel above and below this.

    c. Body wholly covered with small scales, those on the "corselet" and lateral line sometimes larger; vertebræ normal.

      d. Teeth of jaws strong, subtriangular, more or less compressed; teeth on vomer and palatines villiform; gill rakers few; corselet obscure; dorsal spines 14 to 18; body compressed; head short; vertebræ 45. . . . . SCOMBEROMORUS, 125.

     dd. Teeth of jaws subconic, scarcely compressed; gill rakers numerous; corselet distinct.

      e. Vomer toothless; palatines with one row of strong conical teeth; body elongate; vertebræ about 52. . . . . . SARDA, 126.

       ee. Vomer and palatines with sand-like teeth; body robust; vertebræ 40. . . . . ALBACORA, 127.

  cc. Body scaleless, excepting on corselet and about lateral line; abdominal vertebræ with enlarged foramina and a trellis-like structure between the vertebra proper and the hæmapophyses; vertebræ about 38.

    f. Dorsals close together, the interspace about 5 in head; palatine teeth villiform; no teeth on vomer. . . GYMNOSARDA, 128.

    ff. Dorsals well separated, the interspace more than half head; teeth small, on vomer and not on palatines; gill rakers numerous. . . . . AUXIS, 129.

bb. Caudal peduncle without median keel (the two lesser keels present as usual); dorsals well separated, the interspace more than half head; spinous dorsal short; body scaly; corselet obsolete; vertebræ normal, about 31, in number; teeth slender, on jaws, vomer and palatines; gill rakers long, numerous. . . . . SCOMBER, 130.

125. SCOMBEROMORUS Lacépède. (σκόμβρος, Scomber; ὅμορος, near.)

a. Gill rakers short, thick, less than X + 8; dorsal spines 14 or 15.

263. **S. cavalla** (Cuvier). KING-FISH. Lateral line abruptly bent below soft D. Iron gray, nearly plain; spinous D. not black. Head 5; depth 6. D. XV–15–VIII. A. II–15–VIII. L. 6. W. Indies, rarely N., a fine food-fish. (Spanish, horse.)

aa. Gill rakers rather long and slender, more than X + 8; dorsal spines 17 or 18; lateral line wavy, not abruptly bent; teeth strong; spinous dorsal largely black.

  b. Side with one or two narrow blackish stripes breaking up posteriorly into irregular spots; similar spots usually present below these.

264. **S. regalis** (Bloch). SIERRA. PINTADO. Teeth 40 in each jaw; snout bluntish. Head 4¼; depth 4½. D. XVIII–16– VIII. A. II–14–VIII. L. 2½ feet. W. Indies, rarely N.

  bb. Sides with numerous round bronze spots, but never with dark longitudinal stripe.

265. **S. maculatus** (Mitchill). SPANISH MACKEREL. Bluish above, sides silvery; teeth about 30 in each jaw; snout pointed.

Head 4⅔; depth 5. D. XVIII-17-IX. A. II-18-VIII. L. 2⅓ feet. Tropical America, N. in summer, a favorite food fish. (Lat., spotted.)

126. SARDA Cuvier. (Lat. name, from Sardinia, where it abounds.)

266. **S. sarda** (Bloch.) BONITO. Steel-blue, with several blackish streaks obliquely downward and forward from back. Head 3¾; depth 4⅓. D. XXI-1, 13-VII. A. II-13-VII. L. 4 feet. Atlantic, abundant N. to Cape Cod. (*Eu.*)

127. ALBACORA Jordan (gen. nov.)

(*Orcynus* and *Thynnus* Cuvier, both names preoccupied.) (*Albacore*, a word said to be of Moorish origin.)

a. Pectoral fins short, about reaching 9th dorsal spine, and 6 to 7 in body. (*Albacora.*)

267. **A. thynnus** (L.). GREAT TUNNY. ALBACORE. Very robust. Dark blue, dusky below with obscure paler spots. Head 3¾; depth 4. D. XIV-1, 12-VIII. A. II-12-VIII. L. 12 to 15 feet. Atlantic, everywhere, one of the largest of fishes, sometimes reaching 1500 lbs. (θύννος, tunny.) (*Eu.*)

128. GYMNOSARDA Gill. (*Euthynnus* Lütken.)
(γυμνός, naked; Sarda.)

a. Lateral line abruptly curved behind second dorsal.

268. **G. pelamis** (L.). OCEANIC BONITO. Bluish; four brown stripes on each side of belly. Head 3½; depth 4. D. XV-12-VIII. A. 12-VII. Atlantic, scarce, W. (*Eu.*) (πελαμύς, tunny.)

aa. Lateral line without abrupt curve.

269. **G. alletterata** (Rafinesque). LITTLE BONITO. Bluish; no stripes on lower parts; several oblique wavy dark streaks above lateral line, about 5 blackish spots below P. Head 3¾; depth 4⅓. D. XV-12-VIII. A. 12-VII. L. 2½ feet. Warm seas, rarely N. (*Eu.*) (From *alletteratu*, the Sicilian name.)

129. AUXIS Cuvier. (αὖξίς, a young tunny.)

270. **A. thazard** (Lacépède). FRIGATE MACKEREL. Blue, somewhat mottled with darker. Head 4; depth 4⅓. D. X-12-VIII. A. 13-VII. L. 18. Warm seas, occasional schools on our coast. (*Eu.*) (From *tassard*, the French name.)

130. SCOMBER (Artedi) Linnæus. (σκόμβρος, Scomber, mackerel.)

a. Air bladder none; top of head without translucent area. (*Scomber.*)

271. **S. scombrus** L. COMMON MACKEREL. Dark blue above, silvery below, the lower parts unmarked; eye moderate;

back with about 35 dark wavy stripes. Head 3; depth 3½. D.
XI-12-V. A. 12-V. L. 2 feet. Atlantic, everywhere abundant,
one of the best known of food fishes.

   bb. Air bladder small; top of head with a translucent area. (*Pneuma-
     tophorus* Jordan & Gilbert.)

   272. **S. colias** Gmelin. CHUB-MACKEREL. THIMBLE EYE.
Dark blue; sides soiled silvery, in the adult showing dusky cloud-
ings; back with about 30 dark wavy streaks, extending to just below
the lateral line; eye large. Head 3; depth 3¼. D. IX or X-12-V.
A. 12-V. L. 12. Warm seas, not rare N., a food fish of much
less value than the mackerel. (*Eu.*) (κολίας, old name of some
mackerel.)

FAMILY LVI. **CARANGIDÆ.** (THE POMPANOS.)

Fishes closely allied to the Mackerels, but with the vertebræ in
moderate number, about 25. Anal fin always preceded by two
spines, which sometimes disappear with old age; finlets usually
few or none. Teeth all small. Coloration usually metallic silvery.
Genera, 25; species 180; in all warm seas; most of them excellent
as food.

   a. Premaxillaries not proctractile (except in the very young); soft dorsal
     similar to anal, both very long. (*Scombroidinæ.*)
     b. Maxillary without supplemental bone; no pterygoid teeth: scales linear,
       imbedded. . . . . OLIGOPLITES, 131.
   aa. Premaxillaries protractile.
     c. Pectoral fins long, falcate; anal similar to soft dorsal, its base longer
       than abdomen; maxillary with supplemental bone. (*Caranginæ.*)
       d. Dorsal outline not less curved than ventral.
         e. D. and A. each with one free finlet; body slender.
                           DECAPTERUS, 132.
        ee. D. and A. without finlets.
         f. Lateral line with well developed scutes for its entire length.
                            TRACHURUS, 133.
        ff. Lateral line with scutes on its straight posterior portion only
          (these sometimes few and small in species with the body
          compressed).
         g. Shoulder girdle with a deep cross-furrow at its junction with
          the isthmus; body oblong. . . . . TRACHUROPS, 134.
        gg. Shoulder girdle normal.
         h. Body oblong or moderately elevated, not as below.
                            CARANX, 135.
        hh. Body oblong-ovate, very strongly compressed, its out-
          lines all trenchant, the anterior profile vertical; scutes
          almost obsolete. . . . . VOMER, 136.
       fff. Lateral line without any scutes anywhere; body short and
        elevated, strongly compressed. . . . . . SELENE, 137.
     dd. Dorsal outline less strongly curved than ventral; body compressed,
       with trenchant outlines; scutes of lateral line obsolete.
                        CHLOROSCOMBRUS, 138.

    cc. Pectoral fins short, not falcate.

        i. Maxillary without supplemental bone; anal similar to soft dorsal; its base much longer than abdomen; tail unarmed. (*Trachinotinæ.*)

          j. Forehead convex; teeth small, lost with age; membrane of spinous dorsal disappearing with age. . . . . . . TRACHINOTUS, 139.

        ii. Maxillary with supplemental bone; A. shorter than soft D., its base not longer than abdomen. (*Seriolinæ.*)

          k. D. and A. without finlets.

            l. Membrane of D. spines disappearing with age.

                           NAUCRATES, 140.

            ll. Membrane of D. spines persistent. . . . . . SERIOLA, 141.

          kk. D. and A. each followed by a two-rayed finlet. ELAGATIS, 142.

    131. OLIGOPLITES Gill. (ὀλίγος, small; ὁπλίτης, armed.)

    273. **O. saurus** (Bloch & Schneider). LEATHER-JACKET. RUNNER. Bluish, silvery below; fins yellow. Body lanceolate; fins low. Head 5; depth 4. D. V – 1, 20. A. II – 1, 20. L. 18. Warm seas; rarely N. (σαῦρος, old name of some fish that skips like a lizard.)

    132. DECAPTERUS Bleeker. (δέκα, ten; πτερόν, fin.)

    a. Scutes about 40: teeth present.

    274. **D. punctatus** (Agassiz). SCAD. CIGAR-FISH. ROUND ROBIN. Bluish; a dark opercular spot; about twelve small black spots on lateral line anteriorly. Head 4⅓; depth 5. D. V III – 1, 30-I. A. II, – 1, 24-I. L. 12. W. I., etc.; occasional N.; common S. (Lat., dotted.)

    aa. Scutes about 25; teeth obsolete.

    275. **D. macarellus** (Cuv. & Val.). Lateral line unspotted; D. soft rays 33. A. 27; depth 5¾. W. I., rarely N.

    133. TRACHURUS Rafinesque. (τράχουρος, ancient name, from τραχύς, rough; οὐρά, tail.)

    276. **T. trachurus** (L.). HORSE-MACKEREL. SAUREL. Scutes all large, about 72 (35 + 37) in number; depth about 4. D. VIII – 1, 29. A. II – 1, 28. L. 12. S. Europe, etc., occasional on our coast. (*Eu.*)

    134. TRACHUROPS Gill. (*Trachurus*; ὤψ, appearance.)

    277. **T. crumenophthalmus** (Bloch). BIG-EYED SCAD. CHICHARRO. GOGGLER. Eye very large, 3 in head, with very large adipose eyelid; scutes 40. Head 3½; depth 3⅓. D. VIII – 1, 26. A. II – 1, 22. L. 12. Warm seas. N. to Cape Cod. (Lat., *crumena*, purse; ὀθφαλμός, eye.)

    135. CARANX Lacépède. (A corruption of the Portuguese *Acarauna*, French *Carangue.*)

    a. Teeth in jaws in few series, unequal, those above enlarged, those below uniserial; teeth on vomer, palatines and tongue; soft dorsal and anal falcate in front; maxillary broad. (*Caranx.*)

b. Body subfusiform, the depth less than ⅓ the length; breast scaly; no canines; scutes numerous, 40 to 50.

278. **C. chrysos** (Mitchill.) HARD-TAIL. YELLOW MACKEREL. COJINERA. Greenish, yellow below; a black blotch on opercle; none on P; breast scaly; arch of lateral line about half straight part. Head 3¾; depth 3¼. D. VIII – 1, 24. A. II, 1, 19. Scutes 50. L. 18. Cape Cod, S., rather common. (χρυσός, gold.)

  bb. Body oblong-ovate, the depth more than ⅓ the length; outer teeth stronger; scutes larger, 25 to 30; silvery species.

    c. Breast entirely scaly; opercular spot inconspicuous; lower jaw without distinct canines.

279. **C. latus** Agassiz. JUREL. Pectoral spot usually wanting. Head 3⅔; depth 2¾. D. VIII – 1, 22. A. II – 1, 16. Scutes, 30. L. 18. Warm seas, rarely N. (Lat., broad.)

    cc. Breast naked, except a small rhombic scaly area before V.; lower jaw with two small canines; adult with a large black spot on opercle, and one towards base of P.

280. **C. hippos** (L.). CREVALLÉ. CAVALLA. Head large and deep, especially in adult, mouth large. Head 3¼; depth 2½ to 3. D. VIII – 1, 20. A. II – 1, 17. Scutes 25. L. 36. Warm seas, N. to Cape Cod; common S. (ἵππος, horse.)

aa. Teeth of jaws equally small; breast naked, spinous dorsal disappearing with age; soft dorsal and anal with 3 to 6 anterior rays produced in long filaments. (*Alectis* Rafinesque.)

281. **C. gallus** (L.). THREAD-FISH. Body very deep, broadly ovate, its edges trenchant: scales minute; scutes very feeble; silvery, darker above; a dark blotch on opercle; changes greatly with age. Head 3; depth 2 (young as deep as long). D. VI – 1, 19. A. 16. Scutes 9 to 12. L. 2 feet. Warm seas, N. to N. Y. (The American fish, called *Caranx crinitus* Mitchill, seems to be the same as the East Indian *C. gallus.*) (Lat., cock.)

136. VOMER Cuvier. (Lat., ploughshare.)

282. **V. setipinnis** (Mitchill). MOON-FISH. HORSE-FISH. Body oblong, excessively compressed, but less elevated than in *C. gallus* or in *Selene vomer;* fins in adult all very low, none filamentous; head very gibbous above eye; scutes minute. Head 3¼; depth 2 (deeper in young). D. VIII – 1, 21 to 25; A. II – 1, 18 to 20. L. 18. Tropical America, N. to Maine. (Lat., *seta,* bristle; *pinna,* fin.)

137. SELENE Lacépède. (σελήνη, the moon.)

a. D. with 22 soft rays; A. with about 18; anterior profile of head from base of snout to occiput almost straight, the bones of the head being much distorted.

283. **S. vomer** (L.). Moon-fish. Horse-head. Look-down. Adult with soft rays of D. and A. much produced; young with dorsal spines and V. variously elongate, these fins short with age. Silvery. Head 3; depth 1½. L. 12. Warm seas, frequently N. to Cape Cod.

138. CHLOROSCOMBRUS Girard. (χλωρός, green ; σκόμβρος. mackerel.)

284. **C. chrysurus** (L.). Bumper. Casabe. Greenish; sides and below golden ; a dark blotch on back of tail; head deep; mouth very oblique; P. very long; chord of arch of lateral line 1¼ to 1¾ in straight part; no scutes. Head 3¾; depth 2⅓. D. VIII–1, 26. A. II–1, 26. L. 9. W. Indies, rare N. (χρυσός, gold; οὐρά, tail.)

139. TRACHINOTUS Lacépède. Pompanos. (τραχύς, rough ; νῶτος, back.)

a. Dorsal with 19 to 20 soft rays; anal with 17 to 19.

b. Body broadly ovate, its depth at all ages more than half the body; sides without black bars.

285. **T. falcatus** (L.). Round Pompano. Palometa. Body deep; profile from nostril to dorsal everywhere about equally convex ; lobes of D. and A. high, reaching in adult beyond middle of fin ; bluish, sides silvery ; lobes of D. black in young; no axillary spot. Head 3¾; depth 1⅜. L. 20. Warm seas, occasional N. to N. Y. (*T. ovatus* (L.); *T. rhomboides* Bloch.) (Lat., scythe-shaped.)

aa. Dorsal with 25 soft rays; anal with 22; body oblong, rather robust.

286. **T. carolinus** (L.). Common Pompano. Bluish, golden below; changes greatly with age, the young deeper, with conspicuous fin-spines, and with teeth in jaws; D. and A. lobes about reaching middle of fins. Head 4; depth 2⅔. L. 18. Gulf Coast, etc., N. to Cape Cod, common S.; a famous food fish.

140. NAUCRATES Rafinesque. (ναυκράτης, pilot.)

287. **N. ductor** (L.). Pilot-fish. Romero. Bluish with about 6 broad dark vertical bars. Head 4; depth 4. D. IV–1, 26. A. II–1, 16. Pelagic ; occasional on our coast. (*Eu.*) (Lat., guide.)

141. SERIOLA Cuvier. Amber-fishes. (An Italian name.)

288. **S. zonata** (Mitchill). Rudder-fish. Bluish, with 6 broad black bars, which fade or disappear with age; an oblique dark band from eye to spinous dorsal; V. mostly black. Head longer than deep; occiput compressed; tail keeled. Head 3½; depth 3¼. D. VII–1, 38. A. II–1, 21. L. 30. Cape Cod to W. I., not rare. (Lat., banded.)

142. **ELAGATIS** Bennett. (ἠλακάτη, spindle.)

289. **E. bipinnulatus** (Quoy & Gaimard). Blue, yellow below; side with 3 longitudinal bluish stripes. Head $3\frac{5}{6}$; depth $3\frac{2}{3}$. D. VI - 1, 27 - II. A. II - 1, 17 - II. L. 18. Warm seas, rarely N. to L. I. (*Meek.*) (Lat., *bis*, two; *pinnula*, little fin.)

FAMILY LVII. **POMATOMIDÆ.** (THE BLUE-FISHES.)

Closely allied to the *Carangidæ* but with the scales larger and weakly ctenoid. Mouth large, oblique, with very strong, compressed, unequal teeth; premaxillaries protractile; caudal peduncle stout, the fin forked, with broad lobes; preopercle serrate; lateral line unarmed. First dorsal of about 8 fragile spines; second D. and A. long; anal spines minute. A single species, in most warm seas.

143. **POMATOMUS** Lacépède. (πῶμα, opercle; τομός, cutting.)

290. **P. saltatrix** (L.). BLUE-FISH. SKIP-JACK. Bluish, silvery below; a black blotch at base P.; body robust, somewhat compressed; P. inserted low, nearly 2 in head. Head $3\frac{1}{2}$; depth 4. D. VIII - 1, 25. A. II - 1, 26. Lat. l. 95. L. 3 feet. Warm seas, common on our Atlantic coast; an excellent and gamy fish, but very destructive to other species. (Lat , leaper.)

FAMILY LVIII. **STROMATEIDÆ.** (THE BUTTER-FISHES.)

This family is also very close to the *Carangidæ*, differing chiefly in the presence of numerous horny, barbed or hooked teeth in the œsophagus, and in the greater number of vertebræ (30 or more). There are no free anal spines, and the spinous D. is very much reduced or even wanting. Some of the species differ from other mackerel-like fishes in having the gill membranes attached to the isthmus, while still others have no ventral fins. Genera 5; species about 30; of the warm seas.

a. Ventral fins I, 5, well-developed; premaxillaries protractile; gill openings wide; caudal peduncle stout. (*Centrolophinæ*.)
  b. Preopercle finely serrate; dorsal spines short and stout; anterior rays of D. low; scales moderate. . . . . LEIRUS, 144.

aa. Ventrals minute or absent; premaxillaries not protractile; caudal peduncle slender; the fin widely forked; opercles entire; scales minute; spinous D. almost obsolete. (*Stromateinæ*.)
  c. Gill membranes free from isthmus. . . . . . STROMATEUS, 145.

144. **LEIRUS** Lowe. (λειρός, thin.)

291. **L. perciformis** (Mitchill). BLACK RUDDER-FISH. Blackish-green everywhere; eye large; snout blunt. Head $3\frac{1}{3}$; depth $2\frac{1}{4}$. D. VII - 1, 20. A. III, 16. Lat. l. 75. L. 12. Maine to N. J., not rare N. (Lat., *perca*, perch; *formis*, shape.)

145. STROMATEUS (Artedi) Linnæus.   (στρωματεύς, ancient name.)

a. Pelvis ending in a small spine; V. wanting.

b. D. and A. little falcate, their lobes shorter than head; a row of conspicuous pores along side of back above lateral line. (*Poronotus* Gill.)

292. **S. triacanthus** Peck.   DOLLAR-FISH.   BUTTER-FISH.
Bluish; silvery below; body oval, compressed; snout very blunt. Head 4; depth 2⅖. D. III, 45. A. III, 37. L. 10. Maine to Florida, common N. (τρίς, three; ἄκανθα, spine.)

bb. D. and A. falcate, their lobes longer than head; back without evident pores. (*Rhombus* Lacépède.)

293. **S. paru** L.   HARVEST-FISH.   Bluish, yellow below; body almost round, with vertical snout. Head 4; depth 1⅖. D. III, 45. A. II, 43. L. 8. Cape Cod to S. A., rare N. (Brazilian name.)

FAMILY LIX. **CORYPHÆNIDÆ**. (THE DOLPHINS.)

Body elongate, compressed, with small, cycloid scales; mouth wide, with moderate teeth; opercles entire; occipital crest extending well forward, becoming very high in the adult ♂. D. continuous from nape nearly to C., without distinct spines; A. similar, shorter; V. I, 5 : P. short; C. widely forked. Gill openings wide. No pseudobranchiæ nor air-bladder. Vertebræ more than 24. One genus, with 2 or 3 species; large vigorous fishes of the open seas. The bright coloration grows pale at death, but the accounts of this change have been much exaggerated.

146. CORYPHÆNA (Artedi) Linnæus.   (κόρυς, helmet ; φ. ίνω, to show.)

294. **C. hippurus** L.   COMMON DOLPHIN.   DORADO.   Very bright olive-green, with small round blue spots; V. inserted slightly behind upper ray of P. Head 4⅔; depth 4½; V. 1¼ in head; P. 1½. D. 59 to 63. A. 29. L. 3 to 5 feet. Open sea, N. to Cape Cod, abundant S. (ἵππος, horse; οὐρά, tail.)   (*Eu.*)

With the Dolphins, we close the series of fishes having Scombroid affinities, and begin the equally important series of PERCIFORMES, those related in some degree to the common Perch. The Perch-like fishes have usually larger and rougher scales than the Scombroids, and the development of the spinous armature of the fins is in general more pronounced. We begin with one of the most aberrant forms, the small

FAMILY LX. **APHREDODERIDÆ**. (THE PIRATE PERCHES.)

Body oblong, with thick, depressed head and compressed tail; mouth moderate, the chin projecting; teeth in villiform bands on jaws, vomer, and palatines; premaxillary not protractile; maxillary simple; preopercle and preorbital serrate; opercle with a spine; bones of skull somewhat cavernous; gill rakers tubercle-like; gill membranes slightly joined to isthmus; no pseudobranchiæ; gills complete. B. 6. Scales strongly ctenoid; no lateral line. Vent anterior, below the preopercle in adult, farther back in young, its position changing by a lengthening of the rectum. Dorsal small, with 3 or 4 spines; anal with 2; ventrals without spine and with *seven* soft rays (all other perch-like fishes having one spine and five rays); C. rounded. Vertebræ 29. Air bladder large. Pyloric cæca 12. One species, a small fish of nocturnal habits, abounding in sluggish grassy lowland streams throughout the Eastern U. S.

147. APHREDODERUS Le Sueur. (ἄφοδος. excrement; δέρη, the throat.)

295. **A. sayanus** (Gilliams). PIRATE PERCH. Dark olive, profusely dotted with black; two dusky bars at base of C. Head 3; depth 3. D. III, 11. A. II, 6. Lat. l. 48 to 58. L. 6. N. Y. to La., and N. to Minn. and Lake Erie; variable. (To Thomas Say, the entomologist.)

FAMILY LXI. **ELASSOMATIDÆ.** (THE TINY PERCHES.)

Body oblong, compressed, with large cycloid scales; mouth small; teeth conic, strong, on jaws, a few on vomer; upper jaw very protractile; opercles entire; gill membranes broadly united, free from the isthmus; gill rakers tubercle-like; lower pharyngeals narrower, with sharp teeth. B. 5. No lateral line; pseudobranchiæ rudimentary. V. normal (I, 5). Dorsal small, with 4 spines; anal with 3; C. rounded. Vertebræ 24. One genus, with two species, *E. evergladei* Jordan, of Florida, and the following. They inhabit sluggish, lowland waters of the E. U. S., and they are among the smallest of all fishes.

148. ELASSOMA Jordan. (ἐλάσσωμα. a diminution.)

296. **E. zonatum** Jordan. Olive green, finely speckled; sides with 11 dark bars; a round black spot on side behind shoulder; fins spotted; a bar at base of C. Eye large; mouth small. Head 3; depth 3½. D. V, 9. A. III, 5. Scales 40–19. L. 1 to 1½. S. Ill. to Ark. and La., in grassy brooks. (Lat., banded.)

FAMILY LXII. **CENTRARCHIDÆ.** (THE SUN-FISHES.)

Body more or less shortened and compressed, so that the regions above and below the axis of the body are nearly equal and correspond to each other. Mouth terminal; teeth small; premaxillary protractile; maxillary with a supplemental bone which is sometimes

minute or obsolete: preopercle entire or nearly so; preorbital deep,
not sheathing the maxillary; gill membranes separate, free from
isthmus; pseudobranchiæ small, concealed. B. usually 6; lower
pharyngeals separate; scales usually large; lateral line present.
Dorsal continuous, with 6 to 13 spines; anal spines 3 to 8. Vertebræ about 30. Intestines short, with a few cæca. Sexes similar,
but the changes in form due to age often considerable. Genera 10;
species about 25; carnivorous fishes especially characteristic of the
Mississippi Valley, — all but one (*Archoplites interruptus* of Cal.)
confined to the waters of the E. U. S. Some species build nests,
and all are voracious and gamy.

a. Dorsal fin scarcely larger than anal; gill rakers very long and slender.
  b. Spinous dorsal longer than soft, its spines 12; anal spines about 8.
                                                    CENTRARCHUS, 149.
  bb. Spinous dorsal shorter than soft, with 6 to 8 spines; anal spines 6.
                                                    POMOXIS, 150.
aa. Dorsal fin much larger than anal; gill rakers shorter.
  c. Body comparatively short and deep, the depth usually more than $\frac{2}{5}$
    the length; dorsal fin not deeply divided.
    d. Tongue and pterygoids with teeth; mouth large (the maxillary
      reaching past middle of eye).
      e. Scales ctenoid; caudal concave behind.
        f. Opercle emarginate behind; anal spines usually 6; branchios-
          tegals 6. . . . . AMBLOPLITES, 151.
        ff. Opercle ending in a black convex process or flap: anal spines 3.
                                                    CHÆNOBRYTTUS, 152.
      ee. Scales cycloid; caudal convex.   . . . ACANTHARCHUS, 153.
    dd. Tongue and pterygoids toothless; mouth small (the maxillary barely
      to middle of eye).
      g. Caudal convex; opercle emarginate, without flap.
        h. Dorsal fin continuous, normally with 9 spines; anal nor-
          mally with 3 spines. . . . . ENNEACANTHUS, 154.
        hh. Dorsal fin angulated, some of the median spines elevated;
          dorsal spines 10; anal 3. . . . MESOGONISTIUS, 155.
      gg. Caudal margin concave; opercle prolonged behind in a con-
        vex process or flap which is always black; dorsal spines
        normally 10; anal 3. . . . . LEPOMIS, 156.
  cc. Body comparatively elongate, the depth in adult about $\frac{1}{4}$ the length;
    D. low, deeply emarginate, with 10 spines; mouth large; C. lunate.
                                                    MICROPTERUS, 157.

149. CENTRARCHUS Cuv. & Val.   (κέντρον, spine ; ἀρχός, anus.)

297. C. macropterus Lacépède. Body ovate; fins high. Green,
with rows of dark brown spots along sides; fins reticulated; young
with a black ocellus on D. behind. Head $3\frac{1}{4}$; depth 2. Scales
5–44–14. D. XI or XII, 12. A. VII or VIII, 15. L. 6. N. C.
to Ill., and S., in lowland streams. (μακρός, long; πτερόν, fin.)

150. POMOXIS Rafinesque. (πῶμα, opercle ; ὀξύς, sharp.)

a. Dorsal spines 7 or 8; A. reticulate, like soft D.

298. **P. sparoides** (Lacépède). CALICO BASS. GRASS-BASS. BAR-FISH. STRAWBERRY BASS. Body oblong, compressed, the profile comparatively even; fins very high. Silvery olive, much mottled with clear green ; vertical fins with green reticulations around pale spots. Head 3 ; depth 2. D. VII, 15. A. VI, 17. Lat. l. 41. L. 12. N. J. to Minn. and La.; commonest N.

aa. Dorsal spines 6; A. fin whitish, nearly plain.

299. **P. annularis** Rafinesque. CRAPPIE. BACHELOR. NEW LIGHT. CAMPBELLITE. SAC-A-LAI. Profile more or less distinctly S-shaped, the nape gibbous, the head depressed, the snout projecting; mouth very large. Silvery olive, mottled with dark green. Head 3 ; depth 2¼. D. VI, 15. A. VI, 18. Lat. l. about 40. L. 12. Variable. Miss. Valley, in quiet waters, common S. (Lat., ringed.)

151. AMBLOPLITES Rafinesque. (ἀμβλύς, blunt ; ὁπλίτης, armed.)

300. **A. rupestris** (Rafinesque). ROCK BASS. RED EYE. GOGGLE-EYE. Body oblong ; eye very large. Olive green, sides brassy, much mottled with dark green ; young with blackish bars; adult with rows of dark spots along sides ; iris red. Head 2¾ : depth 2. D. XI, 10. A. VI, 10. Scales 5–40–12. L. 12. Vt. to Manitoba, S. to La. and N. C., common W. (Lat., living among rocks.)

152. CHÆNOBRYTTUS Gill. (χαίνω. to yawn : *Bryttus* i. e. *Lepomis.*)

301. **C. gulosus** (Cuv. & Val). WAR-MOUTH. RED-EYED BREAM. Body oblong, robust ; eye moderate. Olive green, sides brassy with blotches of bluish, greenish, and copper-red ; cheeks with 3 or 4 dark bands; fins dusky, mottled ; a dark spot on last D. rays ; young barred ; some specimens with rows of dark spots on sides. Head 2⅔ ; depth 2¼. D. X, 10 ; A. III, 9. Scales 6–40–12. L. 10. L. Michigan to Va. and Texas, abundant S. in sluggish waters. Northern specimens are deeply colored, the adult with blue and copper-red; the D. is usually a trifle farther forward, over opercular spot ; this is var. *antistius* McKay. (Lat. big-mouthed.)

153. ACANTHARCHUS Gill. (ἄκανθα, spine ; ἀρχός, anus.)

302. **A. pomotis** (Baird). MUD SUN-FISH. Form of the Rock Bass. Dark-green, with 2 or 3 faint dusky longitudinal stripes ; cheeks with dark oblique bands; fins plain. Head 2⅔ ; depth 2. D. XI, 10. A. V, 10. Scales 6–43–12. L. 6. Hudson R. to N. C. in sluggish streams coastwise. (*Pomotis* = *Lepomis.*)

154. ENNEACANTHUS Gill. (*ἐννέα*, nine; *ἄκανθα*, spine.)

a. Depth usually more than half length; opercular spot large, more than half eye.

303. **E. obesus** (Baird). Olivaceous, with 5 to 8 distinct dark cross-bars; spots on body and fins golden or purplish; cheek with lines and spots; a dark bar below eye; cheek with 4 rows of scales; lateral line usually incomplete; fins moderate, spine of V. not reaching vent. Head 2⅔; depth 1¼. D. IX, 10. A. III, 10. Scales 4-32-10. L. 4. Mass. to Fla., common coastwise. (Lat., fat.)

aa. Depth usually less than half length; opercular flap small, bordered with pearly and blue.

304. **E. simulans** (Cope). Dark olive, young faintly barred; a dark bar below eye; ♂ with head, body and vertical fins with round sky-blue spots; ♀ duller, with lower fins and larger, faint spots; lateral line usually complete. Head 2¾; depth 2¼. D. IX, 10. A. III, 9. Scales 3-30-9. L. 5. N. J. to S. C., common coastwise; (number of spines sometimes variable). (Lat., resembling.)

305. **E. eriarchus** (Jordan). Olivaceous; vertical fins with round pale spots; lateral line incomplete; fins very large, especially A., which is reached by the ventral spines; scales on cheek, in 3 rows. Head 2¾; depth 2¼. D. X, 9. A. IV. 8 (in typical example probably abnormal). Scales 4-33-10. L. 3. Wis. to Mo.; two specimens known. (*ἔρι*, an intensive particle; *ἀρχός*, anus.)

155. MESOGONISTIUS Gill. (*μέσος*, middle; *γωνία*, angle; *ἱστίον*, sail.)

306. **M. chætodon** (Baird). Body suborbicular, the mouth very small, the fins high. Straw-color, with dark clouds; 6 to 8 irregular, sharply-defined black bars across body and fins, the first bar through eye. Head 3; depth 1⅜. D. X, 10. A. III, 12. Scales 4-28-10. L. 3. N. J. to Md., in sluggish streams; handsomest of the sun-fishes. (A genus of fishes.)

156. LEPOMIS Rafinesque. SUN-FISHES. (*Ichthelis, Pomotis,* and *Apomotis* Rafinesque.)

(A large genus, one of the most difficult in our fauna, as the species are subject to great individual variations, especially with age. On the other hand the numbers of scales and fin-rays are essentially alike in all, and nearly all the distinctive characters are subject to intergradation. The spines are generally higher in the young, while the "ear-flap" is fully developed only in the adult.) (*λεπίς*, scale; *πῶμα*, opercle.)

a Lower pharyngeals narrow, the teeth not paved.

   b. Pharyngeal teeth all, or nearly all, slender and acute.

      c. Supplemental maxillary well developed; palatine teeth present; gill rakers comparatively stiff and strong. (*Apomotis* Rafinesque.)

         d. Scales rather small, more than 40 in lateral line.

            e. D. and A. in adult, with a conspicuous black spot at base of last ray.

307. **L. cyanellus** (Rafinesque). GREEN SUN-FISH. Body oblong, the back not elevated; mouth large, the maxillary nearly to middle of eye; dorsal spines low, about equal to snout; opercular flap short, with pale margin. Green, with brassy lustre, each scale with a blue spot and gilt edging; fins largely blue, A. edged with orange; iris red; cheeks with blue stripes. Head 3; depth 2½. D. X, 11. A. III, 9. Lat. l. 48. L. 7. Great Lakes to Ga. and Mexico; very abundant in small brooks, especially S.; very variable. (κύανος, dark-blue.)

            ee. D. and A. without black spot.

308. **L. phenax** (Cope & Jordan). Body rather deep; mouth small, the maxillary to middle of eye; opercular spot longer than eye. Plain olive green; scales 6–43–14. L. 16. N. J. (φέναξ, false.)

         dd. Scales rather large, less than 40 in lateral line.

309. **L. symmetricus** Forbes. Body short, deep; mouth moderate. Dark green, sides with 10 vertical bars; dorsal in ♀ with black ocellus on last ray; cheek not striped; opercular spot higher than long; spines low. Head 2⅔; depth 1¾. Scales 6–34–14. L. 2½. Ill. to La., not rare; a neat and very small species.

      cc. Supplemental maxillary reduced to a slight rudiment; the mouth small, the palatine teeth few or none.

         d. Gill rakers stiff, not very short. (*Lepomis.*)

            e. Opercular flap short, little larger than eye, even in adult.

310. **L. ischyrus** Jordan & Nelson. Body robust, mouth large, the maxillary to middle of eye; profile depressed above eye; scales on cheek in 6 rows; opercular flap broad, with a broad pale edge. Dusky, mottled with blue and orange; cheeks with wide blue bands; a dark spot on D. and A. behind. Head 2⅔; depth 2½. Scales 5–46–14. L. 7. Ill. R.; only the type known. (ἰσχυρός, robust.)

311. **L. macrochirus** Rafinesque. Steel-blue with bronze orange spots, so arranged as to form series of vertical chain-like bars; fins with bronze and orange; no blue stripes on cheek; P. long, reaching A.; gill rakers slender, 11; 7 rows of scales on cheeks. Head 3; depth 2½. Scales 6–42–15. L. 5. Ohio valley, rare. (μακρός, long; χείρ, pectoral.)

            ee. Opercular flap in adult becoming more or less elongate and conspicuous.

f. Scales large, 5–34–11; opercular spot wholly surrounded by a very broad red margin.

312. **L. humilis** (Girard). Body oblong; spines high; cheeks with 5 rows of scales. Olive, with greenish specks, posteriorly; sides with round orange spots; belly and lower fins red. Head $2\frac{3}{4}$; depth $2\frac{1}{4}$. L. $2\frac{1}{2}$. Ky. to Neb. and Texas; very abundant S. W. (Lat., humble.)

ff. Scales rather small, 42 to 50 in the lateral line.

g. Opercular flap in the adult, very broad, without pale edge; D. and A. in adult with a large black spot on the last rays.

313. **L. pallidus** (Mitchill). BLUE SUN-FISH. COPPER-NOSED BREAM. DOLLARDEE. Body deep, compressed, the young slender, the adult very deep; tail slender; head small; mouth quite small, the maxillary barely to eye; gill-rakers slender, about 10; D. spines higher than in related species. Olive green; young purplish silvery, with greenish cross-bars; no blue stripes on check; no red on fins; old specimens often dusky, with the belly coppery red. Head 3; depth 2. Lat. l. 44. L. 10. Great Lakes to N. Y., Kans., Fla. and Mexico; very abundant. Very variable, but usually known by the black dorsal spot, which it shares with L. *cyanellus.* (Lat., pale.)

gg. Opercular flap in the adult, very long and narrow, not wider than eye, its lower margin pale; dorsal and anal usually without dark spot.

314. **L. auritus** (Linnæus). LONG-EARED SUN-FISH. Body rather elongate; mouth moderate, the maxillary past front of eye; gill rakers quite short, but stiff and rough; scales on check in 7 rows. D. spines low. Olive, belly and lower fins largely red; scales on sides with bluish spots; bluish stripes on head, especially before eye. Head without flap, 3; depth $2\frac{1}{2}$. Lat. l. 47. L. 8. Me. to La., only E. of the mountains; very abundant; usually known at sight by the long, narrow ear-flap. S. replaced by var. *solis* Cuv. & Val., with larger scales on check and belly, the former in 5 or 6 rows. (Lat., long-eared.)

dd. Gill rakers very short, weak and flexible; no palatine teeth; opercular flap in adult extremely long, with or without pale margin, variously shorter in young; head with blue streaks. (*Xenotis* Jordan.)

315. **L. megalotis** (Rafinesque). Body short and deep, the profile steep; mouth small, the maxillary to middle of eye; scales on check in 5 rows. Brilliant blue and orange, the former color predominating below, the blue in wavy streaks, the orange in spots; head with conspicuous blue stripes; fins mostly with membranes orange, the rays blue; V. dusky; no black spot on D. or A. Head without flap, 3; depth $1\frac{2}{3}$ to $2\frac{1}{4}$. Scales 5–38–14. L. 6. Mich. to Dakota, S. to S. C. and Mexico; very abundant, especially in

small brooks. The adult is readily recognized; the young may be known by the small gill-rakers and blue on head. (μεγάλος, large; οὖς, ear.)

316. **L. garmani** Forbes. Body rather deep; mouth moderate; maxillary not to front of pupil; eye large; cheeks with 5 rows of scales. Dusky; sides with rows of bronze spots, one to each scale, and about 7 rows below lateral line; opercular flap $\frac{2}{3}$ eye. Head $2\frac{4}{5}$; depth $2\frac{1}{4}$. Scales 5-34 to 41-14. L. 4. Wabash Valley. (To Harry Garman, of Champaign, Ill.)

bb. Pharyngeal teeth mostly bluntly conic; gill-rakers stout, rather short. (*Xystroplites* Jordan.)

317. **L. euryorus** McKay. Body very robust, the back high; gill rakers about 8; eye small; scales on cheek in 6 or 7 rows; opercular flap nearly as long as snout, with a very broad paler margin; spines low; P. short. Greenish, nearly plain. Head $3\frac{4}{5}$; depth $2\frac{2}{5}$. Scales 6-43-14. L. 7. Fort Gratiot, L. Huron; one specimen known. (εὐρύς, wide; ὄρος, margin.)

aa. Lower pharyngeals very broad, the teeth paved, almost spherical, and truncate at tip; gill-rakers small; opercular flap rather short and broad; its lower posterior edge always bright scarlet; no distinct black spot on D. (*Eupomotis* Gill & Jordan)

h. Body compressed, the back elevated; a considerable angle formed above the eye by the projecting snout; sides silvery-olive, scarcely spotted with orange; cheek without distinct blue lines.

318. **L. holbrooki** (Cuv. & Val.). Eye large, the maxillary reaching its front; cheeks with 5 rows of scales; spines high; P. long, longer than head; opercular spot large. Dusky olive, silvery below; somewhat mottled; belly yellow; fins nearly plain, the lower yellow. Head 3; depth 2. Scales 6-45-14. L. 8. S. Ill., to S. C. and S., in lowland streams. (The western form, var. notatus Agassiz, has perhaps the scales larger, 4-35-13, and 4 rows on cheek.) (To John Edwards Holbrook, author of Ichth., S. C.)

hh. Body robust, the back elevated, but not much compressed; the profile steeper, scarcely forming an angle above eye; the short snout little projecting; sides bluish, profusely spotted and blotched with orange; cheeks orange, with blue wavy streaks.

319. **L. gibbosus** (L.) COMMON SUN-FISH. BREAM. POND-FISH. PUMPKIN-SEED. SUNNY. Eye large, the maxillary reaching its front; cheeks with 4 rows of scales; spines moderate; P. scarcely longer than head; opercular spot moderate. Greenish olive, the sides bluish, the belly and lower fins orange; the sides profusely mottled with orange; D. bluish, orange-spotted. Head $3\frac{1}{4}$; depth 2. D. X, 11. A. III, 10. Scales 6-17-13. L. 8. Minn. and Great Lakes to Me., and S. to S. C.; exceedingly abundant N. and E., but in Western rivers rarely coming south of the latitude of

Chicago. A familiar and active inhabitant of clear brooks, defending its nests with great spirit. "A very beautiful and compact fish, perfect in all its parts, looking like a brilliant coin fresh from the mint." (Lat., gibbous.)

157. MICROPTERUS Lacépède. BLACK BASS. (μικρός, small; πτερόν, fin.)

a. Mouth moderate, the maxillary in adult not extending beyond eye; scales small, about 11-74-17; young more or less barred or spotted, never with a black lateral band.

320. **M. dolomieu** Lacépède. SMALL-MOUTHED BLACK BASS. Body ovate-oblong, growing deep with age; scales on the cheek small, in about 17 rows; D. less deeply notched than in the next; the ninth spine about half as long as the longest. Coloration variable, the young dull golden-green, with darker spots on sides which tend to cluster in short vertical bars; 3 bronze bands across cheeks; C. yellowish, next black, with a white tip; D. with bronze spots. Adult nearly uniform olive-green. Head 3¼; depth 3¼. D. X, 13. A. III, 10. Scales 10 or 11-72 to 75-17. L. 1 to 2 feet; weight 2 to 7 pounds. St. Lawrence River to Dakota, S. to S. C., Ala., and Ark., preferring clear and running streams; hence less common S. than the next, and for the same reason usually considered the better game-fish. "The Black-bass is eminently an American fish; he has the faculty of asserting himself and of making himself completely at home wherever placed. He is plucky, game, brave, unyielding to the last, when hooked. He has the arrowy rush and vigor of a trout, the untiring strength and bold leap of a salmon, while he has a system of fighting tactics peculiarly his own. I consider him inch for inch and pound for pound the gamest fish that swims." (J. A. Henshall.) (To M. Dolomieu, a scientist of Paris.)

aa. Mouth very large, the maxillary in the adult extending beyond the eye; scales rather large, about 7-68-16; last spines of D. very short, so that the fin is almost divided into two ; young with a blackish lateral band.

321. **M. salmoides** (Lacépède). LARGE-MOUTHED BLACK BASS. GREEN BASS. OSWEGO BASS. BAYOU BASS. Body rather deeper and more compressed than in the preceding, growing deeper with age; scales on cheek large, in about 10 rows; 9th D. spine not half length of longest. Color dark green, silvery below; sides with a broad blackish band in young, with some dark spots above and below it ; three dark stripes across cheeks ; C. pale at base and tip, mesially dusky. Adult dull green, nearly plain. Head 3¼; depth 3. D. X, 13. A. III, 11. Scales 8-68-16. L. 1 to 2½ feet; weight 3 to 8 pounds. Dakota to N. Y., S. to Florida and Mexico; everywhere abundant, preferring lakes, bayous, and sluggish waters. Variable. (Lat., Salmo, salmon ; εἶδος, like, which it is not.)

Body elongate, with rather small ctenoid, adherent scales; lateral line usually present, not extending on caudal fin; mouth various, the teeth usually villiform; no supplemental maxillary; opercle with a flat spine; B. 6 or 7; gills 4, a slit behind the fourth; gill membranes free from isthmus; gill rakers slender, toothed; pseudobranchiæ small, often concealed by skin; lower pharyngeals separate, with sharp teeth; air-bladder usually small or wanting, adherent to abdominal walls. Fins usually large; dorsal fins separate, the first with 6 to 15 spines; anal spines 1 or 2; V. thoracic, I, 5; intestinal canal short; pyloric cæca few; vertebræ more numerous than in *Serranidæ*, 30 to 45. Genera about 7; species about 100, in the fresh waters of the Eastern United States, Europe and Northern Asia. The great majority of the species belong to the singular genus or subfamily, *Etheostoma*, including the Darters, a most singular group of dwarfed perches, peculiar to the waters of Eastern America.

a. Pseudobranchiæ imperfect or wanting; preopercle entire or nearly so; branchiostegals 6; anal papilla usually present; pyloric cæca 2 or 3; supraoccipital crest low; fishes of small size (*Etheostomatinæ*).

ETHEOSTOMA, 158.

aa. Pseudobranchiæ well developed; preopercle serrate, the teeth on its lower margin retrorse; branchiostegals 7; no anal papilla; premaxillaries protractile; size large. (*Percinæ*.)

b. Canine teeth none; body oblong. . . . . PERCA, 159.

bb. Canine teeth on jaws and palatines; body elongate.

STIZOSTEDION, 160.

158. ETHEOSTOMA Rafinesque. DARTERS.

This group comprises a great variety of forms, and it has been usually divided into 10 to 16 genera. It is, however, impossible to maintain most of these subordinate groups as genera on account of intergradations of all sorts. There is no considerable variation in the osteology[1] of the species, except in regard to the numbers of the vertebræ. The group is apparently one of comparatively recent origin, and the differential characters do not seem to have become very firmly fixed. On the other hand, the extremes of the group (as *E. pellucidum* or *E. microperca*) have diverged very far from their perch-like ancestors.

The relations of the Darters to the Perches have been aptly expressed by Dr. Stephen A. Forbes: "Given a supply of certain kinds of food nearly inaccessible to the ordinary fish, it is to be expected that some fishes will become especially fitted for its utiliza-

---

[1] For an account of the osteology of this group, see Jordan & Eigenmann, Proc. U. S. Nat. Mus. 1885, 68. For a popular account of the habits of the species, see Jordan & Copeland on "Johnny Darters," in "Science Sketches."

tion. Thus *Etheostoma* is to be explained by the hypothesis of the progressive adaptation of the young of certain *Percinæ* to a peculiar place of refuge and a peculiarly situated food supply. These are the mountaineers among fishes. Forced from the populous and fertile valleys of the river beds and lake bottoms, they have taken refuge from their enemies in the rocky highlands, where the free waters play in ceaseless torrents, and there they have wrested from stubborn nature a meagre living. Although diminished in size by their constant struggle with the elements, they have developed an activity and hardihood, a vigor of life and a glow of high color, almost unknown among the easier livers of the lower lands. Notwithstanding their trivial size, they do not seem to be dwarfed so much as concentrated fishes."

Their colors are often very brilliant, the males of some species being among the most brilliant fishes known. The sexes are usually unlike; the females being generally dull and speckled. They usually prefer clear running water, where they lie on the bottom concealed under stones, darting, when frightened or hungry, with great velocity for a short distance, by a movement of the large pectorals, then stopping as suddenly. They rarely leave the bottom, and are never seen suspended in the water. A few species prefer a sandy bottom, where they lie buried in the sand, with only the eyes visible. The Darters feed chiefly on the larva of *Diptera*. The largest reach a length of 8 inches, but the average is about $2\frac{1}{2}$ inches. (The name *Etheostoma* is said by Rafinesque to mean " various mouths " ($\check{\epsilon}\tau\epsilon\rho\sigma\varsigma$, various ; $\sigma\tau\acute{o}\mu\alpha$, mouth ?), the three species known to him *caprodes, blennioides,* and *flabellare,* differing much in this respect.)

a. Body extremely elongate, hyaline, subterete, the belly mostly naked; lateral line complete; head, long, pointed; gill membranes somewhat united.

    b. Premaxillaries protractile; dorsal spines 7 to 11.

        c. Anal spine single; A. nearly as large as 2d D. **(Ammocrypta**[1] Jordan = *Pleurolepis* Baird.)

        d. Cheeks and opercles scaly.

322. E. pellucidum Baird. SAND DARTER. Scales of body not very rough, only those along lateral line and on tail well imbricated ; nape thinly scaled, becoming usually wholly naked on median line ; belly naked; maxillary barely reaching the large eye ; P. short. Translucent, finely dotted above ; a series of small square olive blotches along back, and another along lateral line, the latter connected by a gilt band; fins pale. Head $4\frac{1}{2}$; depth 7. D. X – 10. A. I. 8. Scales 6 – 75 – X. Vert. 44. L. $2\frac{1}{2}$. Ohio Valley and N. W., abounding in clear sandy streams, where it buries itself in the sand by a sudden plunge, and lies with only the eyes uncovered.

        [1] $\check{\alpha}\mu\mu\sigma\varsigma$, sand ; $\kappa\rho\upsilon\pi\tau\acute{o}\varsigma$, concealed.

From Ind. W. and S. occurs var. **clarum** (Jordan & Meek). Differs from var. *pellucidum* in having no scales along nuchal region, and none on sides anteriorly except the 5 or 6 rows along lateral line. Cheeks with few scales. From S. Ill., S., and W. is found var. **vivax** (Hay), better scaled than var. *pellucidum*, the region before dorsal being more or less closely covered with scales: scales firmer and rougher; a dusky bar across base of soft dorsal.

cc. Anal spines two; anal small. (**Ioa**[1] Jordan & Brayton.)

323. **E. vitreum** (Cope). Side of head closely covered with large, rough-ctenoid scales; middle and lower part of side with rough scales, breast and part of belly naked as is front of back; fins low; P. long. Translucent, with small dark spots on back and sides; fins plain. Head 4½; depth 7. D. VII to IX – 11 to 13. A. II, 7. Scales 60. L. 2. Va. and N. C., common in Neuse R. (Lat., glassy.)

bb. Premaxillaries not protractile; dorsal spines 14; anal fin large. (**Crystallaria**[2] Jordan & Gilbert.)

324. **E. asprellus** (Jordan). Eyes very large; mouth moderate; cheeks and opercles well scaled; nape scaly; throat and belly naked; fins large; C. lunate; hyaline olive; sides with 10 dark quadrate blotches, small and far apart; body sometimes with 4 or 5 broad dark cross-bands; fins plain; a dusky shade through eye. Head 4½; depth 7. D. XIV – 13. A. I, 12. Scales 7–93 – X. L. 4. S. Ind. (Rising Sun; O. P. Jenkins) to Ill., Ala., and Ark., in clear water, much the largest of the *hyaline* or "sand" Darters, approaching the type of *E. aspro*. (Diminutive of *Aspro*.)

aa. Body less elongate, not hyaline, almost entirely covered with scales.

e. Premaxillaries protractile.

f. Anal spine single, obscure; dorsal spines usually 9; anal smaller than soft dorsal.

g. Lateral line complete or very nearly so. (**Boleosoma**[3] DeKay.)

h. Soft dorsal with 12 to 14 rays.

i. Cheeks and opercles scaly; D. IX – 14.

325. **E. olmstedi** Storer. Body rather slender; fins very high; the spines weak; nape and breast usually naked (closely scaled in var. **atromaculatum** Girard, Cayuga L. and S.); olivaceous; sides with blotches and zigzag markings; fins speckled; head black in males in spring. Head 4; depth 5½. A. I, 9. Lat. l. 50. L. 3½. Mass. to W. N. Y., S. to Ga., abundant; probably a variety of the next. (To Mr. Olmsted who discovered the species in the Conn. Valley.)

ii. Cheeks almost always naked; opercles scaly; breast naked.

---

[1] ἰός, arrow.   [2] κρύσταλλος, crystal.   [3] βολίς, dart; σῶμα, body.

j. Scales about 5-50-9; D. IX-12; lateral line often incomplete behind; fins moderate.

326. **E. nigrum** Rafinesque. "JOHNNY." Body slender, fusiform; snout somewhat decurved; mouth small, sub-inferior; pale olive, back speckled with brown; sides with numerous W-shaped blotches; males in spring dusky anteriorly, sometimes entirely black. Head $4\frac{1}{5}$; depth 5. D. IX-12. A. I, 8. Vert. 15 + 22 = 37. L. $2\frac{1}{2}$. Dakota to W. Penn. and Mo., very abundant in small brooks. (*Boleosoma maculatum* Agassiz.)

yy. Scales about 5-40-6; dorsal rays IX-13; lateral line complete; fins very high.

327. **E. effulgens** (Girard). Snout much decurved; brown, with 9 spots on side; fins black; 2d D. and C. with white specks. Head $4\frac{1}{5}$; depth $5\frac{1}{2}$. L. $2\frac{1}{2}$. Penn. to N. C., probably a variety of *B. nigrum*. (*? B. æsopus* Cope; D. VII-14.) (Lat., shining.)

hh. Soft dorsal with 10 or 11 rays.

i. Cheeks naked; opercles scaly; scales 4-35-6.

328. **E. vexillare** Jordan. Body rather stout; nape naked; snout decurved; fins very high; ♂ dusky olive, faintly barred; 2d D. and C. with pale spots; other fins mostly black. Head 4; depth 5. D. VIII-10. A. I, 7. L. $2\frac{1}{4}$. Rappahannock R., Va., one specimen known. (L., bearing a standard.)

ii. Cheeks and opercles wholly naked; scales in lateral line 45.

329. **E. susanæ** (Jordan & Swain). Very slender; head short and small, the snout decurved; head, nape, breast, and middle of belly naked; fins low. Color of *E. nigrum*. Head $4\frac{1}{2}$; depth $6\frac{1}{2}$. D. VIII-10. A. I, 8. L. 2. Cumberland R., abundant in S. Ky. (To Mrs. Susan Bowen Jordan.)

gg. Lateral line ceasing near middle of body. (**Vaillantia**[1] Jordan.)

330. **E. chlorosoma** (Hay). Body slender, with long tail; back somewhat elevated; mouth small, inferior, the snout strongly decurved; cheeks, opercles and breast scaly, nape naked; fins small. Olivaceous, the back spotted; about ten dark spots on sides; a dark opercular spot; head spotted above; D. and C. barred. Head $4\frac{1}{3}$; depth $5\frac{1}{2}$. D. X-10. A. I, 8. Scales 5-56-10. Vert. 38. L. $2\frac{1}{4}$. Ill. to Ala. and Ark., common S. W. (*Boleosoma camurum* Forbes). (χλωρός, green; σῶμα, body.)

ff. Anal spines two, well developed, the first usually the longer.

m. Gill membranes more or less broadly united; belly with ordinary scales.

n. Maxillary normal, free from the preorbital. (**Ulocentra**[2] Jordan.)

[1] To Léon Vaillant, author of a monograph of the Darters.

[2] οὖλος, complete; κέντρον, spine.

331. **E. simoterum** (Cope). Body short and deep; head small, the snout very obtuse; cheeks, opercles and breast scaly. Olivaceous; back and sides each with a series of quadrate, blackish green blotches; belly saffron; upper parts with red spots; 1st D. with red spots and orange-red edge; 2d D. largely red; C. brown, barred; male in spring with head and fins largely dusky. Head 4⅔; depth 4. D. X–11. A. II, 7. Scales 10–52–12. Vert. 38. L. 3. Tennessee and Cumberland basins. (*Ulocentra atripinnis* Jordan.) (σιμός, snub-nosed.)

> *nn.* Maxillary adnate to the preorbital for most of its length, and therefore nearly immovable; mouth very small, inferior; no teeth on vomer. (**Diplesion** [1] Rafinesque.)

332. **E. blennioides** Rafinesque. GREEN-SIDED DARTER. Body elongate, little compressed, the head thick, its profile very convex; eyes large, high up, close together; cheeks, opercles and neck scaly; breast naked; spines strong. Olive green, mottled above; sides with 8 double transverse bars, each pair forming a Y-shaped figure of a deep green color; sides with orange dots; fins blue green, marked with orange red; ♀ duller. Head 4½; depth 4¾. D. XIII–13. A. II, 8. Lat. l. 65 to 78. Vert. 42. L. 5. Penn. to Ala. and Kans., common, one of the prettiest of the darters. (Blennius, εἶδος, like.)

> *mm.* Gill membranes scarcely connected; anal usually not smaller than second D.
> *o.* Belly with enlarged scales on middle line; these falling off, leaving a naked strip. (**Cottogaster** [2] Putnam.)

333. **E. copelandi** (Jordan). Body slender; head large, narrowed in front; mouth small, subinferior, the snout decurved; cheeks and breast naked; opercles and nape with few scales. Pale olive, speckled above, a series of horizontally oblong black blotches along lateral line; fins somewhat barred, dusky in ♂; a black spot on front of first D. Head 4¼; depth 5¼. D. XI–10. A. II, 9. Lat. l. 56. Vert. 18 + 20 = 38. L. 2⅓. White R., Ind., to Ark. (To the late Herbert Edson Copeland, the discoverer of the species, and one of the most careful and enthusiastic students of these fishes.)

334. **E. putnami** (Jordan & Gilbert). Close to the preceding, but with larger scales; lateral spots quadrate; spinous D. with a dusky band. Head 4; depth 6. D. XI–11. A. II, 8. Lat l. 44 to 48. L. 2½. L. Champlain to L. Huron. (To Frederick Ward Putnam.)

> *oo.* Belly with ordinary scales posteriorly, its anterior part naked. (**Imostoma** [3] Jordan.)
> *q.* P. extremely long, 1¼ times length of head, reaching front of A.

---

[1] δίς, two ; πλησίον, near, *i. e.* nearly two dorsals.
[2] Cottus, γαστήρ, belly.   [3] εἶμι, to move ; στόμα, mouth.

335. **E. longimane** Jordan.   Body moderately slender; head long. bluntish anteriorly, profile of snout steep and nearly straight; mouth moderate, included; maxillary to front of eye; cheeks nearly or quite naked; opercles somewhat scaly, nape naked; dorsals very high; A. spines small.   Olivaceous, with 5 dark cross-shades; a dark spot at base C., fins nearly plain.   Head 4; depth 5.   D. IX or X - 12 or 13.   A. II, 8.   Scales 6–43–7.   L. 2½.   James R., Va. (Lat., *longus*, long; *manus*, hand.)

> *qq.* P. moderate, not reaching A.

336. **E. shumardi** (Girard).   Body robust; head broad and thick; mouth large, scarcely inferior; cheeks, opercles and nape scaly; breast naked; fins all large.   Dark olive, blotched with darker; sides with 8 to 10 vague bars; a small black spot on front of spinous D.; a large one behind; fins barred; suborbital stripe large, black.   Head 3⅔; depth 5.   D. X, 15.   A. II, 11.   Scales 6–56–11.   L. 3.   Wabash R. to Ark., in larger streams.   (To Dr. George C. Shumard.)

> *ee.* Premaxillaries not protractile (the skin of the middle of the lower jaw continuous with that of the forehead).
>
> *p.* Cranium broad between the eyes; snout conic, pig-like, projecting beyond the inferior mouth; ventral line with a series of larger scales which fall off, leaving a naked strip; dorsal spines 13 to 15; gill membranes separate; scales small; vertebræ $23 + 21 = 44$. (**Percina** [1] Haldeman.)

337. **E. caprodes** Rafinesque.   LOG PERCH.   HOG-FISH. CRAWL-A-BOTTOM.   Body elongate; fins rather low; cheeks and opercles scaly.   Yellowish green with about 15 dark cross-bands, these usually alternating with shorter and fainter bands; a black spot at base of C.; fins barred.   Head 4; depth 6.   D. XV - 15. A. II, 9.   Lat. l. 92.   L. 6 to 8.   Great Lakes to Va., Ala., and Texas, abundant, the largest of the darters, and the one most nearly allied to the Perch and similar fishes.   N. and E. occurs var. *zebra* Agassiz, with nape naked, etc.; the ordinary form is scaly.   (κάπρος, the wild boar; εἶδος, like.)

> *pp.* Cranium not broad between the eyes: mouth less inferior, the snout usually not projecting much beyond it.
>
> *r.* Ventral line with the median series cf scales more or less enlarged or (if these are fallen) with a naked strip; anal fin large; lateral line complete.
>
> *s.* Palatine teeth present; dorsal spines 11 to 15.
>
> *q.* Preopercle strictly entire; gill membranes scarcely united across isthmus. (**Alvordius** [2] Girard.)
>
> *t.* Cheeks and opercles wholly naked ; head large and long.

338. **E. macrocephalum** Cope.   Body slender; head eel-like; maxillary reaching eye.   Brown, back with dark quadrate spots;

sides with 9 blackish oblong spots, alternating with smaller ones; fins mottled; 1st D. with median dark band. Head $3\frac{1}{2}$; depth 7. D. XV – 13. A. II, 11. Scales 11–77–15. L. 3. Ohio Valley, rare W. (μακρός, large; κεφαλή. head.)

tt. Checks naked; opercles scaly above only; nape and breast naked; muzzle blunt.

339. E. peltatum Stauffer. Body rather stout ; mouth moderate, maxillary reaching eye; ventral shields large. Olive, the back with short cross-bars ; sides with broad brownish shades; a dark blotch on neck and opercle ; snout and space below eye with the usual bars ; fins barred ; 1st D. with a black band. Head 4 ; depth 5. D. XII, 12. A. II, 8. Scales 7–53–9. L. 4. Penn. to S. C.. E. of Mts. (*E. nevisense* Cope ; *Alv. crassus* Jor. & Brayton.) (Lat., having shields.)

ttt. Cheeks usually with small scales : opercles with larger ones.

u. Head not very slender, the muzzle moderate, the lower jaw included.

340. E. aspro (Cope & Jordan). BLACK-SIDED DARTER. Body fusiform, rather elongate ; maxillary reaching just past front of eye ; breast naked; nape scaly or not. Greenish yellow with dark tessellations and marblings above and about 7 large dark blotches along side, more or less confluent; fins barred ; a small spot at base C. Head 4 ; depth 6. D. XIII to XV, 12. A. II, 9. Scales 9–65–17. Vert. 42. L. 3 to 4. W. Penn. to Dakota and Ark., abundant, one of the most elegant of the darters. (*Aspro*, a related genus of European *Percidæ*, from Lat. *asper*, rough.)

uu. Head very slender, with long-acuminate snout : jaws subequal.

341. E. phoxocephalum Nelson. Body slender ; mouth large, maxillary reaching eye ; nape scaly ; breast naked. Yellowish brown, the lateral spots smaller than in *E. aspro* and more numerous, quadrate in form ; a small dark spot at each end of lateral line. Head 4 ; depth 5¼. D. XII – 13. A. II. 9. Scales 12–68–14. Vert. 39. L. 4. Ind. to Kans. and Ark., common S. W. (φοξός, tapering ; κεφαλή. head.)

qq. Preopercle more or less distinctly serrate, especially in the young; gill membranes broadly united. (Serraria [1] Gilbert.)

342. E. scierum (Swain). Body rather stout ; head short, bluntish ; mouth small, the lower jaw shorter ; maxillary not reaching eye ; cheeks and opercles scaly ; breast partly scaled ; scales of median line of belly slightly enlarged, probably deciduous; fins very large. Yellowish olive, everywhere vaguely blotched with black, especially along sides : ♂ with head and most fins blackish ; ♀ paler. Head 4 ; depth 5. D. XIII, 14. A. II, 9. Scales 7–65–11.

[1] Lat., *serra*, saw.

Vert. 40.   L. 4.   Ind to Ark. and Texas.   S. W. occurs var. *serrula* J. & G., with preopercle more sharply serrate ; markings more definite; lat. l. 68 to 71 ; breast naked.   Resembles *E. aspro.* (σκιερός, shaded.)

ss. Palatine teeth obsolete ; dorsal spines 10 to 12 ; ♂ with the lower fins tuberculate in spring. (**Ericosma** [1] Jordan.)

343. **E. evides** (Jordan & Copeland). Body rather stout, compressed : head heavy, rather blunt forward; eye large ; mouth smallish, the maxillary reaching eye ; lower jaw included ; cheeks, nape and breast naked; ventral scales moderate; fins large. Dark olive, tessellated above; back with 7 broad transverse bars which extend below lateral line ; these bars are black in ♀, with yellowish interspaces : in ♂ deep blue-green, the interspaces yellow with copper-red blotches ; throat, cheeks, upper fins, and two spots at base C., largely orange ; A. and V. chiefly blue-black ; fins not barred : a black spot on last D. spines; ♀ with paler colors. Head 4⅓; depth 5⅓. D. XI - 11. A. II, 8. Scales 9–65–9. Vert. 40. L. 3. Ind. to Iowa and Ark; one of the most brilliant of the darters. (εὐειδής, comely.)

rr. Ventral line covered with ordinary scales, which are never shed in life.
   v. Lateral line complete [2] (with rare exceptions; see *E. nianguæ*).
   w. Anal fin large,[2] little if any smaller than the soft dorsal.
      x. Gill membranes nearly [2] separate from each other. (**Hadropterus** [3] Agassiz.)
      y. Scales very small, lat. l. about 85.

344. **E. aurantiacum** (Cope).   Elongate ; snout longer than eye ; lower jaw included ; cheeks and opercles scaly; throat smooth. Golden brown, a series of small round brown spots traversed by a black lateral band which extends around snout ; yellow below ; fins plain. Head 4⅓; depth 6. D. XV - 15. A. II, 11. L. 4⅓. Upper Tenn. R. (Lat., orange.)

yy. Scales moderate, lat. l. 55 to 75.

345. **E. cymatotænia** Gilbert & Meek. Body robust; head short. the snout short and slender ; mouth small, oblique, included ; maxillary nearly to front of eye ; eye large, 4 in head; cheeks, opercles, nape and breast with large scales; preopercle entire ; gill membranes narrowly joined, the degree of union variable, usually very slight; 1st A. spine long and strong; P. short. Greenish, with fine dark points; two pale streaks along sides, below the lower a broad dusky wavy band; a small black spot at base C.; fins trans-

---

[1] ἦρ, spring-time : κοσμέω, to adorn.
[2] These characters are none of them of high importance and are subject to some variations.
[3] ἁδρός, strong ; πτερόν, fin.

lucent, with dark lines. Head 4⅓; depth 5. D. XII to XIV – 13.
A. II, 10. Scales 7–64 to 70–12. L. 4. Ozark region, S. Mo.
(κῦμα, wave : ταινία. band.)

346. **E. nianguæ** Gilbert & Meek. Body elongate, terete; head
very long and slender, the snout deep and narrow, vertically rounded
at tip; mouth large. maxillary beyond front of orbit; eye shorter
than snout, 5⅓ in head; cheeks with a few rudimentary scales;
opercles and breast naked; nape scaled; A. rather smaller than
2d D.; 1st A. sp. short. Olivaceous, the back with 8 to 10 wide
dusky cross-bars, which extend on sides; ♂ with the dark bars
encircling body; back and sides with carmine-red spots in the pale
interspaces, most numerous in ♂ ; two black spots at base C.; 1st
D. dusky, spotted with red, and with red edge; other fins mostly
mottled with red. Head 3⅓; depth 6. D. XI or XII – 13 or
14. A. II, 11 or 12. Scales 11–74–16. L. 3¾. Niangua R., S.
Mo. Var. **spilotum** Gilbert, from Kentucky R. is similar, but with
the scales much larger (lat. l. 58 to 60), and the lateral line
incomplete.

　　xx. Gill membranes more or less broadly united.
　　z. Scales very small, 10-82-18; preopercle entire.

347. **E. squamatum** Gilbert & Swain. Body elongate, the head
long and slender, the snout long-acuminate; mouth long and narrow, the lower jaw included; maxillary to front of eye: eye moderate; 1st D. low; A. high, its spines strong; cheeks, breast, nape,
and opercle scaly: an enlarged black humeral scale. Yellow-olive,
with 10 broad dusky bars on back, and 10 dark blotches along sides;
a small black spot at base C.; 1st D. pale, with broad, orange band;
2d D. and C. barred with dusky and orange. Head 3½: depth 5⅓.
D. XIV – 13. A. II, 10. L. 4. French Broad R. (Lat., scaly.)

　　zz. Scales large, 8-51-9 ; mouth small, low, horizontal (transition to
　　"*Nanostoma* "). (**Pœcilichthys** [1] Agassiz.)

348. **E. variatum** Kirtland. Body rather robust, the head short
and thick, with short blunt snout, the anterior profile convex; eyes
large, 3¾ in head; maxillary to front of eye (4 in head); top of
head rugose; head almost naked; nape and breast scaled; fins all
very large; A. large, a little smaller than soft D.; P. reaching front
of A.; ♂ dusky greenish, finely punctate; belly and sides orange
yellow; posterior part of body with 5 orange bands; 1st D. with
dark blue band; 2d D., A. and P. blue-black, shaded with orange;
♀ paler. Head 3⅓; depth 4¼. D. XIII – 13. A. II, 9. L. 4.
Ohio Valley, scarce. (Lat., variegated.)

　　ww. Anal fin rather small, notably smaller than soft dorsal.
　　a. Gill membranes broadly united across the isthmus; mouth small, sub-
　　　inferior. (**Nanostoma** [2] Putnam.)

　　　　　[1] ποίκιλος, variegated : ἰχθύς. fish.
　　　　　[2] νανός, small : στόμα, mouth.
　　　　　　　9

349. E. zonale (Cope). Body slender; head small and short,
the snout obtusely decurved; cheeks and opercles scaly; breast
scaly, or naked (var. *arcansanum* Jordan & Gilbert); teeth feeble;
dorsals separate. Olivaceous; 6 brown quadrate spots on back,
connected by alternating spots with a broad, brown lateral band,
from which 8 narrower dark bluish bands nearly or quite encircle
the belly; P., A. and C. golden, speckled with brown; middle half
of 1st D. crimson; base of 2d D. with round red spots; a black
spot on opercle and one at base P.; ♀ duller, with V. barred.
Head 4¼; depth 5. D. XI - 12. A. II, 7. Scales 6–43 to 50–12.
Vert. 39. L. 2½. W. Penn. to Kans. and Miss., in clear streams;
variable. (Lat., belted.)

aa. Gill membranes scarcely united. (**Nothonotus** [1] Agassiz.)
  b. Head short, the muzzle abruptly decurved; scales 7–53–8; 2d D., A. and
       C. black edged.

350. E. camurum (Cope). BLUE-BREASTED DARTER. Body
stout; mouth somewhat inferior; C. truncate. ♂ blackish olive, breast
and throat deep rich blue; sides profusely sprinkled with crimson
dots; faint dark lines along rows of scales; 1st D. with a black spot
in front, above and behind which is a crimson one; 2d D., A., and C.
crimson, bordered with yellow and then by blue-black; P. and V.
crimson-edged; ♀ greenish, faintly barred. Head 4; depth 4½.
D. XI -13. A. II, 8. Scales, 7–53–8. L. 2½. Ind. to Tenn., in
clear streams, perhaps the most beautifully colored of all our fresh-
water fishes. (Lat., blunt-headed.)

  bb. Head rather long and pointed, the snout not decurved.
    c. Dorsal spines 10 or 12.
      d. Scales 9–58–10; vertical fins without black border.

351. E. maculatum Kirtland. Body moderately elongate, with
very deep tail; eye large, maxillary to front of eye; fins short;
1st A. spine large. Olive black, with a wavy leather-colored dorsal
band; throat blue; back and sides with crimson spots; 1st D. with
a black spot at base in front; 2d D. blood-red; C. with two confluent crimson spots at base; P. and V. without red border; ♀
dull, the fins speckled and without red. Head 4; depth 5¼. D.
XII -12. A. II, 9. Scales 9–58–10. Vert. 39. L. 2½. W. Penn.
to E. Tenn., scarce. (*Pœc. sanguifluus* Cope.)

  dd. Scales 6–45–7 : A. and C. narrowly black-edged.

352. E. rufolineatum (Cope). Stout, the back elevated; snout
short, as long as the small eye; tail deep. Olive, with numerous
narrow longitudinal streaks including irregular quadrate spots of
brick-red; breast blue; belly orange; head with 2 brown bands, and
5 red spots on each side; fins all broadly bordered with crimson;
two orange spots at base C.; A. scarlet-yellow at base, edge black;

[1] νόθος, prominent; νῶτος, back

♀ olivaceous, barred ; fins speckled. Head 4 ; depth 4½. D. XI -12. A. II, 8. Scales 6-45-7. L. 3. French Broad R. (Lat., *rufus*, red ; *linea*, line.)

    cc. Dorsal spines 14; scales 8-53-9.

    353. E. vulneratum (Cope). Body stout, fusiform ; tail deep; form of *E. maculatum.* Light olive with 8 dark bars, interrupted above, and a few crimson spots; fins mostly plain; 1st D. with a series of red spots ; C. orange, with narrow black edge, as has also 2d D. Head 4 ; depth 4½. D. XIV - 13. A. II, 8. L. 2. French Broad R. (possibly the young of *E. camurum*). (Lat., wounded.)

    vv. Lateral line incomplete or wanting (sometimes nearly or quite complete in *E. jessiæ.*)
    e. Lateral line developed anteriorly.
    f. Gill membranes broadly united (**Etheostoma**).
    g. Head entirely naked; lower jaw prominent; lateral line developed about half way.

    354. E. flabellare Rafinesque. Body long and low, the back not arched; head long and pointed; fins low; 1st D. in ♂ half as high as 2d ; the spines with fleshy tips, spines higher in ♀ : nape and throat naked. A conspicuous black humeral scale. Dusky olive, with dark longitudinal streaks, ♂ with dark cross-bars ; 2d D. and C. sharply barred ; 1st D. in ♂ tipped with orange. Head 4 ; depth 5. D. VIII - 12. A. II, 8. Scales, 7-50-7. Vert. 13 + 20 = 33. L. 2½. W. N. Y. to N. C., and W. in clear streams, abundant and variable ; the typical form from Ind. E. Var. **lineolatum** Ag., from Ind. N. W., has a black spot on each scale, these marks forming very conspicuous stripes along side. Var. **cumberlandicum** Jordan & Swain, from Cumberland R., has thicker head, and the adult is almost plain olivaceous, except for the black humeral spot and the barred fins. This is the most active and wary of the darters, and the most hardy in the aquarium. (Lat., *flabellum*, a fan ; *i. e.*, fan-tailed.)

    gg. Head more or less scaly.
    h. Cheeks, opercles, nape, and breast scaly; jaws equal; lateral line nearly complete.

    355. E. squamiceps Jordan. Body less elongate than in the preceding, the head shorter and thicker. Dusky olive, without well-defined marks; ♂ mottled, with 6 cross-blotches and the lower fins black ; vertical fins cross-barred ; a black humeral spot. Head 3½ ; depth 5. D. IX - 12. A. II, 7. Scales 5-50-6. L. 3. S. Ind. to W. Fla. (Lat., *squama*, scale ; *ceps*, head.)

    hh. Cheeks naked, or with embedded scales; opercles scaly.

    356. E. whipplei (Girard). Body rather deep, compressed, with deep tail; mouth terminal, oblique ; maxillary to eye, 3½ in head. Grayish, mottled with darker and with faint bars; sides with small round scarlet spots; two orange blotches at base of C.; a black humeral spot. Dorsals barred with dusky and orange; A. similar, more orange; C. barred, its margin black. Head $3\frac{1}{2}$; depth $4\frac{3}{4}$. D. IX to XII – 12 to 14. A. II. 7. Scales 8–60 to 70 – X. Vert. 36. L. $2\frac{1}{2}$. Ozark region and S., abundant. (To Lieut. A. W. Whipple, U. S. A.)

ff. Gill membranes little if at all connected.
   i. Lateral line nearly straight, not arched above P. **(Oligocephalus[1] Girard.)**
      j. Humeral region with a small, black, scale-like process.
        k. Cheeks, opercles, and nape naked, or very nearly so.
          l. Scales small, 63 to 73 in a longitudinal series; mouth large, terminal.

357. **E. sagitta** (Jordan & Swain). Body slender, with long tail; head long, very slender, the snout sharp; mouth very large, oblique; maxillary reaching front of pupil, $3\frac{1}{4}$ in head; jaws subequal; fins high. Green, with faint olive cross-bars; a dark spot at base of C.; sides with orange spots; fins with orange shades. Head $3\frac{1}{5}$; depth $4\frac{1}{5}$. D. X – 13. A. I, 10. Lat. l. 68 (48 tubes). L. $2\frac{1}{2}$. Cumberland R., Ky. (Lat., arrow.)

358. **E. punctulatum** (Agassiz). Body slender, the snout sharp, the mouth vertical; eye large; fins rather low. Dark slaty green, with faint dark bars; belly orange red; body and fins profusely dusted with black specks. 1st D. with black band; other fins with wavy bars of dark specks. Head $3\frac{1}{4}$, depth $5\frac{2}{3}$. D. X or XI – 14. A. II, 8 or 9. Lat. l. 63 to 73; pores 50. L. 2. Ozark region, S. Mo. (Lat., dotted.)

   ll. Scales rather large, about 53 in lat. l.

359. **E. virgatum** (Jordan). Form and appearance of *E. fla-bellare;* head long-pointed; jaws subequal; maxillary reaching pupil; preopercle crenulate. Greenish, each scale with a dusky spot, these forming lengthwise stripes; sides with faint bars. D. and C. barred. Head $3\frac{2}{3}$; depth 5. D. IX – 10. A. II, 8. L. $2\frac{1}{2}$. Rock Castle R., Ky. (Lat., striped.)

   kk. Cheeks, opercles, and nape scaly.

360. **E. boreale** (Jordan). Body rather elongate, the head heavy, the snout bluntly decurved; mouth small, horizontal, the lower jaw included; humeral scale small; lateral line very short; dorsals short and small. Gray (in spirits) with 11 or 12 very distinct (blue?) cross-bands, each alternate one meeting its fellow below; 1st D. with a median dark band; 2d D. barred; ♀ paler. Head $3\frac{2}{3}$; depth $5\frac{2}{3}$. D. VIII – 9. A. II, 7. Scales 4–53–10. L. $2\frac{1}{2}$. Montreal.

   jj. Humeral region without black scale-like process.

      [1] ὀλίγος, small; κεφαλή, head.

m. Cheeks naked, or very nearly so; opercles scaly.

361. **E. cœruleum** Storer. RAINBOW DARTER. SOLDIER-FISH. Body rather stout; head large; mouth moderate, the lower jaw the shorter, the maxillary to front of orbit; neck and breast usually naked. ♂ olivaceous, blotched above with darker; sides with about 12 oblique bars of indigo-blue running downwards and backwards, the interspaces bright orange; cheeks blue; breast orange; fins chiefly orange and deep blue; ♀ duller, with little blue or red, the vertical fins barred. Head $3\frac{3}{4}$; depth $4\frac{1}{4}$. D. X – 12. A. II, 7. Scales 5–45–8; pores 33. Vert. $15 + 21 = 36$. L. $2\frac{1}{4}$. W. Penn. to Iowa and Ky., extremely abundant; one of the gaudiest of fishes. Var. **spectabile** Agassiz, Ind. to Kans., has distinct dark streaks along the rows of scales on back. (Lat., blue.)

mm. Cheeks and opercles more or less scaly.

362. **E. jessiæ** (Jordan & Brayton). Body fusiform, rather stout, compressed, with rather deep caudal peduncle; head moderately pointed; mouth terminal, the lower jaw included; cheeks scaly or partly naked; brownish, with cross-bars or blotches of greenish; sides with dark blue quadrate cross-bars; fins speckled with golden. Head 4; depth 5. D. XII – 12. A. II, 9. Scales 6–47–7, with tubes on 35 scales. Tenn. to Wabash Valley,[1] (Evermann) Ills. and E. Texas. (To Mrs. Jessie Dewey Brayton.)

363. **E. iowæ** Jordan & Meek. Similar to preceding, but slenderer and with notably smaller scales. Green, blotched with darker; first D. shaded with red, its edge very dark. Head $3\frac{5}{8}$; depth $5\frac{1}{4}$. D. IX – 11. A. II, 7. Scales 5–59–9. L. 2. S. Iowa.

364. **E. saxatile** (Hay). Form of *E. nigrum*; snout slender and sharp, profile gently decurved; mouth terminal: lower jaw included; body slender, the caudal peduncle also slender; gill membranes narrowly united; P. as long as head. Dorsals well separated; the spines very slender. Olivaceous with 6 dark cross-shades on back, and with dark marks, much as in *E. nigrum;* from the N-shaped marks on sides light blue bands pass down around belly and tail; two black spots on base of C; a black spot behind eye. Head 4; depth 6. D. XI to XIII – 11 or 12. A. II, 9. Lat. l. 50 to 55. L. 2. Tenn. to Ark. and S. (Lat., pertaining to rocks.)

365. **E. luteovinctum** Gilbert & Swain. Compressed, the back elevated, the tail very slender; head compressed, with short, high snout, its profile strongly decurved; mouth low, horizontal; gill membranes narrowly connected; dorsals low, well separated. Very

---

[1] The form from Ind. and Ill. is probably not different from *E. jessiæ.* It may be called Var. **asprigene** (Forbes). Body rather stout; head somewhat pointed; eye large, longer than snout; mouth terminal: dorsals separate. Dark greenish, much mottled; 1st D. dusky behind with a broad band of blue and crimson: soft fins speckled. Head 4; depth $4\frac{1}{4}$. D. XI - 12. A. II, 8. Lat. l 49; tubes on 34 to 41. L. $2\frac{1}{4}$.

pale olive, with 7 dark cross-bars on back; sides with 9 dark greenish blotches, between which are orange-yellow cross-bars; small black spots at base of C.; spinous D. with a median orange band, and a dark blotch behind; 2d D. and C. barred. Head 3¾; depth 4⅔. D. IX or X - 13. A. II, 7. Scales 6–49 to 55–11. L. 2. Stone R., Tenn. (Lat., *luteus*, yellow; *vinctus*, banded.)

ii. Lateral line forming a slight curve above P.; body fusiform; dorsals separate, subequal. (**Boleichthys** Girard.)

n. Cheeks scaly.

366. **E. fusiforme** (Girard). Body slender, terete; snout short, bluntish; mouth small, oblique; maxillary reaching beyond front of eye; eye large, longer than snout; C. rounded; A. spines small. Olivaceous, mottled with brownish; back with 12 green cross-shades; sides with similar shades, sometimes with red spots; 1st D. black below, with reddish spots above. Head 3¾; depth 6. D. IX or X - 10. A. II, 6. Lat. l. 50 (10 to 20 pores). Vert. 16 + 20 = 36. L. 2. Mass. to S. C., Ind., Ark. and Texas, in lowland streams and mud-holes, variable. Southeastward occurs a form or variety, **barratti** Holbrook, similar but without red or blue. (Lat., spindle-shaped.)

367. **E. eos** (Jordan & Copeland). Body elongate, slender, the tail very long; head long, the snout decurved; mouth small, little oblique, the lower jaw slightly included. Dark olive, with darker markings; about 12 dark-blue cross-bars on back, with as many short dull-blue bars between them on sides; interspaces more or less marked with red; lower parts with irregular dark specks and short lines; 2d D., C. and P. barred; 1st D. with blue and red. Head 4; depth 5⅓. D. IX – 11. A. II, 7. Lat. l. 58; about 25 pores. L. 2½. Ind. to Minn.; abundant. (Probably a variety of the preceding.) (ἠώς, sunrise.)

cc. Lateral line wholly wanting. (*Microperca* Putnam.)

368. **E. microperca** Jordan & Gilbert. LEAST DARTER. Body rather short and deep, compressed; snout somewhat decurved; cheeks naked; opercles scaly; nape and breast naked; fins small; anal spines strong. Olivaceous much speckled, and with zigzag markings; 2d D. and C. barred; a dark humeral spot. Head 3¾; depth 4½. D. VI or VII - 10. A. II, 6. Lat. l. 34. Vert. 14 + 16 = 30. L. 1 to 1½. Smallest of the darters, and one of the smallest of fishes; common from N. Ind. to Minn. (*Microperca punctulata* Putnam.) (μικρός, small; πέρκη, perch.)

159. **PERCA** (Artedi) Linnæus. (Latin name from πέρκη, originally from πέρκος, dusky.)

369. **P. flavescens** (Mitchill). YELLOW PERCH. RINGED PERCH. Body oblong, somewhat compressed, the back elevated; cheeks scaly; opercles mostly naked, striate; premaxillaries pro-

tractile, preorbital serrate; snout projecting; maxillary reaching middle of pupil; top of head rugose; gill rakers stout, the longest but 3 times as high as broad. Dark olivaceous, sides golden yellow; 6 to 8 broad dark cross-bars from back to below middle of sides; lower fins orange, upper olivaceous; spinous D. without distinct black spot. Head 3¼; depth 3¼. D. XIII – 1, 14. A. II, 7. Scales 5–55–17. Vert. 21 + 20 = 41. Pyloric cœca 3. L. 15. Minn. to N. Ohio and Quebec, S. to S. C. E. of Alleghanies, not in Ohio Valley or S. W.; abundant. (Lat., growing yellow.)

160. STIZOSTEDION Rafinesque. (στίζω, to prick; στηθίον, little breast; "the name means pungent throat." *Raf.*)

a. Pyloric cœca 3, subequal, all about as long as stomach; D. XIII – 1, 21. (*Stizostedion.*)

370. **S. vitreum** (Mitchill). WALL-EYE. GLASS-EYE. PIKE PERCH. JACK SALMON. Body elongate, growing deeper with age, the back more arched than in the next; head sub-conic, long; cheeks, opercles and top of head more or less scaly; opercle with radiating striæ, ending in spinules; D. spines high, soft D. nearly as long as spinous. Dark olive, mottled with brassy; sides of head vermiculated; 1st D. with a large jet-black blotch posteriorly, otherwise nearly plain dusky; 2d D. and C. mottled olive and yellowish; base of P. without black spot. Head 4⅔; depth 4 to 6. A. II, 12. Lat. l. 90. L. 1 to 3 feet. Great Lakes, Miss. Valley, E. to Va.; commonest N., where it is one of the leading food-fish. Absurdly called "Salmon" in parts of the South.

aa. Pyloric cœca 4 to 7, unequal; D. XIII–1, 18. (*Cynoperca* Gill & Jordan.)

371. **S. canadense** (C. H. Smith). SAUGER. SAND PIKE. GRAY PIKE. HORN-FISH. Body elongate, more terete than in the preceding, the flesh more translucent; head depressed, pointed; opercular spines variable. Eye small, 5 in head. Olive gray, sides brassy or orange, with dark mottlings, more distinct in young; 1st D. with 2 or 3 rows of round, black spots; no black blotch on last spines; 2d D. with 3 irregular rows of dark spots; a large black blotch on base of P.; C. dusky and yellowish. Head 3½; depth 4¼ to 5. A. II, 12. Lat. l. 95. Vert. 23 + 22 = 45. L. 18. Great Lake region to Ohio Valley and Dakota; common N. (Var. *canadense* in St. Lawrence region has bones of head especially rough, the head more scaly and about 4 opercular spines; Var. *griseum*, of the Great Lakes, etc., with smoother head, and Var. *boreum*, of the Upper Miss., etc., with slenderer and more "snake-like" head.)

FAMILY LXIV. **SERRANIDÆ**. (THE SEA BASS.)

Body oblong, with adherent, mostly ctenoid scales; mouth usually large, with villiform teeth and sometimes with canines; teeth on

vomer and palatines ; maxillary broad, not slipping for its whole
length beneath the preorbital; gill rakers stiff, toothed; gills
normal ; pseudobranchiæ large ; lower pharyngeals separate, with
pointed teeth : gill membranes separate; B. normally 6 or 7.
Preopercle usually serrate ; opercle with flat points or spines. Lateral line present, not on the caudal. Dorsal variously developed ;
anal shortish, with three spines (these wanting in one genus). V.
normal. Tail stout, its fin not deeply forked. Vertebræ usually
$10 + 14 = 24$. Intestine short, the stomach cæcal, with pyloric
appendages. Carnivorous fishes, chiefly of the warm seas, often of
large size, most of them valued as food. Genera about 40. species
200. This group may be regarded as the most typical among
the *Percoid* fishes, and it is perhaps the one nearest the parent
stock from which the others have sprung. (*Serran*, the French
name, from Lat. *serra*, saw.)

a. Anal spines 3; dorsal spines 8 to 14.

  b. Dorsal fins separate, or joined at base only; the rays VII to XI - 1, 12 to
     14; maxillary without supplemental bone; teeth all villiform, without
     canines. (*Latinæ*.)

    c. Caudal lunate; tongue with teeth; preopercle without horizontal spine
      or antrorse hooks.

      x. Dorsal fins separate; spines weak; anal rays about III, 12, the
        spines graduated; lower jaw projecting; base of tongue with
        teeth. . . . . Roccus. 161.

      xx. Dorsal fins joined; spines strong; anal rays III, 9, the spines
        not graduated; jaws subequal; base of tongue toothless.
                                 Morone, 162.

  bb. Dorsal fin continuous, not deeply notched.

    d. Maxillary without supplemental bone; canine teeth, if present, on
      sides of jaw as well as in front; no depressible teeth; temporal
      crests on cranium small; gill rakers rather short; lateral line not
      very high; dorsal spines X; anal rays III, 7, supraoccipital
      crest not extending far forward on top of skull, leaving a
      smooth area before it. (*Serraninæ*.)

    e. Smooth area on top of cranium very short and small, the supra-
      occipital crest long: C. not lunate, usually ending in 3 points;
      teeth small; head naked above; dorsal rays X, 11.
                             Centropristis, 163.

    dd. Maxillary with a supplemental bone; canine teeth usually present,
      in front of jaws; inner teeth of jaws depressible; scales small,
      firm; head more or less scaly above; supraoccipital crest en-
      croaching on top of skull, so as to leave no smooth area at vertex;
      temporal crests distinct; scales small. (*Epinephelinæ*.)

    f. Dorsal rays about XI, 16; anal rays III, 8; preopercle without
      antrorse spines; canine teeth in front of both jaws; temporal
      crests moderate; scales of lateral line simple, without radiat-
      ing ridges. . . . . Cerna, 164.

161. ROCCUS Mitchill. (From vernacular *Rock-Fish.*)

  a. Teeth on base of tongue in two patches; body elongate, little compressed.
    (*Roccus.*)

SERRANIDÆ. — LXIV. **137**

372. **R. lineatus** (Bloch). STRIPED BASS. ROCK-FISH. ROCK. Body slender, growing deep with age; spines slender, the 2d anal spine 5 to 7 in head. Olivaceous silvery, sides with 7 to 9 blackish lengthwise stripes. Head 3½; depth 3½. D. IX - 1, 12. A. III, 11. Lat. l. 65. L. 3 to 5 feet. Nova Scotia to La., entering rivers to spawn, one of our finest game fishes. (Lat., striped.)

aa. Teeth on base of tongue in one patch; body deep, compressed. (*Lepibema* Rafinesque.)

373. **R. chrysops** (Rafinesque). WHITE BASS. Back arched; 2d A. spine 3 in head. Silvery, greenish above, sides with several dusky longitudinal streaks, those below lateral line more or less interrupted. Head 3½; depth 2½. D. IX - 1, 14. A. III, 12. Lat. l. 55. L. 15. Great Lakes and Upper Miss. Valley, rather common. (χρυσός, gold; ὤψ, eye.)

162. **MORONE** Mitchill. (Name unexplained.)

a. Sides striped with black.

374. **M. interrupta** Gill. YELLOW BASS. Body oblong, ovate, the back elevated; anterior profile concave; 2d A. spine about 2 in head; spines very strong. Brassy, sides with 7 very distinct black stripes, those below lateral line interrupted behind, and beginning lower down. Head 3; depth 2⅔. D. IX - 1, 12. A. III, 9. Lat. l. 50. L. 12. Lower Mississippi, N. to Brookville, Ind. (A. W. Butler) and S. Ill. (*M. mississippiensis* Jordan and Eigenmann, the name *interrupta* being preoccupied in the genus *Roccus*, from which *Morone* is scarcely distinct.)

aa. Sides with faint pale streaks.

375. **M. americana** (Gmelin). WHITE PERCH. Body oblong, not strongly compressed; 2d A. sp. about 3 in head; spines strong. Olivaceous, sides silvery. Head 3; depth 3. D. IX - 1, 12. A. III, 9. Lat. l. 50. L. 10. Nova Scotia to S. C., common, ascending streams.

163. **CENTROPRISTIS** Cuvier. (κέντρον, spine; πρίστης, saw.)

376. **C. striatus** (L.). BLACK SEA BASS. BLACK-FISH. BLACK-WILL. Body robust; head large; mouth moderate; teeth small, in broad bands; dorsal spines strong, with short filamentous appendages; P. long Blackish, more or less mottled, with traces of pale streaks along the rows of scales; D. with rows of whitish spots; young with dark cross shades and dusky lateral band. Head 2⅘; depth 2¾. D. X, 11. A. III, 7. Lat. l. 50. L. 14. Cape Cod to Fla., common. (*C. atrarius* (L.); the name *striatus* is still earlier.) (Lat., striped.)

164. CERNA Bonaparte. GROUPERS. (*Epinephelus* authors, not of Bloch.)   (Italian name for the genus.)

a. Second dorsal spine high, not lower than third or fourth; C. lunate.

377. **C. morio** [1] (Cuv. and Val.) RED GROUPER. Preopercular angle little salient, without enlarged teeth. Brown, clouded with whitish; lower parts flushed with orange-red; small dark spots about eye; vertical fins broadly edged with black. Head $2\frac{1}{2}$; depth 3. D. XI, 17. A. III, 9. Lat. l. 106. L. 3 feet. West Indies, sometimes N. to N. Y.   (French, *mérou?*)

FAMILY LXV. **LOBOTIDÆ.** (THE FLASHERS.)

This family is closely allied to the *Serranidæ*, from which it differs chiefly in the absence of teeth on the vomer and palatines. The lips are thick, the upper jaw very protractile, the lower longer, and the bases of the high soft dorsal and anal thickened and scaly. The single species is a large fish, found in most warm seas.

165. LOBOTES Cuvier. (λοβότης, lobed.)

378. **L. surinamensis** (Bloch). FLASHER. TRIPLE-TAIL. Head small, the anterior profile concave, the back elevated. Blackish above, sides grayish, often blotched with yellowish. Head 3; depth $2\frac{1}{2}$. D. XII, 18. A. III, 11. Lat. l. 47. L. 3 feet. Tropics. frequently N. to N. Y.

FAMILY LXVI. **SPARIDÆ.** (THE PORGIES.)

Body oblong or elevated, with adherent scales which are usually scarcely ctenoid. Mouth various, usually terminal, the teeth of various forms. Premaxillaries protractile; maxillary for its whole length slipping into a sheath formed by the edge of the preorbital; gills and gill membranes normal; pseudobranchiæ large. Preopercle serrate or not; opercle unarmed. Dorsal fin usually continuous, with 8 to 13 spines; anal spines 3. V. normal, usually with an enlarged scale at base; lateral line continuous, not extending on C. Air-bladder present. Fishes of the warm seas, some carnivorous, others herbivorous, the latter with very long intestines. As here understood, a rather heterogeneous group of some 60 genera and nearly 500 species. distinguished as a whole from the *Serranidæ* chiefly by the sheathed maxillary. Probably the group needs further subdivision. (σπάρος, *Sparus*, ancient name.)

a. Species carnivorous, with short intestines and few pyloric cæca; teeth not all incisor-like.

   b. Vomer with teeth; no incisors or molars; jaws with canines; D. continuous. (*Lutjaninæ.*)

---

[1] Numerous related species of *Cerna* and *Epinephelus* occur off our Southern Coast, and come to the northern markets. For an account of these, see Jordan & Swain, Proc. U. S. Nat. Mus. 1884.

   c. Interorbital area not flat; fronto-occipital crest not continued forward
to snout; no pterygoid teeth; C. lunate. . . . . . LUTJANUS, 166.
bb. Vomer without teeth.
   d. Teeth all pointed ; no incisors or molars ; preopercle serrate. (*Hæmu-
linæ.*)
   e. Mouth small ; chin with a large pore; anal fin long, its rays III,
11 to III, 14, its spines small, graduated. ORTHOPRISTIS, 167.
   dd. Teeth on sides of jaws molar ; preopercle entire. (*Sparinæ.*)
   f. Second interhæmal spine normal, not "pen-shaped;" front
teeth broad, incisor-like; no canines; first spine-bearing in-
terneural developed as an antrorse spine before D.
   g. Occipital and temporal crests of skull nowhere coalescent;
interorbital area not swollen, its bones thin, transversely
concave; incisors deeply notched. . . . LAGODON, 168.
   gg. Occipital and temporal crests coalescent anteriorly, both
merging into the gibbous interorbital area, the bones of
which are honeycombed; incisors entire or nearly so.
ARCHOSARGUS, 169.
   ff. Second interhæmal spine enlarged, hollowed anteriorly, pen-
shaped, receiving the posterior end of the air-bladder in its
anterior groove; front teeth incisor-like, but very narrow; an
antrorse spine before D.; lateral crest not coalescing with
occipital crest; interorbital area flattish. . STENOTOMUS, 170.
   aa. Species herbivorous, with long intestines and many pyloric cæca; front
teeth all incisor-like. (*Kyphosinæ.*)
   g. Vomer with teeth; soft fins densely scaly; incisors entire at
tip, with horizontal backward projecting roots, the bands
of small teeth behind them narrow. . . KYPHOSUS, 171.

166. LUTJANUS Bloch. SNAPPERS. (*Ikan Lutjang,* Japanese
name of the typical species.)

a. Anal fin low, rounded; color chiefly greenish.

   379. **L. griseus** (L.). MANGROVE SNAPPER. GRAY SNAPPER.
LAWYER. CABALLEROTE. Snout pointed; mouth large; lower
jaw not projecting; canines strong; vomerine teeth in a ⋀-shaped
patch; preorbital deep; rows of scales of sides of back becoming
oblique and irregular behind. Dark green, reddish below; young
with dusky streaks; vertical fins blackish, tinged with red in life.
Head 2¾; depth 2¾ to 3. D. X, 14. A. III, 8. Scales 6–50–12.
L. 18. West Indies, N. to N. J.; very common S. especially along
shore among mangroves. (Lat., gray.)

   aa. Anal fin high, angulated, the middle rays elevated; color chiefly red.

   380. **L. aya** (Bloch). RED SNAPPER. Body robust; upper
canines strong, lower small; teeth and scales much as in *L. griseus.*
Rose-red, nearly uniform, young with a blotch on lateral line.
Head 2¾; depth 2¾. D. X, 14. A. III, 9. Scales 7–60–15.
L. 30. West Indies, etc., rarely N. to Block Island (*Goode*),
abundant on the Gulf Coast, in rather deep water. (*L. blackfordi*
Goode & Bean.) (A Brazilian name.)

167. ORTHOPRISTIS Girard. (ὀρθός, straight; πρίστης, saw.)

381. **O. chrysopterus** (L.). PIG-FISH. SAILOR'S CHOICE.
Body compressed, the head long; mouth low, with small teeth;
spines slender. Grayish, sides with many yellow spots, forming
series along the rows of scales, those above lateral line oblique,
those below parallel with lateral line; fins and head spotted.
Head 3½ : depth 3. D. XII, 16. A. III, 12. Lat. l. 57 (rows).
L. 12. N. Y. to Texas and Cuba, common S. (Numerous species
belonging to the allied genus *Hæmulon* are found S. of Cape Hatteras.) (χρυσός, gold; πτερόν, fin.)

168. LAGODON Holbrook. (λαγώς, hare; ὀδών, tooth.)

382. **L. rhomboides** (L.). CHOPA-SPINA. PIN-FISH. BREAM.
Body elliptic-ovate, compressed; head pointed; upper molars in
two rows. Olive, sides silvery, with faint stripes of blue and golden ; 6 dark vertical bars growing faint with age; a large dark
blotch above P.; fins streaked with yellowish; 2d A. spine scarcely enlarged. Head 3⅓; depth 2⅓. D. XII, 11. A. III, 11.
Scales 8-68-18. L. 6. N. Y. to Cuba and Texas; very common S.

169. ARCHOSARGUS Gill. (ἄρχων, ruler; σάργος, Sargus.)

383. **A. probatocephalus** (Walbaum). SHEEPSHEAD. Body
deep, robust, the back arched; occipital crest strong; 2d A. spine
much enlarged. Gray, with 7 broad black cross-bars; no gilt
streaks or shoulder spot. Head 3⅛; depth 2. D. XII, 11. A.
III, 10. Scales 7-48-15. L. 30. Cape Cod to Texas, one of the
best of our food-fishes. (πρόβατον, sheep; κεφαλή, head.)

170. STENOTOMUS Gill. (στενός, narrow; τομός, cutting.)

384. **S. chrysops** (L.). SCUP. SCUPPAUG. PORGEE. Body
ovate, compressed, the back elevated; incisors very narrow, resembling canines; third dorsal spine elevated; 2d A. spine slightly
enlarged. Purplish gray; sides silvery; vertical fins somewhat
mottled; young faintly barred. Head 3¼; depth 2. D. XII, 12.
A. III, 11. Scales 8-49-16. L. 12. Cape Cod to S. C., abundant
N.; a valuable food-fish. (Farther S. occur numerous species of
the related genus *Calamus*, with the front teeth conical.) (χρυσός,
gold; ὤψ, eye.)

171. KYPHOSUS Lacépède. (*Pimelepterus* Lacépède.)
(κυφός, gibbous.)

385. **K. sectatrix** (L.). RUDDER-FISH. "CHUB." Body
ovate, compressed; mouth small; interorbital space gibbous, the
snout truncate; fins all very low; C. forked; head, body and fins
all closely scaled. Dusky-gray; sides with many pale stripes; pre-

orbital with a silvery streak. Head 4⅓; depth 2¼. D. XII, 12.
A. III, 11. Scales 10-66-20. L. 24. West Indies, rarely N. to
Cape Cod. (*Pimel. bosqui* Lacépède.) (Lat., one who cuts.)

FAMILY LXVII. MULLIDÆ. (The Surmullets.)

Body elongate, with large, ctenoid scales; head with large scales;
profile of head blunt; mouth small, the teeth various; premaxillaries protractile; maxillary simple, partly hidden by the broad
preorbitals; *throat with two long barbels.* Dorsals two, well separated, the first of about 7 high spines, the second short; A. short,
with two small spines; V. and gill structures normal. Tropical
seas, 5 genera and 35 species, rather small, carnivorous fishes
mostly valued as food.

   a. Teeth in lower jaw and on vomer and palatines; none in upper jaw; in-
      terorbital space flat and broad; opercle without spine. MULLUS, 172.

   172. MULLUS (Artedi) Linnæus.   (Ancient name from μύλλος,
                       lip.)

   386. **M. surmuletus** L. SURMULLET. Red: sides with three
yellow stripes; barbel 1⅓ in head, reaching beyond lower anterior
angle of opercle; eye smallish, 5 in head. Head 3¼; depth 4.
D. VII–1, 8. A. II, 6. Lat. l. 36. L. 10. Europe, one of the
most esteemed of food fish, very rarely taken on our coast. (Wood's
Holl; N. Y.; Pensacola.) Our form (var. **auratus** Jordan & Gilbert) differs slightly from the European. (*Eu.*) (Low Lat.,
"above mullets.")

FAMILY LXVIII. SCIÆNIDÆ. (The Drums.)

Body elongate, more or less, with weakly ctenoid scales. Lateral line continuous to the end of caudal fin. Head covered with
scales; cranium cavernous, the muciferous system highly developed,
surface of the skull very uneven; chin with pores; mouth and
teeth various; maxillary without supplementary bone, slipping beneath preorbital; premaxillaries protractile; gills and gill structures normal. D. deeply notched, its soft part long; A. short, with
1 or 2 spines; V. normal. Ear bones very large. Vertebræ about
24; air-bladder usually large and complicated, its structure enabling
the fish to make grunting or drumming sounds. Carnivorous fishes,
most of them valued as food. Genera 25; species 130, in all warm
seas, some genera confined to fresh waters.

   a. Vertebræ typically 14 + 10, the number in the abdominal region always
      greater than that in the caudal; lower jaw prominent; teeth not villi-
      form; preopercle entire; anal spines very weak. (*Otolithinæ.*)

    b. Anal moderate of 7 to 13 rays, its length not half that of soft D.; tip of
      upper jaw with (usually) 2 pointed canines; none at tip of lower.

                                           CYNOSCION, 173.

aa. Vertebræ typically 10 + 14; second anal spine well developed. (*Sci-œninæ.*)

x. Lower pharyngeals separate.
  y. Lower jaw without barbels.
    c. Teeth well developed, permanent in both jaws; lower pharyngeals narrow, with sharp teeth.
      d. Gill rakers slender, rather long; mouth oblique; A. inserted rather posteriorly; preorbital narrow; slits and pores of upper jaw little developed; preopercle serrate, its lowest spine enlarged, turned downward; head not very broad, not spongy above.
        BAIRDIELLA, 174.
      dd. Gill rakers rather short and thick; anal further forward; snout with large pores, and 2 to 4 slits on its edge; preorbital broad; mouth inferior. . . . . SCIÆNA, 175.
    cc. Teeth very small, subequal, those in lower jaw lost with age; lower pharyngeals broad, with paved teeth; gill rakers short, but slender, otherwise as in *Sciæna*. . . . . LEIOSTOMUS, 176.
  yy. Lower jaw with one or more barbels (otherwise essentially as in *Sci-œna*).
    e. Lower jaw with several slender barbels at its rami; preopercle serrate. . . . . MICROPOGON, 177.
    ee. Lower jaw with one thickish barbel at its tip; no air-bladder; anal spine single; body long and low; preopercle crenulate.
        MENTICIRRHUS, 178.
xx. Lower pharyngeals very large, completely united, with coarse paved teeth; snout, etc., as in *Sciæna*.
    f. Lower jaw with numerous barbels along the rami; preopercle nearly entire. . . . . POGONIAS, 179.
    ff. Lower jaw without barbels; preopercle obscurely serrate.
        APLODINOTUS, 180.

173. CYNOSCION Gill. (κύων, dog; σκίαινα, *sciæna*.)

a. Soft dorsal and anal closely scaled.
  b. Back and sides nearly uniform silvery, without spots.

387. **C. nothus** (Holbrook). WHITE WEAK-FISH. Body rather deep; snout short, bluntish; eye very large, 4 in head. Head 3½; depth 3¾. D. X–1, 28. A. II, 9. Scales 6–60–7. L. 12. Va. to Fla. (Lat., spurious.)

  bb. Back and sides with irregular dark spots in undulating streaks.

388. **C. regalis** (Bloch & Schneider). WEAK-FISH. SQUE-TEAGUE. Silvery, brownish above, and with bright reflections; fins without distinct spots; snout sharp; eye moderate, 5 to 7 in head. Head 3½; depth 4¼. D. X–1, 29. A. I, 13. Lat. l. 80. L. 2½ feet. Cape Cod to Fla., an abundant and most excellent food-fish.

  aa. Soft dorsal scaleless; back and upper fins with many conspicuous round black spots.

389. **C. nebulosus** (Cuv. & Val.). SPOTTED WEAK-FISH. Silvery, back bluish. Head 3½; depth 5. D. X–1, 25. A. I, 10.

Lat. l. 85. L. 2 feet. N. J. to Texas, common S. All these species are absurdly called "Trout" in the Southern States, — a name also applied in the same regions to the Black Bass.

174. BAIRDIELLA Gill. (To Spencer Fullerton Baird.)

390. **B. chrysura** (Lacépède). SILVER PERCH. YELLOW-TAIL. MADEMOISELLE. Jaws subequal; teeth in lower mostly in one series; second anal spine moderate, $2\frac{1}{3}$ in head; eye large. Greenish, sides silvery; scales and fins much punctulate; lower fins yellow. Head $3\frac{1}{2}$; depth $2\frac{3}{4}$. D. X-1, 22. A. II, 9. Lat. l. 50. L. 9. Cape Cod to Texas, abundant S. (χρυσός, golden; οὐρά, tail.)

175. SCIÆNA (Artedi) Linnæus. (*Corvina* Cuvier.) (Old name, from σκιά, shade.)

a. Preopercle serrate in young, the teeth disappearing with age; body elongate, little compressed. (*Sciænops* Gill.)

391. **S. ocellata** L. RED-FISH. CHANNEL BASS. Head long; eye small; mouth large, nearly horizontal; teeth in both jaws in bands, the outer enlarged above; anal spines moderate. Grayish-silvery, dark points on the scales, forming undulating brown streaks; a jet black spot edged with orange on base C. above, this sometimes duplicated. Head $3\frac{1}{8}$; depth $3\frac{1}{4}$. D. X-1, 24. A. II, 8. Scales 4-50-7. L. 4 feet. Cape Cod to Mexico; an important food-fish, S.

176. LEIOSTOMUS Lacépède. (λεῖος, smooth; στόμα mouth.)

392. **L. xanthurus** Lacépède. SPOT. GOODY. LAFAYETTE. Compressed; profile steep; snout blunt, fins low. Bluish, sides with 15 dark oblique bars; a round black spot behind shoulder. Head $3\frac{1}{2}$; depth 3. D. X-1, 32. A. II, 12. Lat. l. 60. L. 12. Cape Cod to Texas. (ξανθός, yellow; οὐρά, tail, but the C. is never yellow.)

177. MICROPOGON Cuv. & Val. (μικρός, small; πώγων, beard.)

393. **M. undulatus** (L.). CROAKER. Body rather elongate, with rather long head and large mouth. Grayish-silvery, back and sides with undulating dark streaks; dorsals with lines of dots. Head $3\frac{1}{8}$, depth $3\frac{1}{2}$. D. X-1, 27. A. II, 8. Lat. l. 60. L. 18. N. Y. to Texas.

178. MENTICIRRHUS Gill. (Lat., *mentum*, chin; *cirrus*, barbel.)

a. Gill rakers obsolete; lower pharyngeals narrow, their teeth slender; outer teeth of upper jaw enlarged; scales on breast large; maxillary reaching beyond front of eye, more than $\frac{1}{3}$ head. (*Menticirrhus.*)

b. Outer teeth of upper jaw very strong; lower lobe of C. not black.

394. **M. americanus** (L.). WHITING. D. a little lower than in the next species, the spines barely reaching soft rays. Silver-

gray, usually with faint oblique bars; snout projecting. Head 3⅓;
depth 4. D. X - 1, 25. A. I, 7. Lat. l. 65. Md. to Brazil, abundant S. (*M. alburnus* L.)

bb. Outer teeth of upper jaw little enlarged; lower lobe of C. mostly black.

395. **M. saxatilis** (Bloch). KING-FISH. BARB. SEA MINK.
D. high. Dusky gray, the back and sides with oblique dark cross-
bands : one at the nape vertical, forming with the next a V-shaped
blotch : a dark lateral streak, extending on C. Head 4 ; depth 4½.
D. X - 1, 26. A. I, 8. Scales 7–53–14. L. 18. Cape Cod to Fla.,
common N. (Lat., living among rocks.)

aa. Gill rakers present, small; lower pharyngeals broad, their teeth mostly
molar; outer teeth scarcely enlarged; scales on breast small. (*Umbrula* Jordan & Eigenmann.)

396. **M. littoralis** (Holbrook). SILVER WHITING. SURF
WHITING. Snout projecting, 3½ in head ; maxillary to eye, 3½ in
head. Silver-gray, almost plain ; tip of C. black. Head 3½ ; depth
4⅔. D. X - 1, 24. A. I, 7. Scales 6–53–12. L. 18. Va. to Texas.
(Lat., belonging to the shore.)

179. POGONIAS Lacépède. (πωγωνίας, bearded.)

397. **P. cromis** (L.). DRUM. Robust ; 2d A. spine large.
Grayish-silvery or brassy ; 4 or 5 dark vertical bars lost with age.
Head 3⅓; depth 2½. D. X - 1, 20. A. II. 6. Lat. l. 50. L. 4 feet.
Cape Cod to Brazil. (Old name from χρέμω, to neigh.)

180. APLODINOTUS Rafinesque. (ἁπλόος, simple ; νῶτος,
back.)

398. **A. grunniens** (Rafinesque). FRESH-WATER DRUM.
GASPERGOU. " SHEEP'S-HEAD." WHITE PERCH. CROAKER.
THUNDER-PUMPER. Snout blunt ; back compressed ; 2d A. spine
very strong ; C. rhombic. Grayish-silvery, more or less dotted.
Head 3½ ; depth 3. D. IX - 1, 30. A. II, 7. Lat. l. 55. L. 2 feet
or more. Great Lakes to Texas and Ga., abundant ; a large, coarse
fish of the larger streams and lakes. (Lat., grunting.)

FAMILY LXIX. **GERRIDÆ.** (THE MOHARRAS.)

Body compressed, with large, smoothish scales ; lateral line continuous ; mouth small, the premaxillary excessively protractile, the
spines of the premaxillaries extending backward in a deep groove
on top of head ; maxillary simple, not sheathed by the narrow preorbital ; mandible scaly, with a slit behind it, to permit motion ;
teeth small, in jaws only ; preopercle entire or serrate ; pseudobranchiæ concealed ; gills normal ; gill membranes separate ; lower
pharyngeal bones close together, usually loosely united ; D. single,
with 9 spines ; A. with 3 or 2 ; V. I, 5 ; air-bladder present. Verte-

bræ 10 + 14. Oviparous, carnivorous. One genus, with 30 species; in the warm seas. Silvery fishes, probably allied to the *Sparidæ*, but with no near relatives.

181. GERRES Cuvier. (Old name of some fish.)

a. Preopercle and preorbital entire; body oblong; spines moderate. (*Diopterus* Ranzani.)

b. Premaxillary groove scaled across anteriorly so that the posterior part appears as a naked pit.

399. **G. gula** Cuv. & Val. Silvery. faintly barred ; 3d D. spine not half head; 2d A. spine short. Head 3⅓; depth 2⅓. D. IX, 10. A. III, 8. Scales 5–43–10. L. 6. N. J. to Brazil, common S. (Lat., throat, the fish being called " Petite-Gueule " in W. I.)

PHARYNGOGNATHI. This family closes the series of fishes having Percoid affinities. We now pass to the group or suborder PHARYNGOGNATHI, those forms allied to the *Labroids*, and distinguished especially by the complete union of the lower pharyngeal bones. Of these, the typical forms, *Labridæ*, *Pomacentridæ* have the gills reduced, 3½ in number ; the last gill slit wanting or nearly so. Some of them (*Pomacentridæ*, *Cichlidæ*) differ from other spiny-rayed fishes in having but one nostril on each side; still others (*Embiotocidæ*) are viviparous. The *Pharyngognathi* being chiefly tropical are scantily represented within our limits.

FAMILY LXX. LABRIDÆ. (THE WRASSES.)

Body oblong, covered with cycloid scales; lateral line usually interrupted or angularly bent. Mouth terminal, protractile ; the teeth of the jaws generally strong ; no teeth on vomer or palatines; maxillaries simple, slipping under membranous edge of preorbital ; lower pharyngeals solidly united, with blunt teeth ; D. continuous, with 8 to 20 spines, the number greatest in Northern forms, which, as usual among fishes, have also an increased number of vertebræ ; anal spines 2 to 6, usually 3. V. normal. Pseudobranchiæ present. Gills 3½, usually no slit behind the last; nostrils double; air-bladder present. Genera 65 ; species 450, chiefly of the tropical seas. Many of them are brilliantly colored and some are valued as food. The teeth are adapted for the crushing of shells. (*Labrus*, an old name from *labrum*, lip.)

a. Vertebræ in increased number, 30 to 38; dorsal spines 16 to 20; teeth in jaws distinct, the anterior canine; no posterior canines; lateral line continuous; lips thick. (*Labrinæ.*)

  b. Preopercle serrate; cheeks and opercles scaly; teeth in more than two series, the outer enlarged. . . . . CTENOLABRUS, 182.

  bb. Preopercle entire; cheeks scaly; opercles naked; teeth in about two series. . . . . HIATULA, 183.

182. CTENOLABRUS Cuv. & Val. (κτείς, comb; Labrus.)
a. Interopercle naked; snout not very sharp. (*Tautogolabrus* Günther.)

400. **C. adspersus** (Walbaum). CUNNER. CHOGSET. BER- '
GALL. BLUE PERCH. Brownish blue, with brassy shades; young
with a black dorsal spot. Head 3½; depth 3. D. XVIII, 10.
A. III, 9. Lat. l. 45. L. 10. Newfoundland to Va., common N.
about rocks. (Lat., speckled.)

183. HIATULA Lacépède. (Old name; *hio*, to gape.)
401 **H. onitis** (L.). TAUTOG. OYSTER-FISH. BLACK-FISH.
Blackish; young greenish, irregularly barred. Head 3¼; depth 3.
D. XVI, 10. A. III, 8. Lat. l. 60. L. 16. Maine to S. C., a
common food-fish. (Meaning unknown.)

EPELASMIA. The rest of the *Pharyngognathi* are beyond our
limits, as are also the great bulk of the next group, or suborder, the
Squamipennes, or *Epelasmia* (Cope). Of these only a single species
comes N. of Va. In this group the post-temporal is simple, and the
upper pharyngeals reduced to thin laminæ. The group includes
the *Chætodontidæ*, *Acanthuridæ*, *Teuthididæ*, and the small

FAMILY LXXI. **EPHIPPIDÆ.** (THE ANGEL-FISHES.)

Body compressed and elevated; scales ctenoid densely covering
the body and the soft parts of the vertical fins; lateral line present.
Mouth small, terminal, with bands of setiform (tooth-brush-like)
teeth; premaxillary protractile; maxillary simple, partly slipping
under preorbital; gill membranes broadly attached to the isthmus;
gill rakers very short; pseudobranchiæ present. Dorsal deeply
notched, with 8 to 11 spines, the soft part very high, as is also the
soft anal; A. spines 3 or 4; C. subtruncate; P. short; V. normal.
Air-bladder large. Genera 6; species about 15, in the warm seas.
(ἔφιππος, on horseback, from the long dorsal spine.)

a. Anal spines 3; dorsal spines 8 or 9, the third elevated; profile very steep;
scales small. . . . . CHÆTODIPTERUS, 184.

184. CHÆTODIPTERUS Lacépède. (χαίτοδων, Chætodon;
δίς, two; πτερόν, fin.)

402. **C. faber** (Broussonet). ANGEL-FISH. SPADE-FISH. Grayish, the young with 4 to 7 black cross-bands; soft vertical fins, becoming falcate with age. Head 3; depth 1¼. D. VIII-1, 20.
A. III, 18. Lat. l. 60. L. 24. Warm seas, N. to N. Y.; a
good food-fish. (An old name, meaning blacksmith.)

CATAPHRACTI. We next pass to the group of *Cataphracti* or
Cottoid fishes, an assemblage of families, characterized as a whole
by the development of a "suborbital stay," a bony process extending from the suborbital ring backward across the cheeks to or towards the preopercle. In the extreme forms (*Agonidæ*, etc.), the

check is wholly mailed. In others, as *Cyclopterus*, this stay is little conspicuous. The *Cataphracti* agree with the *Scyphobranchii* in having the third upper pharyngeal large, basin-shaped, but they differ much among themselves, the *Hexagrammidæ* and *Scorpænidæ* resembling the Perciform fishes, while some of the others are widely aberrant.

FAMILY LXXII. SCORPÆNIDÆ. (THE ROCK-FISHES.)

Body oblong, robust, usually covered with ctenoid scales; lateral line present. Head large, with spinous ridges above; opercle with two spinous processes; preopercle with five. Mouth large, the jaws with villiform teeth; premaxillaries protractile; maxillaries broad, simple, not sheathed by preorbital; bony suborbital stay present, usually covered by skin and usually not reaching preopercle. Gill membranes free and separate. Gills 3½, with no slit behind the last. V. normal, I, 5. D. continuous, with 8 to 16 strong spines. Arctic species have more spines and more vertebræ than tropical species. Vertebræ 24 to 32; A. short, with 3 spines; P. broad. Pseudobranchiæ and air-bladder large. Genera 20; species 200. Carnivorous fishes living about rocks in all seas, often at considerable depths, especially abundant about Cal. and Japan. Non-migratory; excellent as food, and usually red in color. Most are viviparous, the young ¼ inch long when born. (σκορπίος, scorpion.)

a. Dorsal spines 15; vertebræ 12 + 19 = 31; palatine teeth present; head not very rough above. . . . . SEBASTES, 185.

185. SEBASTES Cuvier. (σεβαστός, magnificent.)

403. **S. marinus** (L.). ROSE-FISH. HEMDURGAN. NORWAY HADDOCK. Body ovate; top and sides of head evenly scaled; cranial ridges low and sharp; preocular, supraocular, postocular, tympanic, and occipital ridges present; eye very large; chin prominent. Orange red, some dusky on opercle. Head 3; depth 2¼. D. XV, 14. A. III, 8. Lat. l. 40, tubes 85. N. Atl., S. to Cape Cod: common N. Specimens in shallow water are smaller and brownish. (*Var. viviparus* Kröyer.) (*Eu.*)

FAMILY LXXIII. COTTIDÆ. (THE SCULPINS.)

Body elongate, more or less, the head usually large and depressed; eyes high; bony stay conspicuous, but not covering the cheek; preopercle armed; teeth in villiform bands; maxillary simple; gills 3½ or 4; gill membranes connected, often joined to isthmus. Body naked, or irregularly scaled or warty, never evenly scaled; lateral line present. Dorsals usually separate, the spines slender; A. without spines; P. large, with broad procurrent base, the lower rays simple; V. thoracic, usually I, 3 or I, 4, sometimes wanting, never united. Pseudobranchiæ present. Vertebræ, as usual in

Arctic fishes, numerous, 35 to 50. Genera 40; species 150, mostly
of the springs, rock-pools, and seashores of Arctic regions; a few in
the deep sea. Singular fishes, mostly of small size, and of little
value as food. The fresh-water species are very destructive to
eggs of other fishes.

a. Spinous D. longer than soft part, of more than 14 spines. (*Hemitrip-*
    terinæ.)
    b. Spinous D. deeply notched, the anterior spines highest; skin with prickles
    and warts; teeth on vomer and palatines; gill membranes free from
    isthmus; no slit behind last gill. . . . . HEMITRIPTERUS, 186.
aa. Spinous D. shorter than soft part, of less than 13 spines; dorsal spines not
    concealed; gill openings not very small. V. present. (*Cottinæ.*)
    c. Vomer with teeth.
      d. Slit behind last gill obsolete or reduced to a round pore; skin with-
        out true scales.
        Gill membranes broadly united to the isthmus, not forming a fold
        across it; head feebly armed; palatine teeth few or none.
                            COTTUS, 187
      ee. Gill membranes free from isthmus or else forming a broad fold
        across it; head well armed.
        f. Palatine teeth none; skin naked or prickly.
                        ACANTHOCOTTUS, 188.
        ff. Palatine teeth well developed; skin smooth.
                        ARTEDIELLUS, 189.
      dd. Slit behind last gill small, but evident; no palatine teeth.
        g. Skin smooth; gill membranes not quite free from isthmus;
        preorbital, etc., strongly cavernous. . TRIGLOPSIS, 190.
        gg. Skin with minute prickly scales, and with plates along back
        and lateral line; gill membranes free from isthmus.
                        TRIGLOPS, 191.
    cc. Vomer without teeth; preopercular spine antler-like; a fold across
    isthmus and no slit behind last gill; no scales.
                        GYMNACANTHUS, 192.

186. HEMITRIPTERUS Cuvier. (ἡμι-, half ; τρεῖς, three ;
πτερόν, fin.)

404. H. americanus (Gmelin). SEA RAVEN. Head large,
with many humps and ridges above. Brown, body and fins much
variegated with blackish. Head 2⅔; depth 3¾. D. IV, XII, 1, 12.
A. 13. Lat. l. 40. L. 18. Cape Cod to Arctic Sea.

187. COTTUS (Artedi) Linnæus. (*Uranidea* DeKay.) MILLER'S
THUMB. (Ancient name of *C. gobio*, from κοττός, head.)

a. Palatines with teeth; ventrals I, 4 (the spine obscure).
    b. Preopercular spine large, as long as eye, strongly hooked upward; skin
    above with coarse prickles (*Tauridea* Jordan & Rice).

405. C. ricei Nelson. Head broad, body contracted at base of
tail. Olivaceous, finely speckled. Head 3⅔; depth 5¼. D. VIII –17.
A. 12. L. 2½. Lake Mich. and L. Ontario, rare. (To Frank L.
Rice.)

bb. Preopercular spines small, mostly concealed by the skin; skin smooth or prickly in or behind the axil only. (*Potamocottus* Gill.)

406. **C. richardsoni** Agassiz. MILLER'S THUMB. BLOB. MUFFLE-JAW. Body rather stout, the head very broad; preopercle with a short, sharp, straightish spine, turned upward and backward, with 2 smaller spines below it. Olivaceous, much barred and speckled. Head $3\frac{1}{2}$; depth 4 to 6. D. VI to VIII – 16. A. 12. V. I, 4. L. 3 to 7. Lake Superior to Ark., Ga., Md., and Canada very abundant in springs, caves, cold lakes, and rocky brooks. Very variable. The numerous varieties or nominal species are hardly worthy of recognition by name. (To John Richardson, author of the "Fauna Boreali-Americana.")

aa. Palatine teeth, none; V. I, 3; skin mostly smooth. (*Cottus.*[1])
  c. Anal rays 13 or 14.
    d. Preopercular spine large, hooked upward.

407. **C. pollicaris** (Jordan & Gilbert). Light olive, blotched and spotted with black, but not speckled; upper fins spotted. Eye $5\frac{1}{3}$ in head. Head $3\frac{2}{3}$; depth $4\frac{3}{4}$. D. VII –19. A. 13. L. 5. Lake Michigan. (Lat., thumb-like.)

408. **C. spilotus** (Cope). Olive, everywhere closely speckled with darker except on belly; sides barred with blackish: fins barred and spotted. Eye $4\frac{1}{2}$ in head. Head $3\frac{1}{2}$; depth 5. D. VIII –17. A. 13. L. 3. Grand Rapids, Mich. (σπιλωτός, spotted.)

    dd. Preopercular spine short, acute, turned obliquely upward.

409. **C. viscosus** Haldeman. Stout, with many mucous pores; fins low. Olivaceous, body and fins mottled with dark; 1st D. with red edge. Head $3\frac{1}{2}$; depth $4\frac{3}{4}$. D. VI –18. A. 14. Penn. to Md.

  cc. Anal rays 11 or 12.
    e. Preopercular spine short, scarcely hooked.
    f. Preopercular spine bent upward and backward.

410. **C. gracilis** Heckel. Body rather slender; fins large. Olivaceous, mottled, 1st D. edged with red. Head $3\frac{1}{2}$; depth $4\frac{1}{2}$ to $5\frac{1}{2}$. D. VIII –16. A. 12. L. 4. N. Eng. and N. Y. (Var. **gobioides** Grd., with robust body, and var. **boleoides** Grd., with slender body and long fins, have been described.) (Lat., slender.)

    ff. Preopercular spine directed backward and scarcely upward.

411. **C. hoyi** Putnam. Slender; ♀ prickly above; jaws narrower and mouth smaller than in *C. gracilis;* another spine below it turned downward, and one or two others still lower. Olivaceous, speckled and barred. D. VI –15. A. 11. L. 2. L. Michigan. (To Dr. Philo R. Hoy.)

    ee. Preopercular spine distinctly hooked.

---

[1] The species of this group have never been critically studied ; some of them are doubtful, and most of them may prove to be mere varieties of *Cottus gracilis.*

412. **C. franklini** Agassiz. Short and stout; fins low. Head 3½; depth 4½. D. VIII –17. A. 12. L. 3. L. Superior. (To Sir John Franklin.)

413. **C. formosus** Girard. Slender; head small, 4¼ ; depth 5½. D. VIII –16. A. 11. L. 3¼. L. Ontario. (Lat., pretty.)

188. ACANTHOCOTTUS Girard. SCULPINS. (ἄκανθα, spine; Cottus.)

a. Anal fin short, with 10 rays.

414. **A. æneus** (Mitchill). GRUBBY. Upper preopercular spine shorter than eye, nearly twice length of next. Grayish brown, much variegated with blackish; no large white spots. Head 2¾; depth 4. D. IX –13. A. 10. V. I, 3. L. 6. Maine to N. Y., common. (Lat., brassy.)

aa. Anal fin long, of 14 rays.

   b. Upper preopercular spine about as long as eye, reaching middle of opercular spine, not twice length of the spine below it.

415. **A. scorpius** (L.) DADDY SCULPIN. BIG SCULPIN. Dark brown, with darker bars; belly in ♂ dusky, with round black spots; fins spotted and barred; top of head with spinous tubercles; eye large. Head 2½; depth 4½. D. X –17. A. 14. V. I, 3. L. 25. N. Atlantic, S. to N. Y. The American form is var. **grœnlandicus** C. & V., distinguished by its larger size, broader interorbital, and higher fins; the var. *scorpius* ranges S. to Me. (*Eu.*) (Lat., a scorpion.)

   bb. Upper preopercular spine very long, longer than eye, reaching beyond tip of opercular spine, its length more than 4 times that of the spine below it.

416. **A. octodecimspinosus** (Mitchill). Body slender, with long. narrow head; a strong spine above eye; top of head with ridges; eye very large. Olivaceous, with dark bars; fins mottled. Head 2½; depth 5½. D. IX –15. A. 14. V. I, 3. L. 15. N. Y. to Nova Scotia. (Lat., *octodecim*, eighteen ; *spinosus*, spined.)

189. ARTEDIELLUS Jordan. (Diminutive of *Artedius*, a related genus.)

417. **A. uncinatus** (Kröyer). Eye very large ; spine of preopercle large, hooked upward. Olivaceous, mottled and barred. Head 3; depth 4½. D. VIII –13. A. 11. L. 4. Cape Cod, N. (Lat., hooked.)

190. TRIGLOPSIS Girard. (τρίγλα, Trigla; ὄψις. appearance.)

418. **T. thompsoni** Girard. Body very slender; head long, depressed; eye very large, 4 in head; skull extremely cavernous; preopercle with 4 short, sharp spines; soft D. and A., very high: lat. l. chain-like. Olivaceous, with faint dark blotches. Head 3;

depth 6. D. VII –18. A. 15. V. I, 3. L. 3. Deep waters of L. Michigan and L. Ontario. (To Rev. Zadock Thompson, author of Nat. Hist. of Vermont.)

191. TRIGLOPS Reinhardt. (τρίγλα, Trigla; ὤψ, appearance.)

419. T. pingeli Reinhardt. Head slender; eye large; tail very slender; preopercular spines small, simple; sides with peculiar scales and prickles. Olivaceous, variegated with darker; sides spotted with dark; a black ocellus on spinous D. Head 3½; depth about 5¼. D. IX –21. A. 21. L. 5. Arctic seas, S. to Cape Cod.

192. GYMNACANTHUS Swainson. (γυμνός, naked; ἄκανθα, spine.)

420. G. tricuspis (Reinhardt). Eye very large; skin mostly smooth; preopercular spine broad, shorter than eye, with 3 points; V. very long. Dark brown with darker bars; axils dusky in ♂ with round white spots. Head 3½; depth 4¾. D. XII – 16. A. 18. V. I, 3. L. 12. Arctic, S. to Me. (*Eu.*) (Lat., *tris*, three; *cuspis*, cusp.)

FAMILY LXXIV. **AGONIDÆ.** (THE ALLIGATOR-FISHES.)

Fishes allied to the *Cottidæ* and similar in general structure, but with the body completely covered by a coat of mail composed of about eight series of large bony plates; head entirely bony externally; suborbital stay covering the cheek; gills 3½, no slit behind the last, pseudobranchiæ large. Vertebræ (as in other Arctic fishes) numerous, 35 to 40. Spinous dorsal sometimes wanting. Small fishes, chiefly of the Arctic seas; genera 10, species about 20. (*a.* privative; γωνία, joint; *i. e.*, rigid.)

a. Spinous dorsal obsolete; gill membranes free from isthmus. (*Aspidophoroidinæ.*)

b. Bony plates keeled, without spines; fins very small; teeth on vomer.
ASPIDOPHOROIDES, 193.

193. ASPIDOPHOROIDES Lacépède. (ἀσπίς, shield; φορέω, to bear; εἶδος, form.)

a. Snout with two large diverging spines above; no other spines present.

421. A. monopterygius (Bloch). Body elongate, subterete, resembling that of a pipe-fish; eyes very large. Brownish, obscurely banded. Head 5⅔; depth 9. D. 5. A. 6. Lat. l. 50. L. 6. Cape Cod, N. (μόνος, single; πτέρυξ, fin.)

FAMILY LXXV. **CEPHALACANTHIDÆ.** (THE FLYING GURNARDS.)

Body elongate, with bony keeled scales; head blunt, cuboid, its surface almost entirely bony; nuchal shield with a strong spine on

each side ; preopercle with a very long rough spine ; opercle
small; isthmus very broad, scaly; gill rakers minute; mouth
small, with granular teeth in jaws only ; tail with 2 serrate, knife-
like appendages. Spinous D. short, its first spines free; an im-
movable spine between dorsals; C. small, lunate; P. divided to
base, the anterior part corresponding to the free rays in *Triglidæ*,
of about 6 rays connected by membrane; the posterior part very
long (reaching C. in adult), the rays slender and simple; V. close
together, I, 4. Air bladder complex; vertebræ 9 + 13. Two
species in the warm seas, able to flutter for short distances in
the air.

194. CEPHALACANTHUS Lacépède. (κεφαλή, head; ἄκανθα, spine.)

a. Occiput without filament.

422. C. volitans (L.). FLYING GURNARD. SEA ROBIN.
Greenish and brown, mottled with orange or red, the belly usually
orange ; P. with blue streaks and spots; C. with reddish bars.
Color very variable. Head 4⅓; depth 5¼. D. II, IV – 8. A. 6.
P. 28–6. L. 12. Atlantic, N. to Newf'd. (*Eu.*) (Lat., flying.)

FAMILY LXXVI. TRIGLIDÆ. (THE GURNARDS.)

Body subfusiform, covered with scales or bony plates; head
entirely covered with rough bones, most of them armed with
spines; mouth moderate, with small teeth or none; maxillary
simple; gills 4, a slit behind fourth; pseudobranchiæ present;
V. I, 5, wide apart, separated by a flat area. Spinous D. short;
A. without spines; C. narrow. P. large, with broad base, the
2 or 3 lowermost rays detached from the rest and separate. Air
bladder present. Vertebræ about 24, as in nearly all tropical
fishes. Genera 5 ; species 40, in all warm seas. (τρίγλα, old
name of *Mullus.*)

a. Body scaly; teeth present on jaws, vomer and palatines; free P., rays 3.

PRIONOTUS, 195.

195. PRIONOTUS Lacépède. SEA ROBINS. (πρίων, saw; νῶτος, back.)

a. Mouth rather small, the maxillary not ⅓ head; a cross-groove on top of
head behind eye; a black ocellated spot on spinous D.

423. P. carolinus (L.). Body rather stout; preopercular spine
with no smaller one before it; P. short, not ½ body ; P. appendages
broadened at tip; gill rakers about 10; bones of head comparatively smooth. Olive, back with 4 dark cross-shades; pale oblique
streaks on 1st D. Head 3 ; depth 5. D. X –13. A. 12. Lat. l.
58 (pores). L. 12. Cape Ann to S. C. (S. occurs *P. scitulus*
J. & G., slender, with short head and spotted body.)

aa. Mouth rather large; the maxillary about 2½ in head; no cross-groove on top of head; black spot on 1st D. diffuse; preopercular spine with a smaller one before it.

  b. Cheek bone without distinct spine at centre of radiation; edge of preorbital granular serrate; spines on top of head not knife-like.

424. **P. strigatus** Cuv. & Val. Head not very broad; gill rakers long, 15 to 20; interorbital area flattish. Brownish, side with a distinct bronze band parallel with lateral line, this breaking up in spots behind; head spotted; body and fins with dark clouds; P. finely barred with black. Head 2¾; depth 4. D. X –12. A. 11. Scales 10–60–23. P. 2 in body. L. 12. Cape Cod to Va. (Perhaps a variety of *P. evolans* L., which has scales larger, P. not barred, etc.; N. C., S.) (Lat., striped.)

  bb. Cheek bone with a spine at centre of radiation; bones of head sharply striate; head broad, the spines above compressed and knife-like, especially in young.

425. **P. tribulus** Cuv. & Val. Spines much larger than in others, still larger in young; spines on snout and side of cheek in line with preocular spine; gill rakers thickish, about 10. Brownish, much clouded; no lengthwise stripe. Head 2½; depth 4½. D. X –12. A. 11. Lat. l. 50. L. 12. P. 2 in body. N. Y. to Texas. (Lat., a thistle, or other source of tribulation.)

Family LXXVII. LIPARIDIDÆ. (The Sea-snails.)

Body oblong, covered with lax, naked skin; head broad, obtuse; suborbital stay slender; teeth small, mostly tricuspid; opercles unarmed; gill openings small, the membranes joined to the isthmus; gills 3½, no slit behind last; no air bladder; pseudobranchiæ rudimentary. D continuous, the spines feeble. A. without spines. V. I, 5, the two fully united, forming the bony centre of a broad sucking-disk or else wanting. P. broad, the base procurrent, the lower rays longer than those above them. C. short. Vertebræ 40 to 45. Genera 3; species about 20. Small fishes of the Arctic seas, some of them in deep water. Although very different in appearance, they are closely related to some of the *Cottidæ.*

a. Ventral disk present. (*Liparinæ.*)
  b. Ventral disk well developed; vent well behind head. . . Liparis, 196.

196. **LIPARIS** (Artedi) Fleming. (λιπαρός, sleek-skinned.)

a. Dorsal fin continuous; separated by a notch from caudal.

426. **L. montagui** (Donovan). Snout very broad. Yellowish, the fins dark-edged. Disk not quite half head. Head 3½; depth 4½. D. 28. A. 24. C. 14. P. 30. L. 3. Cape Cod, N. (*Eu.*) (To Mr. G. Montagu, a writer on British fishes.)

aa. Dorsal fin joined to the caudal.

427. L. liparis (L.). Sea Snail. Body thick; yellowish with purplish stripes. Disk 2 in head. Head 4; depth 3½. D. 33. A. 28. P. 34. L. 5. Cape Cod, N. (*Eu.*)

Family LXXVIII. CYCLOPTERIDÆ.
(The Lump Suckers.)

Closely related to the *Liparidldæ*, but with the body short and thick, covered with thick skin, which is often tubercular or spinous. Vertebræ fewer, about 28. Adhesive ventral disk well developed, enabling the fishes to fasten themselves firmly to rocks. Genera 3; species 4. In the Arctic seas.

a. Spinous dorsal present; skin with bony plates and tubercles.
  b. Dorsal spines not disappearing; gill opening a small slit on level of eye; sucking disk large. . . . . Eumicrotremus, 197.
  bb. Dorsal spines in adult enveloped in a fleshy hump; gill openings larger; disk small. . . . . Cyclopterus, 198.

197. EUMICROTREMUS Gill. (εὐμικρός, very small ; τρῆμα, aperture.)

428. E. spinosus (Müller). Shields with small tubercles and slender flexible prickles. Olivaceous, the naked skin punctate. Head 3; depth 2. D. VII–11. A. 10. C. 10. Maine, N. (*Eu.*) (Lat., spined.)

198. CYCLOPTERUS (Artedi) Linnæus. (κύκλος, circle; πτερόν, fin.)

429. C. lumpus L. Lump-sucker. Lump-fish. Shields without spines. Olivaceous, punctulate; young black, with green specks (*Kingsley*). Head 3¾; depth 2. D. VII–10. A. 10. L. 15. Chesapeake Bay, N. (*Eu.*) (English, lump.)

Haplodoci. The next group shows no close relation to any other of our families. On account of the simple post-temporal (bifurcate in most fishes), Professor Cope has made of the Batrachidæ a special suborder, Haplodoci.

Family LXXIX. BATRACHIDÆ. (The Toad-fishes.)

Body depressed anteriorly, with compressed tail; head large, depressed, with well-developed mucous channels; mouth very large, with strong teeth; gills 3, a slit behind the last; no pseudobranchiæ; gill membranes broadly united to isthmus; no bony suborbital stay; post-temporal (suprascapula) undivided; scales cycloid, small or wanting; dorsals separate, the first of 2 or 3 low stout spines, the second, like the anal, very long. V. jugular, I, 2 or I, 3; P. broad, procurrent; no pyloric cæca. Vertebræ 30 to 45. Carnivorous fishes, chiefly of warm seas, some of them very large. The young attach themselves to rocks by means of an adhesive ventral disk, which soon disappears. Some species have poison glands at base of dorsal and opercular spines.

a. Body naked; lateral line indistinct, without shining bodies; dorsal spines 3; a foramen in the axil; no poison glands; teeth strong, blunt.
BATRACHUS, 199.

199. BATRACHUS Bloch & Schneider. (βάτραχος, frog.)

430. **B. tau** (L.). TOAD-FISH. OYSTER-FISH. SAPO. Pores on jaws with cirri; subopercle with a strong spine. Blackish green, with dark markings; fins with dark bars. Head 2⅔; depth 4⅓. D. III - 27. A. 24. L. 18. Cape Cod to W. I., very abundant. (T., from the form of the bones of the top of the head.)

XENOPTERYGII. We pass next to the suborder XENOPTERYGII, a little group, distinguished by the peculiar sucking disk at the breast, formed from the skin of the body and not from the ventral fins. There is no spinous dorsal or suborbital ring, and the palatine arcade is said to be materially modified. The relations of these fishes are obscure, but they are probably descended from Batrachoid or Cottoid forms.

FAMILY LXXX. GOBIESOCIDÆ. (THE CLING-FISHES.)

Body elongate, the head very broad and depressed, the skin smooth, naked; mouth moderate, upper jaw protractile; teeth conical or incisor-like; opercle reduced to a spine; pseudobranchiæ small or 0; gills 2½ or 3; gill membranes broadly united; D. small, posterior, similar to anal, both of soft rays only; V. I, 4 or I, 5; the fins wide apart, and between them a very large sucking disk composed chiefly of folds of skin. No air-bladder. Vertebræ 26 to 36. Small carnivorous fishes of the warm seas, living in tide pools and clinging firmly to stones. Genera 10; species 30.

a. Gill membranes free from isthmus; gills 3; lower jaw with incisors; posterior part of sucking disk without free anterior margin.
GOBIESOX, 200.

200. GOBIESOX Lacépède. (Gobius + Esox.)

431. **G. strumosus** Cope. Lower incisors not serrate. Head very wide, its width 2⅝ in total (with C.); eye small; teeth 24/22; no canine. Plumbeous, fins blackish. D. 11. A. 10. Va. to S. C., scarce. (Lat., swollen.)

SCYPHOBRANCHII. The Blennioid, Gobioid, and Uranoscopoid fishes show more or less definite affinities with each other, and in some degree with the HAPLODOCI and CATAPHRACTI. Like the latter they have the third upper pharyngeal enlarged and basin-shaped, but they have no suborbital stay, unless the bony cheek in URANOSCOPIDÆ be regarded as representing the latter. They

form together a group or suborder called by Professor Cope the
SCYPHOBRANCHII.

FAMILY LXXXI. URANOSCOPIDÆ. (THE STAR-GAZERS.)

Body elongate, tapering behind; scales usually small, cycloid;
lateral line mostly obsolete. Head cuboid, usually mailed above
and on cheeks. Eyes small, on front of top of head. Mouth vertical, the lower jaw prominent, the lips mostly fringed; teeth small;
premaxillaries protractile; maxillary broad, simple, not concealed
by preorbital. Gill openings very wide, the membranes free; gills
3¼. a small slit behind the last. Pseudobranchiæ present. Spinous
D. very short, the fin long; A. long; P. with broad oblique base;
V. jugular, 1, 5. No air-bladder. Carnivorous fishes of the shores
in warm regions. Genera about 7; species 20. (οὐρανός, sky;
σκοπέω, to look.)

a. Dorsal fins two; head without spines.
  b. Head above entirely covered by a rugose coat of mail; a small barbel
       in mouth, before tongue. . . . . ASTROSCOPUS, 201.
  bb. Head above with a Y-shaped bony projection extending forward from
       occipital region; on each side of this shield a trapezoidal naked area;
       mouth without tentacle. . . . . UPSILONPHORUS, 202.

201. ASTROSCOPUS Brevoort.   (ἄστρον, star ; σκοπέω, to
look.)

432. **A. anoplos** (Cuv. & Val.).   Jet black above and on lower
jaw and 1st D.: belly and fins pale; scales minute. Head 2½; depth
3¼. D. IV - 14. A. 13. L. 2½. N. Y. to Key West. (ανόπλος,
unarmed.)

202. UPSILONPHORUS Gill.   (ὔψιλόν, Υ; φορέω, to bear,
from the bones on top of head.)

433. **U. y-græcum** (Cuv. & Val.)   Brownish, everywhere finely
spotted with white ; a dark horizontal band on tail; C. with lengthwise stripes. Head 2½ ; depth 3¼. D. IV - 13. A. 12. Lat. l.
113. L. 10. N. J., S.

FAMILY LXXXII. GOBIIDÆ. (THE GOBIES.)

Body oblong or elongate, variously naked or scaly; no lateral
line; mouth and teeth various; premaxillary protractile; suborbital
without bony stay; skin of head covering eyes; opercles mostly unarmed; pseudobranchiæ present: gills 4; gill membranes united
with the isthmus; spinous dorsal little developed, of 2 to 8 flexible
spines; anal without spine; V. I, 5, close together or usually fully
united into a sort of sucking disk; C. convex; anal papilla evident.

No pyloric cæca or air-bladder. Vertebræ about 25. Small, carnivorous fishes, creeping about on sea-bottoms after the fashion of the Darters, a group which the Gobies much resemble. Genera 70; species about 400, chiefly of tropical seas and ponds. South of Cape Hatteras a multitude of species are found, but only one is at all common N. of that point.

  a. Ventral fins united; dorsals separate, free from caudal. (*Gobiinæ.*)
  b. Ventral disk not adnate to belly; teeth simple; shoulder girdle without fleshy processes.
  c. Body with ctenoid scales; dorsal spines 6. . . . . GOBIUS, 203.
  cc. Body with small, cycloid scales; dorsal spines 7 or 8.
  MICROGOBIUS, 204.
  ccc. Body entirely naked. . . . . GOBIOSOMA, 205.

203. GOBIUS (Artedi) Linnæus. (The old name, from κωβιός, gudgeon.)

434. **G. soporator** Cuv. & Val. Olivaceous, dotted. C. short. Head 3; depth 4⅖. D. VI–1, 9. A. I, 8. Scales 35–13. L. 6. Tropics; N. to Carolina. (Lat., sleeper.)

204. MICROGOBIUS Poey. (μικρός, small; Gobius.)

435. **M. eulepis** Eigenmann & Eigenmann. Yellowish, dotted; 1st D. with black spot. Head 4; depth 5½. D. VII–15. A. 16. Scales 50–14. L. 2. Fortress Monroe. (εὖ, well; λεπίς, scale.)

205. GOBISOMA Girard. (Gobius; σῶμα, body.)

436. **G. bosci** (Lacépède). Body moderately chubby; cheeks tumid. Olive with darker cross-shades. Head 3½; depth 5 to 6. D. VII–14. A. 10. L. 2¼. Cape Cod to S. C. (To M. Bosc, French consul at Charleston.)

FAMILY LXXXIII. **BLENNIIDÆ.** (THE BLENNIES.)

Body oblong or variously elongate, naked, or covered with smooth scales; teeth well developed; suborbital ring without "stay"; D. long, continuous, or divided; the anterior portion, and sometimes the whole fin of spines, either stiff or flexible; anal long; V. jugular, few rayed or wanting; C. present; tail not isocercal; pseudobranchiæ present; air-bladder usually wanting. Vert. 30 to 100. Genera 50; species nearly 300, a varied group mostly inhabiting shallow sea-bottoms and rock-pools. A few are ovoviviparous. (*Blennius*, ancient name, from βλέννα, slime.)

  a. Teeth long, slender, curved, like comb-teeth, in front of jaws only; body naked; soft rays forming about half of D.; V. well developed. Vertebræ 30 to 40. Carnivorous, oviparous, tropical. (*Blenniinæ.*)
  b. Gill membranes broadly united to the isthmus.
  d. Mouth large; head pointed; no canines. . . . CHASMODES, 206.
  dd. Mouth small, the head blunt in profile.

    e. Canine teeth none. . . . . **ISESTHES,** 207.
    ee. Canine teeth in one or both jaws behind the other teeth.
                            HYPLEUROCHILUS, 208.
aa. Teeth conic (not like comb-teeth); D. (in our genera) of spines only;
    vertebræ very numerous; lateral line not bent; body scaly; species
    chiefly Arctic.
    f. Gill openings not continued forward below, the membranes broadly
        united. V. minute or wanting. (*Xiphidiinæ.*)
    g. Lateral line none; V. rudimentary; gill membrane free from isth-
        mus; A. with 2 small spines; no pyloric cæca. **MURÆNOIDES,** 209.
    ff. Gill openings prolonged forward below, separated by a narrow
        isthmus; P. long; V. well developed; oviparous; herbivorous.
        (*Stichæinæ.*)
    h. Lateral line present.
        i. Lateral line forked or duplicated. . **EUMESOGRAMMUS,** 210.
        ii. Lateral line simple. . . . . **STICHÆUS,** 211.
    hh. Lateral line wanting; teeth on jaws only. **LEPTOBLENNIUS,** 212.

206. CHASMODES Cuv. & Val. (χασμώδης, yawning.)

437. **C. bosquianus** (Lacépède). Orbital tentacle minute or
wanting; maxillary reaching beyond eye : ♂ olive green with 9 blue
lines; head and 1st D. with orange; ♀ dark green, reticulated and
barred. Head 3½; depth 3¼. D. XI, 19. A. 20. L. 3. N. Y.
to La. (To M. Bosc, a zealous collector of the fishes of S. C.)

207. ISESTHES Jordan & Gilbert. (ἴσος, equal; ἐσθίω. to eat.)

438. **I. hentz** (Le Sueur). Orbital cirrus bifid at tip, as long as
D. spines. D. high, the spines stiff. Olive, with vague bars; head
with distinct black spots. Head 3⅔; depth 3¼. D. XII, 15. A. 19.
L. 2½. Md. to La. (*Bl. punctatus* Wood; name preoccupied.) (To
Mr. Hentz, an early entomologist.)

208. HYPLEUROCHILUS Gill. (ὑ, upsilon; πλευρόν, side; χεῖλος, lip.)

439. **H. geminatus** (Wood). Orbital cirrus branched, very high
in ♂ ; D. spines slender. Olive brown, back and fins with black
spots in ♂. Head 3½; depth 4. D. XI, 15. A. 18. L. 2¼. Va.
to Texas, with the two preceding and others, among oyster shells
and clusters of tunicates; also about ballast piles. (*Bl. multifilis*
Girard, ♂.) (Lat., twin.)

209. MURÆNOIDES Lacépède. (μύραινα, moray; εἶδος, form.)

a. Ventrals present, I, 1.

440. **M. gunnellus** (L.). BUTTER-FISH. Head naked. Brown
with darker bars; black ocelli along base of D. Head 8; depth 9.
D. LXXVIII. A. II, 38. Vert. 85. L. 12. Labrador to Va.,
common in sea weed, N. (*Eu.*) (From "gunnel," gunwale, wrongly
supposed to be its English name.)

210. EUMESOGRAMMUS Gill. (*εὖ*, well; *μέσος*, middle; *γραμμή*, line.)

441. **E. subbifurcatus** (Storer). Brownish with pale blotches; black bars on head; D. with black dots; lateral line with upper branch only. Head 4½; depth 5. D. XLIV. A. 30. Cape Cod; N. rare. (Lat., *sub*, almost; *bis*, two; *furcatus*, forked.)

211. STICHÆUS Reinhardt. (*στιχάω*, to set in rows.)

442. **S. punctatus** (Fabricius). Scarlet; D. with black spots. Head 4½; depth about 6½. D. XL. A. I, 35. Cape Cod. N.

212. LEPTOBLENNIUS Gill. (*λεπτός*, slender; Blennius.)

443. **L. serpentinus** (Storer). Head small; olive, with pale shades; D. with oblique white bands. Head 9; depth 15. D. LXXV. A. 50. V. I, 3. L. 12. Cape Cod, N.

FAMILY LXXXIV. **CRYPTACANTHODIDÆ.** (THE WRYMOUTHS.)

Fishes allied to the *Blenniidæ*, but with the head cuboid, with vertical cheeks, conspicuous muciferous channels in jaws and preopercle; top of head flat; snout short; lower jaw very heavy, cleft of mouth vertical; teeth conical, on jaws, vomer, and palatines; gill membranes joined to the narrow isthmus; P. short; V. wanting; D. very long, of spines only, enveloped in thick skin: D., A., and C. joined. Body naked or scaly. Vertebræ many. Genera, 2; species 2. Arctic.

a. Body scaleless. . . . . CRYPTACANTHODES, 213.

213. CRYPTACANTHODES Storer. (*κρυπτός*, hidden; *ἀκανθώδης*, spined.)

444. **C. maculatus** Storer. WRY-MOUTH. GHOST-FISH. Brown, with dark spots, rarely immaculate. Head 6½; depth 13. D. LXXIII. A. 50. L. 24. Cape Cod, N.

FAMILY LXXXV. **ANARRHICHADIDÆ.** (THE WOLF-FISHES.)

Fishes similar to the Blennies in most respects, but with the vomer very thick and solid, with two series of coarse molar teeth; palatines with similar teeth; jaws with canines in front, the posterior teeth below molar. Scales rudimentary; no lateral line; gill membranes joined to isthmus. D. high, of flexible spines only. V. wanting. Air-bladder present. Vertebræ numerous. Large fishes of northern seas. Two genera and 5 or 6 species; one of them commonly, others rarely, taken off our coast.

a. Tail not very long, with a caudal fin, distinct from D. and A.

ANARRHICHAS, 214.

214. ANARRHICHAS (Artedi) Linnæus. (Ancient name, from ἀναρρἱχάομαι, to scramble up.)

445. A. lupus L. WOLF-FISH. Vomerine teeth extending much farther back than palatine. Brown, sides with 9 to 12 black bars, continued on D., besides dark spots and reticulations. Head 6; depth 5½. D. LXII. A. 42. L. 4 feet. Cape Cod, N. (*Eu.*) (Lat., wolf.)

FAMILY LXXXVI. **LYCODIDÆ.** (THE EEL-POUTS.)

Body more or less eel-shaped, naked or with small, cycloid scales; mouth large, with conical teeth; head unarmed; gill membranes united to isthmus; pseudobranchiæ present; gills 4. D. and A. very long, of soft rays only, or with a few spines in posterior part of D. P. small. Vertical fins confluent around the tail. V. jugular, imperfect or wanting; lateral line obsolete. Vertebræ in large number. Genera 10; species 35. Cold or deep waters, chiefly Arctic. This group seems most closely allied to the Blennies, but it agrees with the *Anacanthini* in wanting the spinous dorsal. (λυκώδης, wolfish.)

a. D. with some of its posterior rays very short and spine-like; V. small. (*Zoarcinœ.*)

b. Scales present; teeth strong, in jaws only. . . . . . . ZOARCES, 215.

215. ZOARCES Cuvier. (ζωαρχής, viviparous.)

446. Z. anguillaris (Peck). EEL-POUT. MUTTON-FISH. MOTHER OF EELS. Brownish, mottled with olive. Head 6; depth 6. D. 95, XVIII, 17. A. 105. L. 20. Del. to Labrador, common N. (On the Grand Banks occur several species of the related genus *Lycodes*, which is without D. spines.) (Lat., like an eel.)

The next family is in several respects peculiar, and marks the transition from the Blenny-like to the Cod-like fishes.

FAMILY LXXXVII. **PHIDIIDÆ.** (THE DONZELLAS.)

Body eel-shaped, naked or covered with very small scales which are placed in oblique series at right angles to each other; mouth large, with villiform or cardiform teeth. Gill openings wide, the gill membranes narrowly joined to the isthmus behind V.; pseudobranchiæ small or 0. Gills 4. Vertical fins low, of soft rays only, confluent around the isocercal tail. Ventral fins at the throat, each developed as a long forked barbel. Air-bladder present, Genera 5, species 15; carnivorous fishes of the warm seas.

a. Body scaly; palatines with a band of villiform teeth only; opercle without spine; teeth in jaws fixed. . . . . OPHIDION, 216.

216. OPHIDION (Artedi) Linnæus. (Diminutive of ὄφις, snake.)

447. O. marginatum DeKay. Brownish; D. and A. edged with black. Air-bladder short and broad, with foramen below; gill

rakers 4. V. as long as head. Head 6½. D. 7½. N. Y. to Texas, scarce.

ANACANTHINI. This suborder is distinguished chiefly by the total absence of spines in the fins, and also by the absence of any foramen in the scapular bone. The ventrals are jugular, the scales various. There are 2 or 3 families, the best known being the

FAMILY LXXXVIII. GADIDÆ. (THE COD-FISHES.)

Body elongate, ending in an isocercal tail; scales small, cycloid. Mouth large, the teeth various. No pseudobranchiæ. Vertical fins separate. D. and A. long; no fin spines. Gill openings very wide, the membranes free from the isthmus. Gills 4. Air-bladder present. Pyloric cœca numerous. Vertebræ about 50. Genera 30, species about 90. Carnivorous fishes, chiefly of the Northern seas, many of them of great economic value. One species in fresh waters.

a. Chin with a barbel; frontal bone normal; top of head without excavated area. (*Gadinæ.*)

  b. First D. composed of a band of fringes, preceded by a single ray; barbels 4; one on chin, one on each nostril, one on snout; anal fin single.
              RHINONEMUS, 217.

  bb. First D. of distinct rays.

    c. Dorsal fins two; anal fin one.

      d. Ventrals narrow, filamentous, each of 2 or 3 slender rays.
              PHYCIS, 218.

      dd. Ventrals broader, each of about 6 rays; vomer with teeth; no canines. . . . . LOTA, 219.

    cc. Dorsal and anal fins each single; ventrals well-developed.
              BROSMIUS, 220.

    ccc. Dorsal fins three; anals two.

      e. Lower jaw included; barbel well developed.

        f. Vent below second dorsal.

          g. Shoulder girdle with its chief bone or coracoid much swollen; (lateral line black; maxillary not reaching eye)
          MELANOGRAMMUS, 221.

          gg. Shoulder girdle normal; (lateral line pale; maxillary reaching past front of eye) . . . . GADUS, 222.

        ff. Vent in front of second dorsal; (skull peculiar)
          MICROGADUS, 223.

      ee. Lower jaw projecting; barbel minute; teeth of upper jaw subequal. . . . . POLLACHIUS, 224.

aa. Chin without trace of barbel; frontal bone divided; top of head with a large triangular excavated area, bounded by ridges. (*Merlucciinæ.*)

  h. Lower jaw projecting; teeth sharp, unequal, the larger ones movable; dorsals two; anal single; A. and 2d D. deeply notched; scales loose, silvery. . . MERLUCCIUS, 225.

11

217. **RHINONEMUS** Gill. (ῥίν, nose : νεμῆ, barbel.)

448. **R. cimbrius** (L.). FOUR-BEARDED ROCKLING. Head high, compressed; no canines; mouth large. Brownish; D. and A. behind, and C. below, abruptly black; mouth black within. Head 5; depth 6. D. 50. A. 43. V. 5. L. 12. Cape Cod, N. (*Eu.*) (Lat., Welsh.)

218. **PHYCIS** Bloch and Schneider. (φυκίς, old name from Fucus, sea-weed.)

a. First dorsal with one or more filamentous rays. (*Phycis.*)
  b. Filamentous ray of D. more than twice head.

449. **P. chesteri** Goode & Bean. Brownish. Head 4¼; depth 5. D. 10, 56. A. 56. Lat. l. 90. Mass., in deep water. (To Captain H. C. Chester, of the U. S. Fish Com.)

  bb. Filamentous ray of D. not twice head.
    c. Scales moderate; lat. l. 110.

450. **P. chuss** (Walbaum). CODLING. SQUIRREL HAKE. Brownish, punctulate, yellowish below. Head 4½; depth 5. D. 9–57. A. 50. L. 15. Va., N. (Vernacular name.)

    cc. Scales very small; lat. l. 140.

451. **P. tenuis** (Mitchill). WHITE HAKE. CODLING. Brown, yellowish below; fins very dark. Head 4¼; depth 5½. D. 9, 57. A. 48. L. 12. Va., N. (Lat., slender.)

aa. First dorsal without filamentous rays. (*Urophycis* Gill.)

452. **P. regius** (Walbaum). SPOTTED CODLING. Yellowish brown; lateral line dark. interrupted by white spots; sides of head and 2d D. with black spots; 1st D. largely black. Head 4¼: depth 4½. D. 8–43. A. 45. Lat. l. 90. L. 12. Cape Cod to N. C.; said to possess electric powers. (Lat., royal.)

219. **LOTA** Cuvier. (Lota, the ancient name.)

453. **L. lota** (L.). BURBOT. LAWYER. LING. Head depressed; maxillary reaching posterior margin of the very small eye; scales very small. Dark olive, thickly marbled and reticulate with blackish, the adult duller; edges of vertical fins dusky. Head 4⅔; depth 6. D. 13–76. A. 68. V. 7. Vert. 59. Cæca 30. L. 30. Arctic America and Europe, abundant in lakes, S. to Conn. R., Ohio R., etc.; a fish of little value as food, but widely distributed. The Amer. form is var. maculosa Le Sueur. (*Eu.*)

220. **BROSMIUS** Cuvier. (From the Danish name *brosme.*)

454. **B. brosme** (Müller). CUSK. Brownish, usually mottled with yellowish. Head 4¼; depth 5¼. D. 98. A. 71. Cape Cod, N. (*Eu.*)

221. MELANOGRAMMUS Gill. (μελανός, black ; γραμμή, line.)

455. **M. æglifinus** (L.). HADDOCK. Snout long; dorsals pointed ; C. lunate ; skull depressed, the bones thin ; the supra-occipital crest very high, with wing-like projections at base. Dark gray, a large black blotch above P. Head $3\frac{3}{4}$; depth $4\frac{1}{2}$. D. 15–24–21. A. 23–21. L. 30. Va., N.; an important food-fish. (*Eu.*) (Low Lat., haddock.)

222. GADUS (Artedi) Linnæus. (The Latin name, akin to the English Cod.)

456. **G. callarias** L. COD-FISH. Head large; occipital keel not high ; fins not elevated ; C. slightly notched. Brownish, the ground color varying much ; back and sides with round brownish spots ; fins dark. Head $3\frac{1}{2}$ to $4\frac{1}{2}$; depth 4. D. 14–21–19. A. 20–18. L. 3 feet or more. N. Atl. and N. Pac., S. to Va. and Ore.; one of the most important of food-fishes. (*Eu.*) (*G. morrhua* L.) (Lat., *Callarias*, a young cod.)

223. MICROGADUS Gill. (μικρός, small ; *Gadus.*)

457. **M. tomcod** (Walbaum). TOM-COD. FROST-FISH. Snout rounded ; maxillary reaching pupil, $2\frac{1}{2}$ in head. Eye $3\frac{2}{3}$. Olive-brown, spotted and blotched with darker; surface punctulate. Head $3\frac{2}{3}$; depth 5. D. 13–17–18. A. 20–17. L. 12. Va. to Labrador, a diminutive Cod-fish, common N.

224. POLLACHIUS Nilsson. (From Pollack.)

458. **P. virens** (L.) POLLACK. COAL-FISH. P. short, scarcely reaching A. Greenish, somewhat silvery below; fins pale; usually a dark spot in axil. Head 4; depth $4\frac{1}{2}$. D. 13–22–20. A. 25–20. Lat. l. 250. L. 18. Va., N. (*Eu.*)

225. MERLUCCIUS Rafinesque. (*Merlucius* "Sea-Pike," the ancient name.)

459. **M. bilinearis** (Mitchill). SILVER HAKE. STOCK-FISH. WHITING. Top of head with well defined W-shaped ridges; teeth not very large. P. and V. long, $\frac{2}{3}$ head. Grayish, sides dull silvery; axil inside of mouth and peritoneum black. Flesh soft. Head $3\frac{3}{4}$; depth $6\frac{1}{2}$. D. 13–41. A. 40. Lat. l. 105. L. 2 feet. Va., N., not rare. (Lat., *bis* two; *linearis*, lined.)

ORDER XXI. **HETEROSOMATA.** (THE FLAT-FISHES.)

This group seems to be an offshoot from the *Gadidæ.* Its essential feature is in the unsymmetrical character of the bones of the head. The head is twisted about, so that both eyes are on the same side. The body is compressed, and the side without eyes is habitually kept lowermost. The blind side is usually colorless. The very

young are symmetrical, one eye on each side, the body is translucent and the fish is vertical in the water. The processes by which
the eye of the lower side becomes transferred through or over the
head to the other side are very curious and interesting. There is
but one family. (ἕτερός different; σῶμα, body.)

Family LXXXIX. PLEURONECTIDÆ. (The Flounders.)

Body strongly compressed, the cranium twisted so that both eyes
are on the colored side; mouth and dentition various; premaxillaries protractile; maxillary simple; pseudobranchiæ present. Gills
4; no air bladder; vent not far behind head; scales various; fins
without spines. D. very long; A. similar, shorter; P. and V. various. Fishes mostly carnivorous, chiefly found on sandy sea-bottoms,
some of them ascending rivers. Genera 50; species 450. Those
species found in Arctic seas have, as usual, an increased number of
vertebræ; the tropical forms have 30 to 35; the others 40 to 70.

a. *Flounders:* Edge of preopercle free; teeth present; P. and V. well developed
(with rare exceptions).

   b. Mouth nearly symmetrical, the teeth nearly alike on the two sides, the
   gape usually but not always wide.

   c. Ventral fins symmetrical, similar in position and in form of base, the
      ventral of eyed side not extended along the ridge of abdomen.
      (*Hippoglossinæ.*)

      d. Vertebræ and fin rays much increased in number (vertebræ about
         50; D. 100; A. 85); body elongate; C. lunate; lateral line simple;
         no anal spine; eye on right side.

         e. Lateral line without arch.   . . . Platysomatichthys, 226.
         ee. Lateral line with an arch anteriorly. . . Hippoglossus. 227.

      dd. Vertebræ and fin rays in moderate number; (Vert. less than 46;
         D. less than 95; A. less than 75); C. double truncate.

         f. Lateral line without arch; vertebræ 45; eyes on right side; scales
            firm, ciliated; spine before A. strong; D. beginning above eye.
                                                 Hippoglossoides, 228.

         ff. Lateral line arched in front; vertebræ 35 to 41; eyes on left side;
            scales nearly smooth; anal spine weak; D. beginning before
            eye. . . . . Paralichthys, 229.

   cc. Ventral fins unsymmetrical, dissimilar in position or in form; the left
      V. extended along ridge of abdomen; eyes on left side. (*Pleu-
      ronectinæ.*)

      g. Vomer with teeth; lateral line arched in front; vertebræ 31 to
         36; mouth large; teeth in bands; form broad-ovate; scales
         cycloid, small or wanting; interorbital space not concave.
                                                 Pleuronectes, 230.

      gg. Vomer toothless; V. free from A; Vert. 34 to 40.

         h. Lateral line arched in front; teeth small, in 1 or 2 series; in-
            terorbital space broad, concave; scales small, ctenoid.
                                                 Platophrys, 231.

hh. Lateral line straightish; scales thin, deciduous.

   i. Mouth moderate, the maxillary more than ⅓ head.

        CITHARICHTHYS, 232.

   ii. Mouth very small, the teeth equal, the maxillary not ⅓ head.

        ETROPUS, 233.

bb. Mouth unsymmetrical, the teeth chiefly on the blind side; V. nearly symmetrical; eyes on right side. (*Platessinæ.*)

  j. Vertebræ in moderate number (36 to 44); D. 65 to 80; A. 45 to 60.

   k. Teeth in one row; lateral line not branched.

    l. Lateral line with an arch in front; scales ctenoid.

        LIMANDA, 234.

    ll. Lateral line without arch; teeth incisor-like.

     m. Scales regularly imbricate, all on eyed side ctenoid, in both sexes; lower pharyngeals very narrow, with slender teeth, in two rows. . . . . . PSEUDOPLEURONECTES, 235.

     mm. Scales imperfectly imbricate, rough — ctenoid in ♂, smoothish in ♀; lower pharyngeals very large, partly united, with blunt teeth in 5 or 6 rows. . . . . LIOPSETTA, 236.

  jj. Vertebræ in increased number (58 to 65); D. 100 or more; A. 70 to 100; teeth broad; left side of skull with large mucous cavities; anal spine strong; body elongate, compressed.

        GLYPTOCEPHALUS, 237.

aa. *Soles:* edge of preopercle obscured by the scales; mouth very small, strongly twisted towards blind side; teeth rudimentary.

  n. Eyes on right side, separated by a bony ridge. (*Soleinæ.*)

   o. Gill openings moderate, confluent below; vertical fins separate; right V. confluent with A.; vertebræ 28: body ovate; scales ctenoid, those on head enlarged and fringed; P. minute or wanting. . . . . ACHIRUS. 238.

  nn. Eyes on left side, very small, without distinct ridge between them: scales ctenoid; vertical fins confluent. (*Cynoglossinæ.*)

   p. V. of eyed side only present, free from A.; no P.: no lateral line; head without fringes. . . . . SYMPHURUS, 239.

226. PLATYSOMATICHTHYS Bleeker. (*Reinhardtius* Gill.)
(πλατύς. flat ; σῶμα, body: ἰχθύς. fish.)

460. **P. hippoglossoides** (Walbaum). GREENLAND HALIBUT. Brown. Head 4; depth 3. D. 100. A. 75. L. 4 or more. Cape Cod, N.

227. HIPPOGLOSSUS Cuvier. (Old name from ἵππος, horse ; γλῶσσα, tongue.)

461. **H. hippoglossus** (L.). HALIBUT. Dark brown; eyes large, widely separated. Head 3¾; depth 3. D. 105. A. 78. L. 6 feet or more. In all northern seas, the largest and most valuable of the flat-fishes, reaching 400 lbs. (*Eu.*)

228. HIPPOGLOSSOIDES Gottsche.

462. **H. platessoides** (Fabricius). ROUGH DAB. Plain reddish brown; eyes large; teeth uniserial. Head 3¾; depth 2½. D.

80 to 93.   A. 64 to 75.   Lat. l. 90.   N. Atl., S. to N. Y.   (*Eu.*)
(Lat., *platessa*, the plaice ; εἶδος, like.)

229. PARALICHTHYS Girard.   (παράλληλος, parallel;
ἰχθύς, fish.)

a. Gill rakers 5 + 16, rather long and slender; D. 85 to 93; A. 67 to 73.

463. **P. dentatus** (L.).   SUMMER FLOUNDER.   Body ovate;
maxillary half head; canines large.   Brownish olive, always with
many paler and darker spots and obscure ocelli.   Head $3\frac{2}{3}$; depth
$2\frac{1}{2}$.   Lat. l. 95.   L. $2\frac{1}{2}$ feet.   Cape Cod to Fla., the common flounder N.   (Lat., toothed.)

aa. Gill rakers, few, shortish, 2 + 8 to 10.

  b. Body ovate, opaque, the depth about $2\frac{1}{2}$ in length; no definitely placed
    ocelli.

  c. D. rays 85 to 93; A. 65 to 73; lat. l. about 100.

464. **P. lethostigma** Jordan & Gilbert. · SOUTHERN FLOUN-
DER.   Eyes small, well separated.   Dusky olive, nearly plain.
Head $3\frac{2}{3}$.   L. $2\frac{1}{2}$.   N. Y. to Texas, the common Flounder S.   (λήθη,
forgetting ; στιγμή, spot.)

  cc. D. rays 75 to 81; A. 59 to 61; lat. l. 95.

465. **P. albigutta** Jordan & Gilbert.   Grayish brown, with many
roundish pale blotches.   L. 18.   Va. to Texas, common S.   (Lat.,
albus, white ; *gutta*, spot.)

  bb. Body oblong, strongly compressed, semi-translucent, side with four
    large oblong black ocelli, each edged with pinkish, the anterior spots
    just behind middle of body, the four forming a trapezoidal figure.

466. **P. oblongus** (Mitchill).   FOUR-SPOTTED FLOUNDER.
Mouth large.   Head $3\frac{3}{4}$; depth $2\frac{1}{4}$.   D. 77.   A. 62.   Lat. l. 93.
L. 18.   Cape Cod to N. J.   (Another "4-spotted flounder," *Ancylopsetta quadrocellata* Gill, with deep body and very rough scales ;
probably ranges N. to Va.)

230. PLEURONECTES (Artedi) Linnæus.   TURBOTS.
(*Rhombus* Cuvier.)   (πλευρόν, side; νήκτης, swimmer.)

a. Scales cycloid, well-developed; no bony tubercles.   (*Bothus* Rafinesque.)

467. **P. maculatus** Mitchill.   WINDOW PANE.   First rays of
D. much exserted; body much compressed, translucent.   Grayish
brown, profusely spotted and mottled with dark brown.   Head $3\frac{3}{4}$;
depth $1\frac{3}{4}$.   D. 65.   A. 52.   Lat. l. 100.   L. 18.   Cape Cod to S. C.,
a small and valueless representative of the great turbot of Europe
(*P. rhombus* L.).

231. PLATOPHRYS Swainson.   (πλατύς, broad;
ὀφρύς, eye-brow.)

468. **P. ocellatus** (Agassiz.)   Maxillary $3\frac{2}{3}$ in head.   Light
grayish, with small round spots of darker gray, and with lighter

rings enclosing areas of ground color; two black blotches along lateral line; fins spotted; no blue markings. Head 4; depth 1½. D. 85 to 90. A. 65. Lat. l. 72 to 78. L. 12. Variable. L. I. to Brazil, abundant S. (*Pl. nebularis* Jordan & Gilbert.)

232. CITHARICHTHYS Bleeker. (*Citharus*, an allied genus; ἰχθύς, fish.)

469. C. spilopterus Günther. Maxillary 2½ in head; eye small, 5 to 6 ; snout short, forming an angle with preorbital; teeth small, those in front larger. Head 3½; depth 2½. D. 75 to 80. A. 58 to 61. Lat. l. 43 to 45. L. 6. N. J. to Brazil and Panama, a little flounder very common on sandy shores. (Related species occur in deeper water, in the Gulf stream.) (σπίλος, spot; πτερόν, fin.)

233. ETROPUS Jordan & Gilbert. (ἦτρον, abdomen ; πούς, foot.)

a. Body very deep, the depth more than half length.

470. **E.** crossotus Jordan & Gilbert. Maxillary 4 in head ; eye 3¾. Olive-brown with darker blotches; fins finely speckled. Head 4½; depth 1½ to 2. D. 76 to 85. A. 56 to 67. Lat. l. 42 to 48. L. 5. Warm seas, N. to Va.; may vary into the next. (κροσσωτός, fringed.)

aa. Body more elongate, the depth less than half length.

471. **E.** microstomus (Gill). Maxillary 4½ ; eye 3 to 3½ in head. Grayish, with small dark blotches ; two dark spots at base C.; fins specked. Head 4; depth 2¼. D. 77 to 78. A. 57 to 61. Lat. l. 38 to 41. N. J. to Fla., scarce. (μικρός, small ; στόμα, mouth.)

234. LIMANDA Gottsche. (Old name.)

472. **L.** ferruginea (Storer). RUSTY DAB. Teeth conical, close-set, 11 + 30 in lower jaw; snout abruptly projecting, leaving an angle at its base; interocular ridge high and narrow, prolonged and rugose above opercle. Brownish, with rusty spots; blind side yellow. Head 4; depth 2¼. D. 85. A. 62. Lat. l. 100. L. 2 feet. N. Y. to Labrador. (Lat., rusty.)

235. PSEUDOPLEURONECTES Bleeker. (ψευδής, false ; *Pleuronectes.*)

473. **P.** americanus (Walbaum). WINTER FLOUNDER. FLAT-FISH. Body elliptical ; interorbital space broad, convex, scaly; a low ridge above opercle. Dark rusty brown, obscurely mottled; fins plain. Head 4; depth 2¼. D. 65. A. 48. Lat. l. 83. L. 18. Labrador to Chesapeake Bay, common.

236. LIOPSETTA Gill. (λεῖος, smooth; ψέττα, flounder.)

474. **L.** glacialis (Pallas). EEL-BACK FLOUNDER. A coarse rugose ridge above opercle; scales in males ctenoid on both sides,

in ♀ mostly cycloid. Dark gray, mottled with darker ; fins
spotted. Head 3½ ; depth 2. D. 55. A. 40. Lat. l. 70. L. 12.
Arctic regions, S. to Cape Cod, our form or variety (L. putnami
Gill) common from Cape Ann to Nova Scotia; the original *glacialis*
in Alaska. (*Pleuronectes glaber* Storer.)

237. GLYPTOCEPHALUS Gottsche.   (γλυπτός, sculptured;
κεφαλή, head.)

475. **G. cynoglossus** (L.).   CRAIG-FLUKE.   P. short, not half ′ ᴄ
head ; eyes large, 3 in head. Grayish brown, fins spotted. Head
5; depth 2¾. D. 101 to 112. A. 87 to 99. Lat. l. 125. L. 12.
N. Atl., S. to Cape Cod. (*Eu.*) (An old name, from κύων, dog;
γλῶσσα, tongue.)

238. ACHIRUS Lacépède.   (ἄχειρ, without hands.)

a. Pectorals wanting. (*Achirus.*)

476. **A. fasciatus** Lacépède.   SOLE.   HOG-CHOKER.   Eyed side
without black hair-like cilia. Olive-brown, mottled and with about
8 dark vertical streaks; vertical fins with dark spots and clouds;
blind side usually with round dark spots. Head 4 ; depth 1⅔.
D. 50 to 55. A. 37 to 46. Lat. l. 66 to 75. L. 8. Cape Cod to
Texas, abundant, ascending rivers. (Lat., banded.)

239. SYMPHURUS Rafinesque.   (*Plagusia* Cuvier ; *Aphoristia*
Kaup.)   (συμφύω, to grow together ; οὐρά, tail.)

477. **S. plagiusa** (L.).   TONGUE-FISH.   Body broadly lanceolate. Brown, with faint darker longitudinal streaks and with black
cross-bars; C. similarly colored, *never black*. Head 5; depth 3 to
3⅓. D. 86 to 95. A. 75 to 80. Lat. l. 85 to 93. L. 5. Va. to
Texas, common S. (πλάγιος, oblique.)

ORDER XXII. PLECTOGNATHI. (THE PLECTOGNATHS.)

Premaxillaries co-ossified with the maxillaries, and dentary with
the articular; post-temporal undivided, grown fast to the skull;
interopercle rod-like ; upper pharyngeals forming vertical transverse laminæ; skin naked or variously covered with rough scales,
shields or spines. Vertebræ usually in less than normal number,
15 to 30. Ventral fins reduced or wanting.

This group is a modified offshoot of the suborder *Epelasmia* of
Acanthopteri. The relations of the *Balistidæ* with the *Acanthuridæ*
of the latter group are very close. (πλεκτός, joined; γνάθος, jaw.)

FAMILY XC. BALISTIDÆ. (THE TRIGGER-FISHES.)

Body ovate, compressed, covered with scales of varying structure. Mouth small, terminal, low; jaws short, each with one or
more series of separate incisor-like teeth ; eye very high. Gill

openings small, slit-like. Dorsals separate, the first of 1 to 3 spines ; 2d D. and A. long; V. wanting ; pubic bone long, movable, with sometimes a spine at its end. Genera 8 ; species 100 ; carnivorous fishes of the warm seas.

a. Dorsal spines 3; body covered with thick, firm scales; pelvis with a blunt spine. . . . . BALISTES, 240.

aa. Dorsal spine single, or followed by a rudiment; skin with minute rough shagreen-like scales.

   b. Pubic spine present; gill-slit short, nearly vertical. A. 25 to 35.
                         MONACANTHUS, 241.

   bb. Pubic spine wanting ; gill-slit long, oblique. A. 36 to 50.
                         ALUTERA, 242.

240. BALISTES (Artedi) Linnæus. (βαλῶ, to shoot; from the trigger-like 2d spine of D.)

a. A groove before eye; larger plates behind gill opening; teeth white; no spines on tail. (*Balistes.*)

478. **B. carolinensis** Gmelin. LEATHER-JACKET. TRIGGER-FISH. Soft D. high; C. lobes elongate in adult. Brownish; young spotted with darker; 2d D. and A. with interrupted brown streaks; C. mottled ; scales on head similar to those on body. Head 3 ; depth 1¼. D. III – 27. A. 25. Lat. l. 51 to 62. L. 18. Warm seas, rarely N. to Cape Cod. (*Eu.*)

241. MONACANTHUS Cuvier. (μόνος, one; ἄκανθα, spine.)

a. Pubic spine movable; ventral flap moderate, not extending beyond it ; dorsal spine with retrorse barbs.

479. **M. hispidus** (L.). FOOL-FISH. FILE-FISH. No recurved spines on tail ; first soft ray of D. sometimes filamentous. Dull greenish, mottled with darker. Head 3⅔ ; depth 1¾. D. I – 32. A. 32. L. 6. Cape Cod to Cuba, common. (Lat., rough.)

242. ALUTERA Cuvier. (? ἄλουτος, unwashed.)

480. **A. schœpfi** (Walbaum). Dull-greenish, marbled with darker; D. spine slender, not barbed ; C. long in young, shorter with age. Head 3⅔; depth 2¼. D. I, 36. A. 38. L. 18. Cape Cod to Texas. (To Johann David Schöpf, a Hessian surgeon in the Revolutionary War, and an excellent naturalist.)

FAMILY XCI. **TETRAODONTIDÆ.** (THE SWELL-FISHES.)

Body oblong, little compressed, the skin naked and usually prickly; stomach capable of great inflation; teeth in each jaw confluent into two, which form a sort of beak; no fin spines; D. opposite A. ; C. distinct ; V. wanting ; P. short ; pelvic bone moderate. Gill openings small; air-bladder present. Genera 7; species 70, in warm seas. They are noted for their power of swallowing air, by

which the stomach may be greatly inflated and the fish float belly
upward out of reach of its pursuers.

a. Back not carinated, skin without scutes; nostril on each side with two
openings.
   b. D. and A. falcate, of 12 to 15 rays; C. lunate; vertebræ 20; nostrils sessile; mucous tubes on head very conspicuous.   LAGOCEPHALUS, 243.
   bb. D. and A. short, rounded, of 6 to 8 rays; C. rounded; vertebræ 18;
nostrils at tip of a hollow papilla; mucous tubes not conspicuous.
ORBIDUS, 244.

243 LAGOCEPHALUS Swainson.   (λαγώς, hare ; κεφαλή, head.)

481. **L. lævigatus** (L.).   RABBIT FISH.   SMOOTH PUFFER.
TAMBOR. Olive green; silver-white below; belly with large 3-rooted
spines; skin elsewhere smooth. Head 3¼; depth 4½. D. 14. A. 12
L. 2 feet or more. Cape Cod to Brazil. (Lat., made smooth.)

244. ORBIDUS Rafinesque.   (Lat., orbis, a sphere.)

482. **O. maculatus** (Bloch & Schneider).   COMMON PUFFER.
SWELL-FISH. SWELL-TOAD. Sides of head and body always prickly,
as is back from upper lip to D.; prickles all similar, small, close-set,
3-rooted, never obsolete. Dark olive above, marbled and dotted
with black ; black blotches on side forming short cross-bars ; C.
nearly plain. Head 2⅔; depth 3. D. 7. A. 6. C. 7. L. 12. Cape
Cod to S. C., very common. (*Tetraodon turgidus* Mitchill.)

FAMILY XCII. DIODONITIDÆ. (THE PORCUPINE-FISHES.)

Fishes similar to the *Tetraodontidæ*, but having the teeth of each
jaw grown into *one;* body with rooted spines; stomach less extensively inflatable than in the *Tetraodontidæ*. Genera 3; species
about 10, in warm seas. (δίς, two; ὀδών, tooth.)

a. Spines robust, all fixed, and 3-rooted (some of them rarely 4-rooted); nasal
tube simple with two lateral openings.   . . CHILOMYCTERUS, 245.

245. CHILOMYCTERUS (Bibron) Kaup.   (χεῖλος, lip ; μυκτήρ, nostril.)

483. **C. schœpfi** (Walbaum).   BURR-FISH.   SWELL-TOAD.   A
ridge above eye. Greenish, with series of parallel blackish stripes
covering most of the body above ; an ocellated black spot above P.;
a larger one behind it; one at base of D. ; a smaller one below it;
fins unspotted. Head 2¾; depth 3. D. 12. A. 10. L. 6. Cape
Cod to Texas; very abundant S. (*C. geometricus* Bloch & Schneider.) (Farther S. occurs *Diodon hystrix* L., larger, with longer
spines, of which some are 2-rooted and movable.) (To Johann
David Schöpf, a Hessian surgeon in the Revolutionary War.)

FAMILY XCIII. **MOLIDÆ.** (THE HEAD-FISHES.)

Body deep, compressed, truncate behind, so that there is no caudal peduncle; skin scaleless, rough. Mouth very small, the teeth united, without median suture as in *Diodontidæ*. D. and A. of soft rays only, confluent around tail, elevated in front. V. wanting; pelvic bone small; belly not inflatable. Three species, placed in as many genera; large fishes of the open sea, consisting apparently of a huge fish-head to which small fins are attached.

a. Body ovate, not twice as long as deep; skin thick, leathery, without hexagonal plates. . . . . MOLA, 246.

246. MOLA Cuvier. (*Orthagoriscus* Bloch & Schneider.) (Lat., millstone.)

484. **M. mola** (L.). SUN-FISH. HEAD-FISH. MOLA. Dark gray, silver-gray below; a dusky bar along bases of vertical fins. D. and A. very high; form varying greatly with age; a hump or snout above mouth in old specimens. Head 3; depth 1⅔ (in adult). D. 17. A. 16. L. 4 feet or more. Pelagic, N. to Cape Cod; not rare, sometimes weighing 500 lbs. (*Eu.*)

ORDER XXIII. **PEDICULATI.** (THE PEDICULATE FISHES.)

Carpal bones reduced in number and notably elongate, forming a kind of arm which supports the broad pectorals. Gill openings reduced to a small pore in or near the axil, behind the pectoral fins; V. jugular, if present; first vertebra united with skull; post-temporal broad, flat, simple; pharyngeals reduced in number; spinous D. often reduced to isolated tentacles. No scales.

This singular group is probably a modified off-shoot of the *Haplodoci* (*Batrachidæ*) or of some similar form. It may fairly be placed at the end of the fish-series, as having gone farther in its divergence from the original fish-stock than any other of the groups called "orders" among fishes. It is not however in any proper sense the "highest" of the fishes, for some of its peculiarities may be due to degradation. Still less is it the order most closely related to the higher vertebrates. Most of the *Pediculati* belong to the tropics or to the deep sea. (Lat., *pediculatus*, provided with a little foot or peduncle.)

FAMILY XCIV. **MALTHIDÆ.** (THE BAT-FISHES.)

Head broad and depressed, the snout elevated, the trunk short and slender. Mouth small, inferior; gill opening very small, above and behind axil of P. Body and head covered with bony tubercles or spines. Spinous D. a single tentacle on snout, retractile into a cavity beneath a long process on snout. Genera 3; species 10, all American.

247. MALTHE Cuvier. (μάλθη, a name of some soft-bodied fish.)

485. **M. vespertilio** (L.). BAT-FISH. DIABLO. Dark gray, reddish below; forehead produced in a long rough process of variable length. D. I, 4. A. 4. L. 6. Warm seas, rarely N. (Lat., bat.)

FAMILY XCV. **ANTENNARIIDÆ.** (THE FROG-FISHES.)

Head and body somewhat compressed, the mouth nearly vertical, the chin projecting; gill openings small, pore-like, in lower axil of P. Spinous D. of 1 to 3 isolated tentacles. Genera 5; species 40. living in floating seaweed, etc., in warm seas. (Lat., *antenna*, a feeler.)

a. Head compressed; dorsal spines 3; skin smooth with many fleshy tags; V.
long. . . . . PTEROPHRYNE, 248.

248. PTEROPHRYNE Gill. (πτερόν, wing; φρύνη, toad.)

486. **P. histrio** (L.). MOUSE-FISH. Yellowish, much marbled; wrist slender. Head 2¼; depth 1¼. D. III–14. A. 7. V. 5. L. 5. Warm seas, occasional N. (Lat., stage-player.)

FAMILY XCVI. **LOPHIIDÆ.** (THE ANGLERS.)

Head wide, depressed, very large; body contracted, tapering, scarcely longer than head; mouth enormously wide, with a stomach proportionate; teeth very strong, unequal, some of them long, sharp canines and most of them depressible; strong teeth on vomer and palatines. Gill openings large, in lower axil of P. Skin smooth, with many dermal flaps. Spinous D. of 3 isolated tentacles, and 3 spines joined by membrane, the first spine enlarged at tip and extending over the mouth, said to serve as a bait for smaller fishes. One genus with 3 or more species, large fishes of the cool seas, remarkable for voracity.

249. LOPHIUS (Artedi) Linnæus. (Old name from λόφος, crest.)

487. **L. piscatorius** L. GOOSE-FISH. ANGLER. FISHING-FROG. ALL-MOUTH. BELLOWS-FISH. Brownish, mottled; mouth behind tongue, unspotted. D. III–III, 10. A. 9. V. I, 5. L. 3 feet or more. N. Atl., S. to Cape Lookout, common N. The eggs of this fish are remarkable, in ribbon-like bands, pink in color, 30 to 40 feet long and a foot in width. These float near the surface in summer. (Lat., fishing.)

With this monstrous creature, unexcelled for pure ugliness in the class to which it belongs, we may close the long series of fishes.

Next come the *Batrachians*, animals bearing close relations to the "central stem " of the fishes, now represented by the *Dipnoi*. They are decidedly fish-like in their early conditions, but this stage is ultimately outgrown. " The undivided cartilaginous coracoid of *Polyterus* (a *Dipnoan*) has a tubercle articulating with diverging rods; in the one we have the rudiment of the humerus, in the other the representatives of the ulna and radius, while the undifferentiated cartilage between the diverging rods is material for the carpal bones, and in bones radiating from that cartilage are the homologues of the metacarpals. The attempts of a primitive animal of such a type to travel on land might develop the fore-limb, and a hind one would follow in sympathy with the other. Then we would have the first of the quadruped vertebrates," the Batrachians. (*Gill.*)

NOTE.— Page 47. The RED-HORSE or WHITE SUCKER of the Chesapeake region has the anterior rays of the dorsal elevated, the outline of the fin decidedly concave. It is perhaps a distinct species from the common Red-Horse of the West and South described in the text. It may stand as 81. *Moxostoma macrolepidotum* (Le Sueur), and the Western form, N. Y. to Dak. and Ga., as 81 (b). *M. duquesnei* (Le Sueur).

Class F. **BATRACHIA.** (The Batrachians.)

Cold-blooded vertebrates, intermediate between the fishes and the reptiles. They differ from the fishes chiefly in the absence of rayed fins, the limbs being usually developed and functional with the skeletal elements of the limbs of reptiles, and in the reduction or absence of the various bones of the branchial, opercular and suspensory systems.

The Batrachians undergo a more or less complete metamorphosis; the young ("tadpoles") being fish-like and more or less aquatic, breathing by means of external gills. These differ from the gills of fishes in standing on fleshy processes of the branchial bones and not on the bones themselves. In the tadpole, the tail is provided with a more or less distinct fin-like membrane, which usually disappears with age. Later in life, lungs are developed, and in most cases the gills disappear. Skin mostly naked and moist, used to some extent as an organ of respiration. Heart with two auricles and a single ventricle.

Reproduction by means of eggs which are of comparatively small size, without hard shell. These are deposited in water or in damp places. In one salamander the young are born alive. Professor Cope recognizes nine orders of Batrachians, four of these being extinct. (βάτραχος, frog.)

Orders of Batrachia.

a. Body lengthened, with a distinct tail throughout life; hind limbs, if present, not especially enlarged.
  b. External gills and gill-clefts persistent throughout life, the gills 3 on each side; no eyelids; vertebræ amphicœlian; maxillary small or wanting.
    c. Body eel-shaped, without hind legs; teeth on vomer; floor of mouth rough; jaws with horny sheath. . . . Trachystomata, XXIV.
    cc. Body salamander-shaped, the hind limbs present; jaws with teeth.
                              Proteida, XXV.
  bb. External gills normally disappearing in adult life; limbs 4 (or wanting, present in all our species); jaws with teeth; maxillaries and palatines present. . . . . Urodela, XXVI.
aa. Body short, depressed; tail disappearing with age; limbs 4, the posterior much enlarged. . . . . Salientia, XXVII.

ORDER XXIV. TRACHYSTOMATA.

This order contains a single family. (τραχύς, rough ; στόμα, mouth.)

FAMILY XCVII. SIRENIDÆ. (THE SIRENS.)

Body elongated, eel-like, with no posterior limbs, not even a vestige of pelvis; head flattened; snout obtuse; mouth narrow, jaws with horny sheaths; floor of mouth with teeth or asperities; vomer with two large patches; eye very small ; lips thick ; tail compressed, finned. Genera 2; species 2. *Pseudobranchus striatus* (LeC.), of Georgia, a small species with 3 toes and with thickened, functionless gills, and the following : —

a. Gills large, bushy, in function throughout life ; toes 4 ; spiracles 3.

SIREN, 250.

250. SIREN Linnæus.

488. S. lacertina L. MUD EEL. Tail shorter than body, pointed at tip. Blackish, sometimes dotted. L. 36. Lowland streams and swamps, N. Ind. to N. C. and S. (Lat., like a lizard.)

ORDER XXV. PROTEIDA.

This order contains a single family.

FAMILY XCVIII. PROTEIDÆ. (THE MUD PUPPIES.)

Salamanders provided with bushy external gills, and having the branchial clefts remaining open through life; teeth well developed; limbs 4. Genera 2; species 3 or 4. *Proteus* inhabits caves in S. W. Austria, and *Necturus* the fresh waters of the U. S. *Proteus* is blind, nearly colorless, and has the toes 3-2.

a. Toes 4-4; tongue large, free in front ; vomerine teeth in one strong series ; eyes small, not covered. . . . . NECTURUS, 251.

251. NECTURUS Rafinesque (1819). (*Menobranchus* Harlan, 1825.) (νήκτης, a swimmer; οὐρά, tail.)

489. N. maculatus Rafinesque. MUD PUPPY. (N.) WATER DOG. (S.) Brown, more or less spotted; young with traces of a lateral band ; gills large and bushy, bright red, forming 3 tufts on each side ; a strong fold across throat; head broad, depressed ; tail much compressed. E. U. S., chiefly N. and W. of the Alleghanies, abundant in the Great Lake Region. L. 24. (Lat., spotted.)

ORDER XXVI. URODELA. (THE SALAMANDERS.)

Body naked, elongate, subterete; both jaws with teeth; 4 limbs present (wanting in the tropical family Cæciliidæ); tail persistent through life; no external gills in the normally developed adult.

This group is divided by Cope into 7 families, all but one of these (*Salamandridæ*) being represented in our fauna. These families are based chiefly on technical characters, most of which can be ascertained only by a careful study of the osteology. "It may be stated as characteristic of the Batrachia in general that their characters cannot be determined **without a study of the skeleton**." (*Cope.*)   (οὐρά, tail; δῆλος, visible.)

Families of Urodela.

a. Side of neck with a spiracle or rounded opening; no eyelids; vertebræ amphicœlian; teeth on front or outer edge of palatines.

 b. Limbs rudimentary; body eel-shaped. . . . . . AMPHIUMIDÆ, 99.

 bb. Limbs well-developed; body not eel-shaped. CRYPTOBRANCHIDÆ, 100.

aa. Side of neck without spiracle in the adult; limbs well developed; eyelids present; teeth on posterior or inner edge of palatines.

 c. Palatine teeth in a transverse (or posteriorly converging) series, inserted on posterior portion of vomer.

  d. Vertebræ amphicœlian (double concave).

   e. Parasphenoid (behind vomer) without teeth; carpus and tarsus ossified; tongue (in our species) large, thick, with radiating folds, its margin little free; digits 4–5. . AMBLYSTOMATIDÆ, 101.

   ee. Parasphenoid with teeth; tongue small, and largely free.

             PLETHODONTIDÆ, 102.

  dd. Vertebræ opisthocœlian (concave behind only); teeth on parasphenoid; palatine teeth often wanting; tongue moderate, largely free; toes 5. . . . . DESMOGNATHIDÆ, 103.

 cc. Palatine teeth in two longitudinal series diverging behind, inserted on inner margin of two palatine processes; parasphenoid toothless; vertebræ opisthocœlian; skull with a bony post-fronto-squamosal arch; tongue small, laterally free. . . . PLEURODELIDÆ, 104.

FAMILY XCIX. AMPHIUMIDÆ. (THE CONGO SNAKES.)

Body elongate, eel-shaped; limbs rudimentary, with 2 or 3 toes each; a spiracle on each side of neck; tongue indistinct, wholly adherent; a strong series of vomerine teeth parallel with the teeth in jaws. Tail short, compressed. One species, inhabiting the ditches and streams of the S. U. S.

252. AMPHIUMA Garden. (Name unexplained.)

490. **A. means** Garden. CONGO SNAKE. Blackish. L. 3 feet. Ark. to N. C. and S. (Lat., swift-moving.)

FAMILY C. CRYPTOBRANCHIDÆ. (THE GIANT SALAMANDERS.)

Body robust, with well-developed limbs; an orifice on each side of neck usually persistent throughout life; tongue covering floor of mouth; vomerine teeth strong; nostrils very small; no external gills; toes 4-5. Aquatic. Genera 2, species 2. *Megalobatrachus maximus* of Japan and the following.

a. Spiracles persistent; gill arches 4. . . . . . . CRYPTOBRANCHUS, 253.

253. CRYPTOBRANCHUS Leuckart. (κρυπτός, concealed; βράγχος, gill.)

491. C. alleghaniensis (Daudin). HELLBENDER. Blackish; side of body with a thick fold of skin. L. 24. Ohio Valley and S., a very unprepossessing but harmless creature. Var. *fuscus* Holbr., brown, paler below, occurs in Tenn. R.

FAMILY CI. AMBLYSTOMATIDÆ. (THE BLUNT-NOSED SALAMANDERS.)

Vertebræ amphicœlian; carpus and tarsus ossified; toes 4-5, not webbed; tongue thick; a band of teeth across posterior part of vomer; no teeth on parasphenoids (behind vomer). Genera 6; species about 25, mostly North American. The larvæ of *Amblystoma* often reach a large size before the gills disappear, and sometimes breed while in this condition. These were formerly considered as forming a separate genus, *Siredon*, supposed to be allied to *Necturus*.

a. Tongue sub-circular, with radiating folds, its lateral borders free; palatine teeth in a long series, continuous or interrupted; tail compressed; mucous pores before eye.

b. Folds of tongue radiating from behind; palatine teeth extending laterally behind inner nares. . . . . AMBLYSTOMA, 254.

bb. Folds of tongue radiating from the median longitudinal furrow; series of palatine teeth not extending laterally behind inner nares.
CHONDROTUS,[1] 255.

254. AMBLYSTOMA Tschudi. (ἀμβλύς, blunt; στόμα, mouth.)

a. Costal grooves 10.

492. A. talpoideum (Holbrook). Blackish brown, with gray, lichen-like markings; tail short, compressed, 2½ in length; head very broad; body short and squat. Southern, N. to S. Ill. (Lat., like a mole, *talpa*.)

aa. Costal grooves usually 11.

b. Sole with one indistinct tubercle, or none.

c. Body with gray cross-shades.

493. A. opacum (Gravenhorst). Black above, with about 14 bluish gray bars; belly dark blue; no dorsal furrow; no enlarged pores on the head; tail 2½ in total length; body stout. L. 3½. Penn. to Wis., and S.

cc. Body with yellowish spots.

[1] The essential character of this genus lies in the osteology of the tongue and hyoid bones, and cannot easily be explained without figures. See Cope, Amer. Nat., 1887.

494. A. punctatum (L.). Spotted Salamander. Black above with a series of round yellow spots on each side of the back; body broad, depressed and swollen : skin punctate with small pores from which exudes a milky fluid; two or three clusters of enlarged pores on head ; a strong dorsal groove; tail 2¼ in length; costal grooves sometimes 10 ; large. L. 6. Nova Scotia to Nebr. and S., common.

495. A. conspersum Cope. Lead colored, with one or two series of small yellowish spots along sides ; no dorsal groove ; skin smooth; body slender; tail shorter than head and body; tail 2½ in length; small. Penn. to Ga. (Lat., sprinkled.)

bb. Sole with two distinct tubercles.

496. A. bicolor (Hallowell). Olive brown, yellowish below, the yellow rising in blotches on the sides; a few ill-defined yellowish spots above ; limbs banded; tail yellow with brown spots ; body stout and heavy. L. 6. N. J.

497. A. copianum Hay. Dark brown, yellowish below; no distinct spots; limbs not banded; tail not spotted ; body very short and stout, the distance from snout to axil equal to distance from axil to groin : tail long, compressed. Irvington, Ind., one specimen known. (To Edward Drinker Cope.)

aaa. Costal grooves 12.

e. Sole with two distinct tubercles; snout with mucous pores.

498. A. tigrinum (Green). Dark brown, with usually many irregular yellow blotches, sometimes arranged in cross-bands ; body thick and strong; the head comparatively long; tail not much, if any, longer than head and body ; color varying from uniform brown to yellow, but usually spotted. L. 8. N. E. to Minn. and S., common.

499. A. xiphias Cope. Yellow olive, brighter below ; back and sides with brown reticulating bands; head small, blunt; tail very long, much longer than head and body. L. 11. Ohio. (ξιφίας, sword-shaped.)

ee. Sole with one indistinct tubercle or none; palatine teeth interrupted.

500. A. jeffersonianum (Green). Olive brown or blackish, usually with pale or bluish spots, but sometimes uniform plumbeous. Head small, eyes far back; body slender; fore limb not reaching hinder when appressed. L. 5 to 8. Va. to Ind. and N., common, variable. Prof. Cope recognizes the typical variety *jeffersonianum*, Penn. to Ill. and N.; var. *laterale* Hallowell, Canada to Wis., with large white spots on sides and tail ; var. *fuscum* Hallowell, S. Ind. to Va., dark brown, a darker band along sides ; var. *platineum*, Ohio to S. Ill., with narrower head, 5½ to 6 in length to groin : plumbeous, paler below, sometimes with whitish blotches. (To Thomas Jefferson.)

255. CHONDROTUS Cope. (χόνδρος, cartilage ; οὖς, ear.)

501. C. microstomus Cope. Blackish, usually with plumbeous shades and specks ; head small, short, broad ; body slender ; skin very smooth and slippery ; snout very short. the lower jaw projecting beyond it. Costal grooves 14. Ohio to Kansas and S. (μικρός, small ; στόμα, mouth.)

FAMILY CII. **PLETHODONTIDÆ.**

Vertebræ amphicœlian ; carpus and tarsus cartilaginous ; parasphenoid with one or two laminæ which are covered by a coarse brush of teeth which look downwards on roof of mouth. The species with cylindric tails rarely or never enter water. Genera 11 : species 35 ; chiefly North American.

a. Tongue attached by a band running from its central or posterior pedicel to the anterior margin ; premaxillaries two, with fontanelle.

b. Toes 4–4. . . . . HEMIDACTYLIUM, 256.

bb. Toes 4–5. . . . . PLETHODON, 257.

aa. Tongue free all around, attached by its central pedicel only ; toes 4–5, all free.

c. Premaxillaries two, with fontanelle. . . . . GYRINOPHILUS, 258.

cc. Premaxillary single, with fontanelle. . . . SPELERPES, 259.

256. HEMIDACTYLIUM Tschudi. (ἥμι-, half ; δάκτυλος, toe.)

502. H. scutatum (Schlegel). Brown above ; snout yellow ; whitish below, with dots like ink spots ; body short ; tail slender ; skin of back with depressions resembling scales. Costal grooves 13. L. 2½. R. I. to Ill., and S.

257. PLETHODON Tschudi. (πλῆθος, crowd ; ὀδών, tooth.)

a. Costal groove 16 to 18 ; palatine teeth not extending outward beyond inner nares.

503. P. erythronotus (Green). Plumbeous above, often with a broad brownish red dorsal band ; belly marbled ; body very slender ; tail cylindric ; inner toes rudimentary. L. 3½. E. U. S., common under logs, etc. ; nocturnal in habit and very active. Var. **cinereus** Green, found with the other, lacks the red dorsal band. (ἐρυθρός, red ; νῶτος, back.)

aa. Costal grooves 14 ; palatine teeth extending outside of inner nares.

504. P. glutinosus (Green) Black, usually with bluish-white blotches and specks ; stout ; tail rounded ; inner toes well developed. L. 5 to 7. E. U. S., chiefly terrestrial.

258. GYRINOPHILUS Cope. (γυρῖνος, tadpole ; φίλος, lover.)

505. G. porphyriticus (Green). Yellowish or purplish brown above, irregularly blotched with gray ; head broad ; tail rounded at

base, not finned. Costal grooves 14. L. 6. Aquatic. Vt. to Ala.
in the mountains. "The only one of our Eastern Salamanders
which attempts self-defence. It snaps fiercely but harmlessly and
throws its body into contortions." (*Cope.*) (πορφύρα, purple.)

259. SPELERPES Rafinesque. (σπέος, cave; ἑρπετόν, reptile.)
a. Costal grooves 13 or 14: palatine teeth not confluent with sphenoid patches.
    b. Tail about as long as rest of body.

506. **S. bilineatus** (Green). Yellow, with a dark line along each
side of the back : belly unspotted : tail not keeled anteriorly ; costal
grooves 14, rather faint. L. 3. Maine to Wis. and S.
    bb. Tail 1½ to 2 times as long as rest of body.

507. **S. guttolineatus** (Holbrook). Yellow, with black band on
back and one on side : tail black, barred with yellow; belly mottled : tail keeled : costal grooves 13. Ohio to N. C. and S.

508. **S. longicauda** (Green). CAVE SALAMANDER. Orange
yellow : back and sides with many irregular small black spots ; a
median dorsal series : belly spotless ; tail keeled, spotted or barred
with black. L. 5. Maine to Minn. and S., abundant in caves in
Ky. and Ind. (Lat.. *longus ; cauda*, tail.)

aa. Costal grooves 15 to 17; tail rounded at base, not keeled; palatine and
    sphenoid teeth continuous.

509. **S. ruber** (Daudin). Vermilion red, with numerous, crowded
faint dark spots ; head wide ; tail shorter than body. L. 6. Maine
to Neb. and S. Var. **montanus** Cope (Penn. to S. C.) has tail as
long as body, and lacks the dark bar across eye usually present in
var. *ruber.*

FAMILY CIII.   **DESMOGNATHIDÆ.**

Vertebræ opisthocœlian; carpus and tarsus cartilaginous; palatine teeth few, sometimes wanting; no crests or other dermal appendages developed at the breeding season. Genus 1 ; species 3 ;
all of the Eastern U. S., the species aquatic, seldom leaving the
water. In external characters, this family is scarcely distinguishable from the preceding, but the skeletal distinctions are very
strongly marked.

260. DESMOGNATHUS Baird. (δεσμός, band; γνάθος, jaw.)
a. Costal grooves 13 or 14.
    b. Tail sub-terete.

510. **D. ochrophæa** Cope. Brownish yellow with a brown shade
on each side ; a yellowish dorsal band ; back with a few spots ; belly
unspotted ; ♂ with lower jaw toothless behind. L. 3. Scarcely
aquatic. N. Y. to Ga. in mts. (ὠχρός, yellowish ; φαιός, dusky.)
    bb. Tail compressed and keeled.

511. **D. fusca** (Rafinesque). Brown above, with gray or purplish spots or shades, becoming blackish with age; marbled below; eyes prominent; tail as long as head and body. L. 4. Mass. to Ohio and S.; common in springs; remarkable for its activity. Represented from Ind. S. and W., by var. **auriculata** Holbrook, with small red spots on sides and sometimes a dark ear-spot. (Lat., dusky.)

aa. Costal grooves 12; tail compressed and keeled.

512. **D. nigra** (Green). Uniform black : body stout : palatine teeth never wanting. L. 6. Penn. to Ill. and S., in mountain springs.

FAMILY CIV. **PLEURODELIDÆ.** (THE NEWTS.)

Vertebræ opisthocœlian; carpus and tarsus ossified. Palatine teeth in two series diverging backward; no parasphenoid teeth; skull with a bony post-fronto-squamosal arch, a skeletal character which separates this family from the European *Salamandridæ.* Genera 5; species 16; chiefly of Europe and Asia.

a. Tongue small, thick, oval, attached by nearly its whole inferior surface; toes 4–5, outer and interior on hind foot rudimentary; tail compressed.

DIEMYCTYLUS, 261.

261. **DIEMYCTYLUS** Rafinesque. (δίς, two; ἡμι-, half; δάκτυλος, toe.)

513. **D. viridescens** Rafinesque. NEWT. EVET. EFT. Above olive green or reddish of varying shades; lemon yellow below; each side usually with a row of several rather large scarlet spots, each surrounded by a black ring; back with a pale streak; belly, with small black dots; head with three longitudinal grooves; 3 large pores behind eye. L. 3½. E. U. S., abundant N. and N. E.; in ponds.

Var. **miniatus** Rafinesque, the RED EFT, is entirely similar, but bright vermilion red, and with the skin rougher. It is found in the same region but away from water, under stones, etc., coming out after rain. It is probably a form of the preceding, its peculiarities being due to life out of water. (Lat., greenish.)

ORDER XXVII **SALIENTIA.** (THE TAILLESS BATRACHIANS.)

Body short and broad; all four limbs present, the hinder limbs long and strong, adapted for leaping; lower jaw usually toothless; tail wanting in the adult. Young (tadpole) fish-like, with broad head, external branchiæ, a long tail, no limbs and no teeth; the intestinal canal very long, adapted for a vegetable diet; from

this form by degrees it develops into the adult animal, which is
always more or less frog-like. (Lat., *saliens*, leaping.)

Families of Salientia.

a. Tongue present, adherent in front, more or less free behind ; eustachian
tubes widely separated.
  b. Thoracic[1] region capable of expansion : the free and divergent ends of
  the coracoid and precoracoid connected by two longitudinal cartilaginous bands, the cartilage of one side overlapping the other. Toads
  and Tree-toads. (*Arcifera.*)
    c. Upper jaw toothless; toes webbed, not dilated at tip ; paratoids
    (glandular bodies behind ear) generally present; terrestrial.
                                                                BUFONIDÆ, 105.
    cc. Upper jaw with teeth.
      d. Fingers and toes tapering, without viscid disks; ours with a sharp
      flat-edged spur at heel ; paratoids present; subterranean.
                                                                PELOBATIDÆ, 106.
      dd. Fingers and toes more or less dilated at their tips, this dilation
      forming a viscid disk; paratoids none in our species; chiefly
      arboreal. . . . . HYLIDÆ, 107.
  bb. Thoracic region incapable of expansion, the two bands of cartilage
  united in a median mass between the adjacent ends of the nearly
  parallel coracoid and precoracoid bones. Frogs. (*Firmisternia.*)
    e. Upper jaw toothless; diapophyses of sacral vertebræ dilated (tympa-
    num hidden and toes free in our species). ENGYSTOMATIDÆ, 108.
    ee. Upper jaw with teeth ; no paratoids; toes webbed, and usually fin-
    gers also ; tympanum evident ; no viscid disks; sacral diapophyses
    scarcely dilated. . . . . RANIDÆ, 109.

FAMILY CV. BUFONIDÆ. (THE TOADS.)

Jaws toothless; toes webbed, not dilated at their tips; sacra
vertebræ with dilated processes; paratoids prominent. Genera 8;
species 85, in most warm regions.

a. Snout not pointed ; no lateral fold of skin; skin more or less warty.
                                                                BUFO, 262.

262. BUFO Laurenti. (Lat. Toad.)

514. B. lentiginosus Shaw. AMERICAN TOAD. Brownish
olive with a yellowish vertebral line and some brownish spots ;
two black patches below eyes ; tympanum large; adults very
warty ; young nearly smooth ; a bony ridge above and behind eye ;
paratoids elliptical. L., 3½. E. U. S., very common, variable ;
the northern form is var. **americanus** (Le Conte) having the bony
ridges moderate, not swollen behind; var. **fowleri** Putnam, Mass.
and N., has these crests much swollen and coalescent, "forming
an osseous boss on the skull." (Lat., freckled.)

FAMILY CVI. PELOBATIDÆ. (THE BURROWING TOADS.)

Upper jaw with teeth; heel usually provided with a more or less
developed spur. Genera 8, species 18 ; Europe and America.

---

[1] To understand the character of the structure here briefly described, the student
should dissect a toad (*arciferous*) and a frog (*firmisternial.*)

c. Forehead and crown bony, rough; skin slightly tuberculate; sacrum not co-ossified with coccyx; vomer with teeth: heel with a spadelike process covered by a horny sheath; toes more or less webbed.

SCAPHIOPUS, 263.

263. SCAPHIOPUS Holbrook. (σκάφη, spade ; πούς, foot.)

515. S. holbrooki Harlan. SPADE-FOOT. Olive brown, a yellowish band on each side. E. U. S., rare W. of Penn.; burrows in the ground; extremely noisy in spring. " The machinery for producing sounds equal to an ordinary steam whistle is apparently confined to the throat of this rare and curious Batrachian." (*Abbott.*) L. 3. (To Dr. J. E. Holbrook.)

FAMILY CVII. **HYLIDÆ**. (THE TREE FROGS.)

Fingers and toes more or less dilated into viscous disks at their tips; upper jaw and vomer with teeth; lower parts usually covered with small warts; ear well developed. Genera 14; species 170; found in most warm regions, especially abundant in tropical America; noted for their loud and varied voices, some of them being heard at all times from early spring until frost comes.

a. Disks small; fingers not webbed; palustrine.

b. Toes broadly webbed; tympanum indistinct. . . . . . . ACRIS, 264.

bb. Toes scarcely webbed; tympanum distinct. . . CHOROPHILUS, 265.

aa. Disks round, conspicuous; fingers somewhat webbed; skin roughened; arboreal. . . . . HYLA, 266.

264. ACRIS Dumeril & Bibron. (ἀκρίς, locust, from its sharp note.)

516. A. gryllus Le Conte. CRICKET FROG. Hind legs very long. Brownish above; middle of back and head bright green or reddish brown; a dark triangle between the eyes; sides with three oblique blotches ; a white line from eye to arm. L. 1½. E. U. S., in swamps, not on trees; var. *gryllus*, S., N. to S. Ill. The northern form is var. **crepitans** Baird. Its snout is more blunt and the inner surface of thigh not reticulate; its note resembles the rattling of pebbles. (γρύλλος, a pig.)

265. CHOROPHILUS Baird. (χορός, chorus; φίλος, lover.)

517. C. triseriatus (Wied). SWAMP TREE FROG. Bluish ash, a dark dorsal stripe from snout backward, bifurcating above middle of body; a stripe on each side of this and one on side of head and body, the later pale-edged below. L. 1. Variable. In swampy ground, rarely in trees. Its voice is a " rattle with a rising inflection at the end " (*Cope*), or like the scraping of a coarse-toothed comb. (Lat. 3-rowed.)

266. HYLA Laurenti. (ὕλη, forest.)

518. H. versicolor Le Conte. COMMON TREE TOAD. Green, gray or brown, with irregular dark blotches; below yellow, behind white; tympanum ⅔ diam. eye; fingers ¼ webbed; skin with small warts. L. 2. E. U. S., W. to Kan., very abundant and variable. Its "clear, loud trilled rattle" is heard mostly in the evening and in damp weather.

519. H. pickeringii Holbrook. Yellowish brown or fawn-color, with dusky rhomboidal spots and lines, the latter usually arranged in the form of an oblique cross; head with lines; limbs barred; tympanum very obscure. L. 1. E. U. S.

520. H. squirella Daudin. Olive green, with irregular dark blotches ; a dark bar between eyes; a white line along upper jaw to shoulder ; greenish white below, darker behind ; throat with a few dark spots : legs marked with darker above : tympanum half diam. eye. L. 1¼. Ind. (Brookville, A. W. Butler) to S. C. (Eng. squirrel.)

521. H. andersonii Baird. Deep pea-green ; sides with irregular yellow spots: a green spot on throat; a purplish band from eye to arm ; tympanum ⅓ eye. L. 1½. N. J. to S. C., rare.

FAMILY CVIII. **ENGYSTOMATIDÆ.** (THE TOOTHLESS FROGS.)

Frog-like Batrachians with the maxillaries toothless and the diapophyses of the sacral vertebræ dilated. Genera 18; species 54, chiefly tropical.

a. Pupil erect; tongue elliptical ; **tympanum hidden**; toes free; no precoracoids. . . . . ENGYSTOMA, 267.

267. ENGYSTOMA Fitzinger. (ἐγγύς, contracted ; στόμα, mouth.)

522. E. carolinense Holbrook. Snout obtuse, not twice eye; skin smooth, a fold across head behind eyes. Brown, dotted with paler below. L. 1. S. U. S., N. to Mo.

FAMILY CIX. **RANIDÆ.** (THE FROGS.)

Teeth well developed on upper jaw, and usually on vomer also; toes 4-5, all more or less webbed; ear well developed. Genera 18, species 250, chiefly of the Northern Hemisphere and the East Indies. Most of them are aquatic, and similar to our common frogs.

a. Vomerine teeth present; no finger opposable to the others; tongue emarginate behind; hind toes full-webbed. . . . . RANA, 268.

268. RANA Linnæus. (Lat., frog.)

a. Glandular folds on each side of back more or less distinct; web of feet not reaching tip of fourth toe.

b. Tympanum smaller than eye.

c. Back with large distinct dark spots, more or less regularly arranged; vomerine teeth between the inner nares.

523. **R. areolata** Baird. The Northern form is var. **circulosa** Rice & Davis, thus described: Brownish black, divided by narrow clay-colored lines into irregular circular blotches, largest behind; arms and legs barred or blotched; head broad depressed: snout very obtuse; skin coarsely punctate; a deep hollow between nostril and eye; region above and behind ear swollen; glandular folds large; toes narrowly webbed. L. $3\frac{1}{2}$; leg, $5\frac{1}{2}$. N. Ind. and Ill. The typical *areolata*, from Texas, has spots smaller, bordered with white. (Lat., with little areas or spots.)

524. **R. virescens** Kalm. COMMON FROG. LEOPARD FROG. Green, usually bright, with irregular black blotches edged with whitish, these mostly in two irregular rows on back; usually two spots between eyes; legs barred above; belly pale; glandular folds large; head rather elongate. L. $2\frac{3}{4}$. N. Am., W. to Sierra Nevada, very common. (*R. halecina* " Kalm.") (Lat., greenish.)

525. **R. palustris** Le Conte. PICKEREL FROG. Light brown, with two rows of large oblong square blotches of dark brown on back; one or two on sides; a brown spot above eye; a dark line from nostril to eye; upper jaw white, spotted with black; head short, obtuse; toes well webbed; glandular folds low. L. $2\frac{3}{4}$. E. U. S., in mountains, etc. (Lat., in swamps.)

cc. Back with small dark spots or none.

d. Side of head without distinct dark band; vomerine teeth between the inner nares.

526. **R. septentrionalis** Baird. Brown or olive with paler vermiculations; sometimes a few dark blotches behind; pale below; femur and tibia equal, $\frac{1}{2}$ length of body. L. $2\frac{1}{2}$. Canada to Montana. (Lat., northern.)

dd. Side of head with a dark brown band, wider behind, from snout to near shoulder, bordered below by a yellowish white line; usually a black spot at base of arm; vomerine teeth extending beyond level of hinder edge of inner nostril.

527. **R. sylvatica** Le Conte. WOOD-FROG. Pale reddish-brown; arms and legs barred above; head small, pointed; femur and tibia about equal, the latter considerably more than half body; a rounded outer metatarsal tubercle present. L. $1\frac{3}{4}$. E. U. S., W. to the plains; common in damp woods; an almost silent frog.

528. **R. cantabrigensis** Baird. Very similar to preceding, but the tibia half length of body; a narrow pale line along thighs behind; a dorsal line from snout to arms; back sometimes with dark spots; no outer metatarsal tubercle. Mass., to Alaska and N. (Lat., of Cambridge.)

bb. Tympanum as large or larger than eye.

529. R. clamata Daudin. GREEN FROG. Green or brownish,
brighter in front; generally with irregular small black spots; arms
and legs blotched, yellowish or white below; tympanum large;
glandular folds large; toes well webbed; first finger not extending
beyond second; tibia and femur equal, $\frac{1}{2}$ body. L. 3. E. U. S.,
in springs, etc. (Lat., called to.)

aa. Glandular folds on sides of back obsolete or nearly so; dark spots on back
      small; web of feet reaching tip of fourth toe.

530. R. catesbiana Shaw. BULL-FROG. Greenish, of varying
shades, with small faint dark spots above ; head usually bright
pale green ; legs blotched; ear large   toes broadly webbed; femur
equal to tibia, not half body. L. 5 to 8. Largest of the frogs; in
ponds and sluggish rivers, from Kansas E. ; remarkable for its
sonorous bass notes. (To Mark Catesby, who first figured the
bull-frog.)

CLASS G. — REPTILIA. (THE REPTILES.)

THE Reptiles are cold-blooded air-breathing vertebrates, usually scaly or covered with bony plates, never with feathers or hair. The limbs when present are usually adapted for walking, sometimes for swimming. There is an incomplete double circulation of the blood; the septum between the two ventricles being usually wanting or imperfect. There is no metamorphosis after leaving the egg, and the eggs are large and mostly provided with a leathery skin. The skeleton is usually firm, and the nervous system is better developed than in the preceding groups. There are various other anatomical and embryological peculiarities of the Reptiles, too numerous to be noticed here. We may say however that the Reptiles are obviously distinguished from the Birds by the absence of feathers, and from the Batrachians by the presence of scales, and by the absence of gills after leaving the egg. The extinct forms of Reptiles are numerous, and their close relation with the earlier birds show the propriety of uniting the two classes in a single group, *Sauropsida.* The three orders represented in our fauna are well distinguished from each other. A fourth (CROCO-DILIA) is represented by two species (*Alligator mississippiensis* Daudin, and the rare *Crocodilus americanus* Seba,) in the lowlands of the South.

Orders of Reptilia.

a. Body covered with imbricated scales ; vent a cross-slit ; bones of skull separate; jaws with teeth; dorsal vertebræ and ribs movable.
b. Mouth very dilatable; bones of mandible (and of head generally) united by ligaments; limbs wanting or represented by short spurs on sides of vent; no shoulder girdle ; no eyelids ; no tympanum.
OPHIDIA, XXVIII.
bb. Mouth not dilatable; bones of mandible united by a bony suture in front; limbs 4 (rarely obsolete); shoulder girdle present; eyelids and tympanum usually evident. . . . . . . LACERTILIA, XXIX.
aa. Body short, depressed, enclosed between two bony or cartilaginous shields (*carapace; plastron*), from which the head, limbs, and tail may be protruded; jaws with a horny shield and no teeth; vent roundish or longitudinal, plaited. . . . . TESTUDINATA, XXX.

ORDER XXVIII. OPHIDIA. (THE SERPENTS.)

Reptiles with elongate, terete bodies, obsolete limbs, and with an epidermal covering of imbricated scales, which is shed as a whole and replaced at regular intervals; the mouth very dilatable; the

bones of both jaws and of the palato-pterygoid arch freely movable,
united by ligaments only.   Limbs wanting; the shoulder girdle
wanting: the pelvic girdle usually so, rarely rudimentary, and with
the hinder limbs represented by small spurs on the sides of the
vent; vent a transverse slit; tongue forked, capable of protrusion ;
no eyelids, nor external ears.   Various anatomical characters distinguish the snakes, but the elongated form and absence of limbs
separate them at once from all our other vertebrates, excepting the
lizard *Ophiosaurus*, and this is not in any other respect, snake-like.
(ὄφις. snake.)

Families of Ophidia.

a. Maxillary horizontal, not excavated; no trace of hinder limbs; no deep
   pit between eye and nostril; poison fangs wanting, or if present, permanently erect.
  b. Upper jaw with solid teeth only; no grooved nor perforated fangs. (*Non-
   venomous*.) . . . . COLUBRID.E, 110.
  bb. Upper jaw with a permanently erect perforated fang in front. (*Somewhat
   venomous*.) . . . . ELAPID.E, 111.
aa. Maxillary vertical; upper jaws in front with large, erectile perforated
   fangs; fangs not grooved in front; a deep pit on each side behind
   nostril, partly occupying the excavated maxillary. (*Venomous*.)
                                       CROTALID.E, 112.

FAMILY CX. COLUBRIDÆ. (THE COLUBRINE SNAKES.)

Both jaws fully provided with teeth, which are conical and not
grooved; head covered with shields; no poison fangs; no spur-like
appendages to vent; belly covered with broad band-like plates
(ventral plates or gastrosteges); tail conical, tapering ; sub-caudal
plates (urosteges) arranged in pairs.

A very large family comprising 225 genera, and upwards of 700
species, found in nearly every part of the world, but most abundant
in warm regions.   They differ from the *Elapidæ* in the want of
erect poison fangs; from the *Crotalidæ*, in having both jaws fully
provided with teeth, and in the absence of erectile poison fangs; and
from the *Boidæ* and their relatives in the want of the spur-like rudimentary posterior limbs.

a. Head conic, not distinct from the body, which is cylindric and rather
   rigid. (*Calamariinæ*.)
  b. None of the teeth grooved; scales not keeled; anal plate bifid; internasals 2.
   c. Prefrontals 2.
    d. Nasal plate single, pierced by the nostril; lorals present; no pre-
     ocular.
     e. Scales in 13 rows; postorbital single (ventral plates 120 to 135).
                                       CARPHOPHIOPS, 269.
     ee. Scales in 19 rows; postorbitals 2 (V. P. 170 to 185).
                                       ABASTOR, 270.

dd. Nasal plates 2, the nostril between them; a loral plate; no preocular; (scales 15 or 17, V. P. 115 to 125). . . . VIRGINIA, 271.

   cc. Prefrontal single; no preocular; nasal single; (scales 19; V. P. 170 to 205). . . . . FARANCIA, 272.

aa. Head more or less distinct from the body which is not specially rigid (*Coronellinæ,*[1] *Colubrinæ, Homalopsinæ*).

  f. Rostral plate normal, not recurved nor keeled.

    g. Anal plate divided; head not very short.

     h. Dorsal scales keeled more or less.

      i. Nasal plates 2. the nostril between them.

       j. Prefrontals 2.

        k. Loral plate present.

         l. Scales on back and sides all keeled (ventral plates 130 to 170).

          m. Posterior teeth not longer (scales 19 to 21). . REGINA, 277.

          mm. Posterior teeth longer (scales 23 to 31).

              TROPIDONOTUS, 278.

         ll. Scales on sides not keeled; those on back often with the keels obscure (V. P. 200 to 270; scales 25 to 29).

              COLUBER, 279.

        kk. Loral plate absent (scales 15 to 17; V. P. 125 to 130).

              STORERIA, 274.

       jj. Prefrontal single; nasals 2; loral present; no preocular (scales 17; V. P. 120 to 130); body slender, the head distinct.

              HALDEA, 273.

      ii. Nasal single, pierced by the nostril; loral plate present.

        n. Nasal plate grooved; tail short (scales 19; V. P. 130 to 140). . . . . TROPIDOCLONIUM, 275.

        nn. Nasal plate not grooved below nostril; tail very long, about ⅓ of length (scales 17; V. P. 150 to 160).

             CYCLOPHIS, 280.

    hh. Dorsal scales not keeled.

      o. Nasal single, pierced by the nostril; loral present; tail long (scales 15; V. P. 125 to 140). LIOPELTIS, 281.

      oo. Nasal plates 2. the nostril between them; loral present; preoculars 2.

        p. Head depressed; preoculars nearly equal in size (scales 15 to 17; V. P. 140 to 200 or more).

             DIADOPHIS, 284.

        pp. Head not depressed; upper preocular much the larger; lower sometimes wanting; (scales 17; V. P. 170 to 210). . . . . . BASCANION, 282.

  gg. Anal plate entire.

      q. Dorsal scales all or part of them keeled; head rather long.

---

[1] These three sub-families are so vaguely bounded that I cannot use their distinctive characters in the key. Professor Cope gives the following definitions : —

Homalopsinæ. (Genera 273 to 278.) "Hypapophyses spinous to caudal region; anterior teeth not enlarged; body not slender; head distinct."

Colubrinæ. (Genera 279 to 283.) Head more distinct and elongate; body and tail longer; teeth entire, not longer in front.

Coronellinæ. (Genera 284 to 288.) Head slightly distinct, short ; teeth entire, not enlarged in front.

r. Postfrontals 2 pairs ; loral single ; prefrontals 4 or more (scales 25 to 35; V. P. 200 to 250). . . . . Pituophis, 283.

rr. Postfrontals 1 pair; nasals 2; posterior teeth rather larger; viviparous (scales 19 to 21; V. P. 140 to 180). . . . Eutainia, 276.

qq. Dorsal scales not keeled, rather loosely imbricate; head short (V. P. 160 to 240).

    s. Rostral plate not acute.

       t. Loral present (scales 21 to 25). . . . . . . Ophibolus, 285.

       tt. Loral absent (scales 19) . . . . Osceola, 286.

    ss. Rostral plate acute, the snout sharp-pointed (scales 19).

                                   Cemophora, 287.

ff. Rostral plate (at tip of snout) produced, recurved and keeled; dorsal scales keeled; anal plate divided ; head broad and short; some of the posterior teeth enlarged (scales 23 to 27; V. P. 120 to 150).

                                   Heterodon, 288.

The following purely artificial key may aid in finding the names of specimens : —

I. *Dorsal scales not keeled ; anal plate bifid.*

    a. Scales 13; V. P. about 130; color brownish . Carphophiops, 269.

    aa. Scales 15 to 17.

       b. Ventral plates about 120; brownish. . . . . Virginia, 271.

       bb. V. P. about 140.

          c. Blackish, with yellow collar. . . . . . Diadophis, 284.

          cc. Green; no collar. . . . . Liopeltis, 281.

       bbb. V. P. about 185; blue-black, young blotched. Bascanion, 282.

    aaa. Scales 19; V. P. about 180.

       d. Blue-black with three red lines. . . . . . Abastor, 270.

       dd. Blue-black with square red spots on sides. . . Farancia, 272.

    aaaa. Scales 25 to 29 (median dorsal scales faintly keeled).

                                   Coluber, 279.

II. *Dorsal scales not at all keeled ; anal plate entire.*

    e. V. P. about 170 to 205; snout not sharp.

       f. Scales 19; red, black-banded. . . . . . Osceola, 286.

       ff. Scales 21 to 25; black, brown or red, mostly variegated.

                                Ophibolus, 285.

    ee. V. P. 160 to 170; scales, 19; snout sharp; red with black rings.

                                Cemophora, 287.

III. *Dorsal scales more or less keeled ; anal plate entire.*

    g. Scales 19 to 21; V. P. about 155; striped . . . Eutainia, 276.

    gg. Scales 25 to 35; V. P. about 220; blotched . . Pituophis, 283.

IV. *Dorsal scales keeled; anal plate bifid.*

    h. Scales 15 to 17.

       i. Tail about ⅓ of length; V. P. 155; green. . . Cyclophis, 280.

       ii. Tail not ⅓ of length; brownish.

          j. Loral absent. . . . . Storeria, 274.

          jj. Loral present. . . . . Haldea, 273.

    hh. Scales 19 to 21.

       k. V. P. about 135; blotched. . . . . Tropidoclonium, 275.

kk. V. P. about 150; striped. . . . . REGINA, 277.
hhh. Scales 23 to 29.
  l. Snout without recurved keel at tip.
    m. V. P. 130 to 160; brownish, usually with cross blotches.
                                                    TROPIDONOTUS, 278.
    mm. V. P. 200 to 240; brown or black, mostly blotched. COLUBER, 279.
  ll. Snout recurved and keeled; V. P. 125 to 150.   HETERODON, 288.

269. CARPHOPHIOPS Gervais. (κάρφος, dry twig; ὄφις, snake;
                          ὄψ, appearance.)
a. Frontals, two pairs.

531. C. amœnus (Say). GROUND SNAKE. Glossy chestnut
brown; belly salmon-red; head very small; vertical plate broad;
scales 13; V. P. 112 to 131. L. 12. Mass. to Ill. and S. (Lat.,
pleasing.)

532. C. vermis (Kennicott). WORM SNAKE. Purplish-black,
belly flesh color, the color extending on sides; scales 13; larger
than the others. Mo. to Kan. (Lat. worm.)

aa. Frontals, a single pair.

533. C. helenæ (Kennicott). Lustrous chestnut-brown, flesh
color beneath; snout short and narrow; scales 13. S. Ind. to
Miss. (To Miss Helen Tennison.)

270. ABASTOR Gray. (A coined name.)

534. A. erythrogrammus (Daudin.) "HOOP SNAKE." Blue-
back; sides with three red lines; belly flesh color, with black
blotches; eyes very large; nostril in the middle of nasal plate;
scales 19; V. P. 167–185. L. 25. N. C. to S. Ill. and S.; a
harmless snake concerning which many absurd stories have been
told. (ἐρυθρός, red; γραμμή, line.)

271. VIRGINIA Baird & Girard.

535 V. valeriæ Baird & Girard. Grayish, with minute black
dots, often in two rows; yellowish beneath; scales 15; V. P. 120
to 130. L. 12. Md. to Ill. and S. (To Miss Valeria Blaney.)

536. V. elegans Kennicott. Scales very narrow and elongate;
olivaceous above, yellowish beneath; scales 17. S. Ind. to Ark.

272. FARANCIA Gray. (A coined name.)

537. F. abacura (Holbrook). HORN SNAKE. Blue-black with
red, squarish spots on side; belly red, blotched with black; eyes
small; scales 19; V. P. 170 to 203. L. 36. S. C to S. Ill. (*Nelson*) and S. (ἄβαξ, checker; οὐρά, tail.)

273. HALDEA Baird & Girard. (To Prof. Samuel
                          S. Haldeman?)

538. H. striatula (L.). BROWN SNAKE. Head elongated,
on a small neck. Eye large. Reddish gray, salmon red beneath;

scales 17; V. P. 110 to 130.   L. 10.   Va. to Wis. and Texas.
(Lat.. narrowly striped.)

274. STORERIA Baird & Girard.   (To Dr. David Humphreys
Storer).

539. **S. occipitomaculata** (Storer).   RED-BELLIED SNAKE.
Greyish or chestnut brown, usually showing a paler vertebral band
bordered by blackish dots; obscure dots on side; occiput with
three pale blotches (a very constant feature); belly salmon red;
scales 15; V. P. 120 to 125.   L. 12.   Minn. to Mass. and Ga.;
abundant E.   (Lat., occiput-spotted.)

540. **S. dekayi** (Holbrook).   Grayish brown; a clay-colored
dorsal band, bordered by dotted lines; grayish below : a dark patch
on each side of the occiput; scales 17; V. P. 120 to 138.   L. 12.
E. U. S., W. to Rocky Mts.   (To James E. DeKay.)

275. TROPIDOCLONIUM Cope.   (τρόπις, keel ; κλωνίον, a
small twig.)

541. **T. kirtlandi** (Kennicott).   Light reddish brown, with 4
series of round black spots; belly reddish with a row of black
spots on each side; head shining black; head small ; vertical
plate broad ; scales 19, all carinated ; V. P. 115 to 140.   L. 16.
Ohio to Ill.   (To Dr. Jared P. Kirtland.)

276. EUTAINIA Baird & Girard.   GARTER SNAKES.
(εὖ, well ; ταινία, ribbon.)

a. Lateral stripe on 3d and 4th rows of scales.
  b. Scales little or not spotted, in nineteen rows; a dorsal band; body very
    slender.
    c. Stripes alike in color.

542. **E. saurita** (L.).   RIBAND SNAKE.   Chocolate with three
yellow stripes ; light brown below the lateral stripes; tail usually
3½ in length; colors bright ; V. P. 150 to 160.   L. 36.   E. U. S.,
chiefly E. of the Alleghanies, about streams.   (Lat., lizard-like.)

543. **E. faireyi** Baird & Girard.   Blackish, with three greenish
yellow stripes ; body relatively stout ; tail less than ⅓ length ; space
below bands same color as above; V. P. 165 to 180.   L. 30.   Wis.
to La.   (To James Fairie.)

  cc. Stripes not uniform in color.

544. **E. proxima** (Say).   Blackish, dorsal stripe brownish yellow : lateral stripes greenish; tail ⅔ of total length : sides colored
like back ; V. P. 165 to 175.   L. 35.   Wis. to Mexico.   (Lat.,
near.)

  bb. Scales above and below lateral line with subquadrate black spots.

515. E. radix Baird & Girard. Green or black with three narrow yellow stripes ; six series of black spots ; scales very rough, the outer row broad; colors deep; head short; tail short, 5 in length; scales 19 to 21 ; V. P. 150 to 160. L. 25. Wis. to Oregon. (Lat., root, from Root R., Wis.)

aa. Lateral stripe on 2d and 3d rows of scales ; body stoutish, the tail 4 in length.

   d. Scales in 19 rows.

546. E. sirtalis (L.). COMMON GARTER SNAKE. STRIPED SNAKE. Olivaceous, dorsal stripe narrow, obscure ; 3 series of small dark spots on each side, about seventy between head and vent; sides and belly greenish ; lateral stripes rather broad but not conspicuous; colors generally duller than in the other species. V. P. 130 to 160. N. Am., everywhere except in Cal.: our commonest snake ; very variable. (Lat., like a garter.)

Prominent varieties are : Var. **ordinata** (L.) with the stripes obscure or wanting and the spots more distinct, square, 85 in number before anus; V. P. spotted on sides. Chiefly northeastward. Var. **dorsalis** (Baird & Girard) has the dorsal stripe broad, and two rows of small distinct spots on each side. N. Am., everywhere. Var. **obscura** Cope, uniform brown, the spots obscure, the bands distinct. Var. **parietalis** (Say) has the stripes dull greenish and the space between the lateral spots of a more or less vivid brick red. Ind. to Cal.

   dd. Scales in 21 rows.

547. E. vagrans Baird & Girard. Ashy brown, usually a narrow dorsal stripe; each side with about 100 small black spots in two series. V. P. 160 to 180. Ills. (*Nelson*) to Cal.

277. REGINA Baird & Girard.

a. Postorbitals 2.

548. R. rigida (Say). Greenish brown; two brown dorsal bands ; a brown spot at base of each scale on sides; belly yellowish, blotched; outer row of scales smooth; scales, 19 ; V. P. 130 to 170. L. 24. Penn. to Ga., chiefly E. of mts., W. to Central Ill. (*Hay.*)

549. R. leberis (L.). Chestnut brown; a yellow lateral band and three narrow black dorsal stripes ; belly yellow, with two brown bands; scales all keeled; scales 19; V. P. 140 to 150. L. 24. U. S., about streams. (Lat., cast-off snake-skin.)

aa. Postorbitals 3, the lower very small.

550. R. grahami (Baird & Girard). Brown; a pale brown dorsal band; besides this two narrow black streaks on each side ; a straw-color lateral stripe ; belly unspotted; scales all strongly keeled ; head slender ; scales 19 to 21; V. P. 160. L. 20. Mississippi Valley, N. to Mich.

278. TROPIDONOTUS Kuhl. (τρόπις, keel ; νῶτος, back.)
a. Scales in 23 to 25 rows.
b. Rostral plate single.

551. **T. sipedon** (L.). WATER SNAKE. "MOCCASIN." Brownish, back and sides each with a series of large, square, dark blotches alternating with each other; about 80 in each series; rarely uniform brownish; belly with brown blotches. Scales, 23. V. P. 130 to 150. L. 30 to 50. N. Eng. to Kan. and S.; very abundant about streams, feeding on fishes and frogs. Variable ; an unpleasant and ill-tempered, but perfectly harmless snake. Prominent varieties are var. **woodhousei** Baird & Girard : color of *sipedon*, a narrow whitish line between dorsal blotches. Scales 25. S. Ill. to Texas. Var. **erythrogaster** Shaw, uniform red-black above, coppery below ; head long; scales strongly keeled ; Mich. to Kan. and S.

552. **T. fasciatus** (L.). SOUTHERN WATER SNAKE. Dark brown, with transverse black blotches on back and about 35 oblong red spots on sides ; back sometimes with broken rings of yellowdots; belly reddish, usually blotched. Scales 25. V. P. 128 to 135. S. Ind. (*Ridgway*) to Texas and S. E., swarming in the lowland swamps S.

bb. Rostral plate divided into two by a vertical suture.

553. **T. bisectus** Cope. Olive brown, a row of small blackish spots on side ; head and belly nearly plain. V. P. 143. Scales 25. One specimen from Washington, D. C.

aa. Scales in 27 to 33 rows.

554. **T. cyclopium** Duméril & Bibron. Plumbeous, with alternating blackish vertical bars 1 to 1½ scales wide. V. P. 140 to 150. Scales 27 to 33. S. Ill. to Fla. (κύκλωψ, round-eyed.)

555. **T. rhombifer** Hallowell. Brown with about 50 black quadrangular blotches bordered by black lines. Scales 27, all keeled. V. P. 140 to 145. Mich. to Ill. and S. W. (Lat., bearing rhombs.)

279. COLUBER Linnæus. (Old name of some snake.)
a. Body without longitudinal brown stripes.
b. Scales in 25 to 27 rows.
c. Vertical plate longer than broad.

556. **C. guttatus** L. CORN SNAKE. Red brown with a dorsal series of large, red, dark-edged blotches; belly checkered with black. Scales 27. V. P. 210 to 230. L. 50. Va. and S. (Lat., spotted.)

557. **C. obsoletus** Say. PILOT SNAKE. Lustrous black, some scales white-edged ; belly slaty-black ; median scales of back obscurely keeled, the rest smooth. Scales 27. V. P. 235. L. 50 to 75. Mass. to Ill. and Texas; one of our largest snakes, often climbing trees to a great height by following the depressions in

rough bark. Var. **confinis** Baird & Girard is ashy gray, with 45 dark chocolate blotches on back, their edges faintly darker: two smaller series on side; a dark band between eyes; belly blotched. S. Ind. (Brookville, E. R. Quick) to S. C. and S. W.

cc. Vertical plate broader than long.

558. **C. vulpinus** (Baird & Girard). FOX SNAKE. Light brown, with quadrate, chocolate-colored blotches; vertical plate broader than long. Scales 25. V. P. 200 to 210. L. 60. Mass. to Kan. and N. (Lat., fox-like.)

bb. Scales in 29 rows.
d. Vertical plate longer than broad.

559. **C. emoryi** (Baird & Girard). Ashy gray with transverse brown blotches; vertical plate elongated; 6 or 8 median rows of scales only keeled. Scales 29. V. P. 210 to 220. L. 40 to 50. Ill. to Kan. and Texas.

dd. Vertical plate as broad as long.

560. **C. lindheimeri** (Baird & Girard). Back and sides with black blotches, the interspaces paler; scales edged with white : greenish white below; centres of shields slate color ; about 9 rows of scales obscurely keeled. Scales 29. V. P. 225 to 235. S. Ill. to Texas.

aa. Body with 4 longitudinal brown stripes.

561. **C. quadrivittatus** Holbrook. CHICKEN SNAKE. Greenish yellow, with two brownish stripes on each side; straw-color below. Scales 27, only 5 to 8 rows keeled. V. P. 230 to 245. Va. to Fla. (Lat., four-striped.)

280. CYCLOPHIS Günther. (κύκλος, ring; ὄφις, snake.)

562. **C. æstivus** (L.). GREEN SNAKE. Head conical, neck very small; bright clear green, yellowish below. Scales 17. V. P. 150 to 165; tail more than ⅓ of body. L. 30. Southern N. J. to Ind. and S., abundant S. ; a most exquisite little creature, often climbing bushes over water.

281. LIOPELTIS (Fitzinger) Cope. (λεῖος, smooth ; πέλτη, shield.)

563. **L. vernalis** (DeKay.) GRASS SNAKE. Head elongate. neck slender; eyes very large; uniform deep green (bluish in spirits), yellowish below; tail not quite ⅓ of length. Scales 15. V. P. 125 to 140. L. 20. E. U. S., chiefly N.; a beautiful species.

282. BASCANION Baird & Girard. (βάσκανος. malignant.)

564. **B. constrictor** (L.). BLACK SNAKE. BLUE RACER. Lustrous pitch black, greenish below, chin and throat white; young

olive. with rhomboid black blotches; body slender; eye very large.
Scales 17 (rarely 19). V. P. 170 to 190. L. 50 to 60. E. U. S.,
common E. and S. (Lat , one that hugs.)

283. PITUOPHIS Holbrook. (πίτυς, pine-tree ; ὄφις, snake.)

565. **P. melanoleucus** (Daudin). PINE SNAKE. BULL SNAKE.
Whitish, with chestnut brown blotches which are margined with
black. besides 3 series of lateral blotches. Scales 29. V. P. 220 to
230. L. 60. Pine woods; N. J. to Mich. and S. (μέλας, black ;
λευκός, white.)

566. **P. sayi** (Schlegel). WESTERN PINE SNAKE. Chestnut
brown with many orange cross-blotches and spots; sides mottled
with black and orange. Scales 25 to 29. V. P. 220 to 245. L.
40 to 70. Ill. to Kan. and N. W. (To Thomas Say.)

284. DIADOPHIS Baird and Girard. (διά, through;
ὄφις, snake.)

567. **D. punctatus** (L.). RING-NECKED SNAKE. Eye rather
large. Blue-black above, bright pale orange below (yellowish in
spirits) ; each plate usually with a black spot on each side and
sometimes a median one ; a very conspicuous yellowish ring about
neck, 2 scales wide. Scales 15. V. P. 140 to 160. L. 15. E.
U. S. W. to Kan. Represented W. by var. **amabilis** Baird & Girard,
slender, with V. P. 180 to 185 ; below darker and more spotted;
scales on sides considerably larger than those on back. W. U. S.,
E. to Ohio.

568. **D. arnyi** Kennicott. Lead black; belly spotted and mottled
with black; occipital ring narrow, 1½ scales wide. Scales 17. Ill.
to Ariz.

285. OPHIBOLUS Baird & Girard. (ὄφις, snake; βολίς, dart.)
a Dorsal scales in 21 rows.
b. Color chiefly black.

569. **O. getulus** (L.). CHAIN SNAKE. THUNDER SNAKE.
Black with narrow yellowish lines forking on the flanks, each fork
embracing a large black spot ; belly checkered. Scales 21. V. P.
210 to 240. L. 50. Va. to La., E. of the mountains; variable.
Represented westward by var. **sayi** (Holbrook). KING SNAKE.
Lustrous black, many scales with a yellow spot in the centre, these
sometimes forming cross-lines on back ; belly blotched. Alleghany
to Rocky Mts., abundant, N. to Ills.; a handsome snake, said to
be an enemy of the rattlesnake.

bb. Color red or grayish, with dark markings.

570. **O. doliatus** (L.). RED SNAKE. CORN SNAKE. Red,
with twenty to twenty-five pairs of black rings, each set enclosing

a yellowish one; the lines of each pair separate on sides and become confluent with the nearest one of adjacent pair; head red. Scales 21. V. P. 180 to 210. L. 30 to 50. Md. to Kan. and S.; exceedingly variable, running by degrees into the following varieties, extremes of which bear little resemblance to the typical *doliatus*. (Lat., sorrowful.)

Var. **coccineus** Schlegel, the black rings not confluent and usually meeting on belly. S. Ill. to Fla. and W.

Var. **triangulus** (Boie). MILK SNAKE. HOUSE SNAKE. SPOTTED ADDER. Grayish, with three series of brown, rounded blotches bordered with black, about fifty of them in the dorsal row ; an arrow-shaped occipital spot ; belly with square black blotches. Va. to Iowa, and N.; very common.

aa. Dorsal scales in 25 rows.

571. **O. rhombomaculatus** (Holbrook). Light chestnut, back and sides with 3 series of darker rhomboidal blotches, about 50 in dorsal series; belly obscurely blotched. V. P. 200 to 205. Ill. to N. C. and S.

572. **O. calligaster** (Say). Light olive gray, with about sixty quadrate, chestnut colored, emarginate blotches on back and two rows of smaller ones on each side. Ill. to Kansas and S. (καλός, beautiful ; γαστήρ, belly.)

286. **OSCEOLA** Baird & Girard. (Name of an Indian chief.)

573. **O. elapsoidea** (Holbrook). SCARLET SNAKE. Brilliant red, with about 18 pairs of jet black rings on body and three on tail, each pair enclosing a white ring; the black rings tapering towards the sides, the white ones spreading : a yellow collar on upper part of neck, bordered by black lines; rostral plate very broad ; resembles closely *O. doliatus*. Scales 19. V. P. 175 to 180. L. 20. Va. to S. Ill. and S.

287. **CEMOPHORA** Cope. (κημός, muzzle ; φορός, bearing.)

574. **C. coccinea** (Blumenbach). Crimson, with 20 to 26 black rings enclosing yellow ones; yellowish below. V. P. 160 to 170. S., N. to S. Ohio and Ark. (Lat., crimson.)

288. **HETERODON** Beauvais. (ἕτερος, different; ὀδών, tooth.)

a. Vertical plate in direct contact with frontals.

575. **H. platyrhinus** Latreille SPREADING ADDER. BLOWING VIPER. Brownish or reddish, with about 28 dark dorsal blotches, besides lateral ones and half rings on the tail; often (var. *niger*) uniform black. Vertical plate longer than broad, about equal to occipitals. L. 30. V. P. 120 to 150. Scales 23 or 25. E. U. S., abundant. A very variable species; when angry it de-

presses and expands the head, hissing and threatening, but it is
perfectly harmless. (πλατύς, flat; ρίς, nose.)
aa. Vertical plate encircled by 5 to 10 small plates.

576. **H. simus** (L.). HOG-NOSED SNAKE. Dorsal blotches
about 35; ground color usually pale yellowish brown; vertical
plate much longer than occipitals, broader than long. V. P. 115 to
150. Scales usually 25. Ill. and Wis. to S. C., chiefly S. (Lat.,
flat-nosed.)

FAMILY CXI. **ELAPIDÆ.** (THE HARLEQUIN SNAKES.)

Venomous snakes, provided with two or more permanently erect,
perforated fangs in the upper jaw, and usually a series of smaller
teeth behind them; scales not keeled; head usually quadrangular,
with flat crown and short muzzle; no loral plate. Genera 3, species
about 20, chiefly East Indian, a few inhabiting the warmer parts of
America.

a. Anal plate entire; sub-caudal plates two-rowed; two nasal plates; inter-
    nasal plate touching the nasal laterally. . . . . . . ELAPS, 289.

289. ELAPS Schneider. (Old name of some snake.)

577. **E. fulvius** (L.). BEAD SNAKE. Jet black, with about
17 broad crimson rings, each bordered with yellow, and spotted
below with black; a yellow occipital band; tail with yellow rings.
V. P. 200 to 215. U. 32. Scales, 15 rows. L. 30. Va. to Ark.
and S. A beautiful snake, apparently harmless, although provided
with venom-fangs. Resembles *Ophibolus doliatus.* (Lat., reddish-
yellow.)

FAMILY CXII. **CROTALIDÆ.** (THE RATTLESNAKES.)

Maxillary vertical, without solid teeth, but provided with long,
erectile, perforated poison-fang on each side in front; a deep pit
between eye and nostril, extending into the excavated maxillary.
Body stout; head large, flat, triangular, on a slender neck; pupil
elliptical, placed vertically. Tail usually provided with a rattle
composed of horny rings, modified scales. Subcaudal plates generally undivided, at least anteriorly. Scales keeled, in all our
species; anal plate entire. Genera 12; species about 60, all American, renowned for their venom. All are viviparous.

a. Tail short, without rattle, ending in a horny point; top of head with about
    8 symmetrical plates arranged around the vertical plate; tail not pre-
    hensile. . . . . AGKISTRODON, 290.
aa. Tail with a rattle.
    b. Top of head with about 8 plates symmetrically arranged; rattle small.
                                              SISTRURUS, 291.
    bb. Top of head covered with small scales; rattle large. CROTALUS, 292.

290. AGKISTRODON Beauvais. (ἄγκιστρον, hook; ὀδών, tooth.)

a. Loral plate present. (*Agkistrodon.*)

578. **A. contortrix** (L.). COPPERHEAD. COTTON-MOUTH. Hazel brown; top of head coppery-red; back with a series of 15 to 25 V-shaped blotches; belly yellowish, with 35 to 45 dark spots on each side; loral plate present. Scales 23. V. P. 150 to 155. L. 40. N. E. to Wis. and S. in damp places, becoming rare N.; a dangerous reptile. (Lat., one who twists.)

aa. Loral plate wanting. (*Toxicophis.*)

579. **A. piscivorus** (Holbrook). WATER MOCCASIN. BLACK MOCCASIN. Greenish brown with 20 to 30 dark vertical bars, often obscure; belly black and yellow, blotched. Scales 21 to 25. V. P. 138 to 145. L. 50. Aquatic, N. C. to S. Ill., Ark. and S., often resting on overhanging bushes over streams watching for frogs and fishes. The most dangerous of our snakes. (Lat., fish-eating.)

291. SISTRURUS Garman. (σεῖστρον, rattle; οὐρά, tail.)

580. **S. catenatus** (Rafinesque). PRAIRIE RATTLESNAKE. MASSASAUGA. Brown or blackish with about 7 series of about 34 deep chestnut blotches, these blackish exteriorly and edged with yellowish; a yellowish streak from pit to neck; body sometimes all black. Scales 23 to 25. V. P. 135 to 150. L. 30. Prairies, Ohio to Min. and S., abundant in grassy fields where not exterminated. Another species (*S. miliarius* L.) occurs S. (Lat., forming a chain.)

292. CROTALUS Linnæus. (κρόταλον, rattle.)

a. Scales in 23 to 25 rows.

581. **C. horridus** L. COMMON RATTLESNAKE. Yellowish-brown of various shades, with 3 rows of confluent irregular brown spots, forming zigzag-shaped cross-blotches; tail black; a pale line from mouth to eye with a dark patch below. V. P. 165 to 175. L. 60. N. Eng. to Rocky Mts. and S. in rocky places; once common, but nearly exterminated in well-settled regions.

aa. Scales in 27 to 29 rows.

582. **C. adamanteus** Beauvais. DIAMOND RATTLESNAKE. Brown, with 3 series of complete brown yellow-edged rhombs. V. P. 165 to 180. Va. to Miss. and S. (Lat., diamond-like.)

> " I only know thee humble, bold,
> Haughty, with miseries untold,
> And the old curse that left thee cold,
> And drove thee ever to the sun
> On blistering rocks. . . .

Thou whose fame
Searchest the grass with tongue of flame,
Making all creatures seem thy game,
When the whole woods before thee run,
Asked but — when all is said and done —
To lie, untrodden, in the sun!" — BRET HARTE.

ORDER XXIX. LACERTILIA. (THE LIZARDS.)

Reptiles not shielded, with the body usually covered with over-lapping scales: mouth not dilatable; tongue free; jaws always with teeth. Limbs 4, distinct, rarely rudimentary and hidden by the skin ; shoulder girdle developed. Feet usually with 5 digits, the phalanges normally 2, 3, 4, 5, 3, or 4. Tail usually long and in many cases very brittle, readily broken by a slight blow; this is owing to a thin, unossified, transverse septum, which traverses each vertebra. "The vertebra naturally breaks with great readiness through the plane of the septum, and when such lizards are seized by the tail, that appendage is pretty certain to part at one of these weak points." (*Huxley.*) Vent a cross slit; quadrate bone articulated to the skull. The great majority of the numerous species belong to tropical and sub-tropical regions. The few found within our limits give but a slight idea of the whole great group. (Lat., *lacerta*, lizard.)

Families of Lacertilia.

a. Tongue covered with imbricate, scale-like papillæ or with oblique plicæ ; clavicle dilated proximally, often loop-shaped.
   b. Premaxillary double; temporal fossæ roofed over by bone; sternal fontanelle usually wanting; (tongue not deeply bifid). . SCINCID.E, 113.
   bb. Premaxillary single; temporal fossæ not roofed; sternal fontanelle present; (tongue deeply bifid). . . . . TEID.E, 114.
aa. Tongue smooth or with villous papillæ; clavicle not dilated proximally.
   c. Temporal fossæ roofed over by bone; tongue sheathed at tip; body with osteodermal plates; (limbs obsolete in our species).
      ANGUID.E, 115.
   cc. Temporal fossæ not roofed over; tongue thick; (limbs present).
      IGUANID.E, 116.

FAMILY CXIII. SCINCIDÆ. (THE SKINKS.)

Head regularly shielded ; scales smooth, underlaid by bony plates; body fusiform or subcylindrical; nasal plate single, ungrooved, the nostril in the centre; limbs present; toes compressed, 5-5; head usually without posterior vertical plate. Genera about 60; species 200; in most parts of the world.

a. Palate with teeth; two supranasal plates; ear large; its front edge dentate; lower eye-lid scaly. . . . . EUMECES, 293.
aa. Palate toothless; no supranasal plates; ear very large, circular, exposed; lower eye-lid with a transparent disk. . . . . . OLIGOSOMA, 294.

293. EUMECES Wiegmann. (εὐμήκης, of good length.)

583. **E. fasciatus (L.).** BLUE-TAILED LIZARD. "SCORPION."
Blackish olive, with 5 yellowish streaks, middle one forked on the
head; tail usually bright blue; old specimens reddish olive, the
stripes very faint or even wanting; head becoming coppery red with
age. L. 8 to 11. U. S., E. of the Rocky Mts.; abundant N. to
N. Ind.; very variable. (Lat., banded.)

584. **E. obsoletus** (Baird & Girard). Greenish white, the scales
narrowly edged with black. Parieto-occipital and vertical, the
largest plates on head. Ill. (*Forbes*) to Sonora.

585. **E. anthracinus** (Baird). Bronze, with 4 yellow stripes,
between and below which are coal-black lines; tail blue. Penn. to
Texas, in mountains. (Lat., coal-black.)

586. **E. septentrionalis** (Baird). Olive, with 4 dark stripes
above; sides with 2 narrow white lines margined on each side with
black. Minnesota to Nebraska. (Lat., northern.)

294. OLIGOSOMA Girard. (ὀλίγος. small; σῶμα. body.)

587. **O. laterale** (Say). GROUND LIZARD. Chestnut color;
on each side a black lateral band, edged with white; abdomen yellowish; tail blue below; head short; limbs weak; small and slender.
L. 5. Southern States, abundant; N. to S. Ind.

FAMILY CXIV. TEIDÆ.

Tongue flat, elongate, ending in 2 long, smooth points; its surface
mostly covered with imbricate scale-like papillæ; teeth not hollow
at base; premaxillary single; shields of head free from the cranial
ossification; limbs present, rarely rudimentary; clavicle dilated and
perforated proximally. Genera 35; species about 110; all from
tropical America.

a. Scaly portion of tongue arrow-headed, bifid, and not retractile posteriorly;
tail not compressed; shields of head large, regular; eyelids developed;
ear exposed; a double collar-fold; scales small; ventral plates large,
limbs developed; toes 5–5. . . . . CNEMIDOPHORUS, 295.

295. CNEMIDOPHORUS Wiegmann. (κνημιδοφόρος, wearing
leg-armour.)

588. **C. sexlineatus (L.).** Dusky brown, with 3 yellow streaks
on each side; the interspaces jet black; throat silvery; belly blue
in breeding ♂. L. 6 to 9. Conn. to Va., Wis. and Mexico; common S.; very active.

FAMILY CXV. ANGUIDÆ. (THE SLOW WORMS.)

Tongue of 2 parts, the posterior larger, thick, covered with villiform papillæ; the anterior thin, emarginate, covered with scales,

extensible and retractile into a sheath formed by a transverse fold
at anterior extremity of posterior part, this sheath disappearing
when the tongue is drawn out.   Premaxillary single; dermal cranial
ossifications roofing over the temporal fossa; clavicle slender; limbs
present or absent, the shoulder girdle and pelvis always present;
no abdominal ribs; bony plates underlying the scales; vertical plate
on head present.   Genera 7; species 45; in warm regions.

a. Side with a conspicuous fold; limbs wanting or the hinder rudimentary;
  body snake-like, the tail very brittle; scales squarish rhombcidal, forming straight series, in either direction. . . . . . Ophisaurus, 296.

296. OPHISAURUS Daudin.   (ὄφις, snake; σαῦρος, lizard.)

589. **O. ventralis** (L.).   Glass Snake.   Joint-Snake.   Greenish or brownish; sides largely yellow, with narrow black streaks.
Dorsal scales in 14 rows or 120 transverse series; 10 rows on belly;
scales on back obtusely keeled, others smooth; ear much larger than
nostril.   L. 25.   Wis. to Kan. and S.

Family CXVI. IGUANIDÆ.   (The Iguanas.)

Tongue thick, villous, nearly or quite entirely fixed to the floor
of the mouth, and little if at all notched in front; pupil round; eyelids well developed; scales various, those on head usually small;
head generally with an enlarged interparietal scale; teeth subequal.
Habits various, mostly insectivorous.   A very large family of 50
genera and 320 species, swarming in the hotter parts of America;
a very few in the East Indies.

a. Femoral pores absent; toes dilated or depressed, the distal joint narrower,
  cylindrical or compressed, raised above the one before it; scales small
  or granular; ♂ with an inflatable gular sac; tail long, not prehensile;
  lateral teeth tricuspid; no sternal fontanelle; tympanum distinct.

Anolis, 297.

aa. Femoral pores present; fourth toe longer than third; lateral teeth tricuspid.
  b. Head without spines; no dorsal crest; occipital scale very large.
    c. Gular folds 2, the second denticulated; dorsal scales minute, uniform;
      caudal scales small; tympanum concealed. . . Holbrookia, 298.
    cc. Gular folds none; tympanum distinct; scales keeled, equal; no crest.

Sceloporus, 299.

  bb. Head armed with bony spines; body short, depressed; a large sternal
    fontanelle; scales unequal. . . . . Phrynosoma, 300.

297. ANOLIS Daudin.

590. **A. principalis** (L.).   "Chamæleon."   Grass-green; head
brownish, the color changing at times in life to grayish, yellowish,
bronze, and black; gular sac crimson when inflated; head scales
large and rough; scales of body subequal, keeled.   L. 6.   Pine
woods, Tenn. to Cuba; common S.; one of the most beautiful of
lizards.   (*A. carolinensis* Cuvier.)

298. HOLBROOKIA Girard. (To Dr. John Edwards Holbrook, of Charleston, author of " North American Herpetology," etc.)

591. **H. maculata** Girard. Gray, paler above, with a row of large darker spots on sides ; 1 or 2 black spots on side of belly ; scales nearly smooth ; hind leg not reaching eye. Tenn. to Kan. and S. W.

299. SCELOPORUS Wiegmann. (σκέλος, leg; πόρος, pore.)

592. **S. undulatus** (Daudin). COMMON LIZARD. SWIFT. Greenish, bluish, or bronzed, with black, wavy cross-bands above ; throat and sides of belly in ♂ with brilliant blue and black ; dorsal scales rather large, strongly keeled, mucronate similar to lateral scales ; head shields striated or rugose ; body depressed ; tail slender. L. 7. U.S., in forests and along fences, N. to Mich. ; abundant S. ; varies greatly in color.

300. PHRYNOSOMA Wiegmann. (φρῦνος, toad ; σῶμα, body.)

593. **P. douglassi** (Bell). Ventral scales smooth. Head spines small, shorter than eye ; grayish, with large, dark, pale-edged spots. Kan. to Cal. and S.

594. **P. cornutum** (Harlan). COMMON HORNED TOAD. Ventral scales keeled ; head with very long spines; back with spinous scales; gray, with pale dorsal streak and some dark spots. L. 5. Kan. to Cal. and S. ; common S. W.; a most grotesque little creature ; terrestial. (Lat., horned.)

ORDER XXX. **TESTUDINATA.** (THE TURTLES.)

Reptiles with the body enclosed between 2 more or less developed bony shields, which are usually covered by horny epidermal plates, but sometimes by a leathery skin. Upper shield (carapace) and lower shield (plastron) more or less united along the sides. Neck and tail the only flexible parts of the spinal column; these, together with the legs, usually retractile within the box made by the two shields. The bony part of the carapace is formed by the dorsal and sacral vertebræ, and the ribs co-ossified with a series of overlying bony plates, usually accompanied by a marginal row. The dorsal vertebræ have their ends flattened and immovably united by cartilage, and all of them, except the first and last, have their neural spines flattened horizontally so as to form the median line of plates. On either side of this series is a single row of ossified dermal plates overlying the ribs and corresponding in number to the developed ribs, of which there are usually 8 pairs. No true sternum ; plastron consisting of membrane bones, of which there are usually 9 pieces, — 4 pairs and a single symmetrical median piece. The osseous plates, both above and below, correspond neither in number nor position with the overlying dermal plates.

The skull is more compact than that of the other reptiles. There are no teeth, but the jaws are encased in horny sheaths, usually with sharp cutting edges; the eye is furnished with two lids and a nictitating membrane as in the birds; the tympanic membrane is always present, although sometimes hidden by the skin. Respiration is effected by swallowing air. (Lat., *testudo*, tortoise.)

Families of Testudinata.

a. Limbs developed as paddles, not capable of distinct movements at wrist or ankle-joint; digits flattened, elongated, bound immovably together by the integument. (Sea Turtles.)

   b. Feet scaleless, the anterior very large. . . DERMOCHELYDIDÆ, 117.

   bb. Feet scaly; carapace heart-shaped. . . . . . CHELONIIDÆ, 118.

aa. Limbs not in the form of paddles, capable of movement at wrist and ankle-joints. (Land and pond-turtles.)

   c. Carapace leathery, its margins flexible; no dermal plates; toes 5–5, the claws 3–3; head small, the snout pointed; body very flat.

                                                   TRIONYCHIDÆ, 119.

   cc. Carapace firm, ossified; dermal plates present; claws mostly 5–4.

     d. Fingers and toes spreading, not closely bound together, more than one joint being free.

       e. Tail very long and strong, with a crest of tubercles; plastron narrow and small, cross-shaped, with 9 plates (besides the bridge); head large; body highest in front. . . . CHELYDRIDÆ, 120.

       ee. Tail short, not crested; plastron broad.

         f. Lower jaw ending in a long sharp point; carapace highest behind the middle, its edge not flaring outward; plastron with 9 or 11 plates. . . . . KINOSTERNIDÆ, 121.

         ff. Lower jaw without long point at symphysis; carapace highest at about the middle, its edge flaring outward; plastron with 12 dermal plates. . . . . EMYDIDÆ, 122.

     dd. Fingers and toes bound closely together, only the last joint free; plastron very broad. . . . . TESTUDINIDÆ, 123.

FAMILY CXVII. DERMOCHELYDIDÆ. (THE LEATHER-TURTLES.)

Sea turtles with the body covered by a smooth leathery skin; carapace with several longitudinal ridges with deep grooves between them; body highest in front and widest just before bridge; hind legs much exposed; toes without nails; head short, high, very broad behind; upper jaw with 2 pits and 2 tooth-like projections. One species, widely distributed.

301. DERMOCHELYS Blainville. (δέρμα, skin; χέλυς, tortoise.)

595. **D. coriacea** (Vandelli). TRUNK-BACK. LEATHER-TURTLE. Dark brown. L. 6 to 8 feet. Open sea, N. to Cape Ann. (Lat., leathery.) (*Eu.*)

FAMILY CXVIII. **CHELONIIDÆ.** (LOGGER-HEAD TURTLES.)

Sea turtles, with the carapace covered with bony plates; carapace heart-shaped, broad and flat, highest in front, widest near middle; head large, jaws without tooth-like projections. Genera 4; species about 7, of the open sea, coming to shore only to deposit and bury their eggs.

a. Scales around large median plate on top of head 13 to 20; plates of carapace not imbricate; edge of lower jaw not serrate; costal plates 5 on each side; scales on cheeks small, 15 to 20; head broad. . THALASSOCHELYS, 302.
aa. Scales around vertical plate 7; costal plates 4.
  b. Tomia of lower jaw not serrate; shields of carapace imbricated; scales on cheeks large, 7 to 10; head broad. . . . . . ERETMOCHELYS, 303.
  bb. Tomia of lower jaw serrate; shields of carapace not imbricated; scales on cheeks small, 15 to 20; head high and narrow. . CHELONIA, 304.

302. THALASSOCHELYS Fitzinger. (θάλασσα, sea; χέλυς, tortoise.)

596. **T. caretta** (L.). LOGGER-HEAD TURTLE. Scales not imbricate; 2 nails to each foot. Atlantic, N. to Mass.; reaches 450 lbs. (Eu.) (An old name.)

303. ERETMOCHELYS Fitzinger. (ἐρετμός, oar : χέλυς.)

597. **E. imbricata** (L.). TORTOISE-SHELL TURTLE. HAWKS-BILL TURTLE. Jaws produced in a beak; nails two. N. C. to Brazil. Smaller and fiercer than the preceding, its scales used in making combs.

304. CHELONIA Brongniart. (χελώνη, tortoise.)

598. **C. mydas** (L.). GREEN TURTLE. Plates thin: nail single; body oblong. L. I. to Brazil, herbivorous, reaching 850 lbs., and valued as food. (μυδάω, to be wet.)

FAMILY CXIX. **TRIONYCHIDÆ.** (THE SOFT-SHELLED TURTLES.)

Body flat, nearly orbicular; carapace not completely ossified, the ribs projecting freely towards the outer extremities; marginal ossicles rudimentary; carapace and plastron covered by a thick leathery skin which is flexible at the margins. Head long and pointed, with a long, flexible, tubular, pig-like snout; neck long. Feet broadly webbed; toes long, 5–5, but the claws only 3–3.

Aquatic, carnivorous and voracious; species about 30, in both hemispheres.

a. Nostrils rather under the tip of snout; nasal septum without an internal longitudinal ridge on each side; head narrow; edge of upper jaw serrate behind. . . . . AMYDA, 305.

aa. Nostrils terminal, crescent-shaped; a prominent longitudinal ridge projecting from each side of septum; head broad; edge of upper jaw entire.

ASPIDONECTES, 306.

305. AMYDA Agassiz. (Lat., turtle.)

599. **A. mutica** (Le Sueur). LEATHER-TURTLE. A depression along median line of carapace; no spines nor tubercles along anterior margin nor on back. Olive, young spotted; feet not mottled below. L. 12. Canada to Ohio R., and N. W. (Lat., unarmed.)

306. ASPIDONECTES Wagler. (ἀσπίς, shield; νήκτης, swimmer.)

a. Lower parts of body and feet spotted with dark.

600. **A. spinifer** (Le Sueur). COMMON SOFT-SHELLED TURTLE. Carapace olive brown with dark spots; head and neck olive green with light and dark stripes; legs and feet mottled everywhere with dark; ♂ with the tubercles on the front of the carapace smaller than in the ♀, the body also longer and the tail extending considerably beyond the margin of the carapace. Canada to Ky. and Minn., abundant. (Lat., spine-bearing.)

601. **A. nuchalis** Agassiz. A marked depression on either side of the blunt median keel, which is dilated and triangular anteriorly; spines and tubercles prominent in ♂. Cumberland and Upper Tenn. Rivers.

aa. Lower parts of body and feet white.

602. **A. ferox** Wagler. Tubercles on shell largest in ♂; back blotched in adult; young with black spots and ocelli and with 2 or 3 concentric black marginal lines. S. Ind. to Ga. and La. (Lat., fierce.)

FAMILY CXX. CHELYDRIDÆ. (THE SNAPPING TURTLES.)

Shell high in front, low behind; body heaviest forward; head and neck very large, the snout narrowed forward; jaws strongly hooked, and very powerful; tail long, strong, with a crest of horny, compressed tubercles; plastron small, cross-shaped, with 9 plates besides the very narrow bridge. Claws 5–4, strong, the web small.

Large turtles of great strength and voracity, chiefly aquatic; 2 of the 3 species are American, the third (*Platysternum*) is from China.

a. Eyes close together, partly superior; head covered with soft skin; tail with two rows of moderate scales beneath; ridges of carapace becoming obsolete with age; jaws moderately hooked. . . . . . . CHELYDRA, 307.
aa. Eyes distant, lateral; head very large, covered with smooth, symmetrical plates; tail with many small imbricate scales beneath; carapace with 3 large persistent keels; jaws very strongly hooked.

MACROCHELYS, 308.

307. CHELYDRA Schweigger. (χέλυς, turtle ; ὕδωρ, water.)

603. **C. serpentina** (L.). COMMON SNAPPING TURTLE. Dusky brown ; head with dark spots. L. 25 or more. Canada to Equador, everywhere abundant about water.

308. MACROCHELYS Gray. (μακρός, large ; χέλυς.)

604. **M. lacertina** (Schweigger). ALLIGATOR SNAPPER. Blackish ; head with many fleshy slips. Gulf States, N. to Wis. L. 40 or more ; " perhaps the most ferocious, and, for its size, the strongest of reptiles."

FAMILY CXXI. **KINOSTERNIDÆ.** (THE BOX TURTLES.)

Carapace rather long and narrow, the outline usually rising gradually from the front to a point beyond the centre of the shell, then abruptly descending ; the bulk of the body therefore thrown backward ; margin of the carapace turning downward and inward rather than outward ; plastron proportionally large, covered with 7. 9 or 11 horny plates, the anterior pair coalescing into one ; anterior, and sometimes also posterior lobe of plastron, often movable upon the fixed central portion ; head pointed ; jaws usually strong ; eyes far forward ; limbs slender ; feet short.

Turtles of small size, chiefly American.

a. Plastron with its anterior and posterior lobes nearly equal in length, both freely movable and capable of closing the shell; posterior lobe emarginate behind, its angles rounded ; carapace without traces of keel in adult.
KINOSTERNON, 309.

aa. Plastron with its posterior lobe longer than anterior, truncate behind, its posterior angles not rounded; lobes of plastron little movable, incapable of closing the shell; carapace more or less keeled, at least when young; head very large, with strong jaws. . . . . . . AROMOCHELYS, 310.

309. KINOSTERNON Spix. (κινέω, to move ; στέρνον, breast.)

605. **K. pennsylvanicum** (Bosc). MUD TURTLE. Shell dusky brown ; head dark, with light dots. L. 4. N. Y. to Fla.

310. AROMOCHELYS Gray. (ἄρωμα. odor ; χέλυς.)

606. **A. odoratus** (Latreille). MUSK TURTLE. STINK-POT. Shell dusky, clouded, sometimes spotted ; neck with two yellow stripes, one from above eye, the other from below ; head very large with strong jaws ; carapace with traces of a keel, but the plates not imbricated in the adult ; no point at symphysis of upper jaw ; odor strong, musky. L. 6. E. U. S., abundant, W. to N. Ill. (*Rice & Davis.*)

607. **A. carinatus** Gray. Plates of carapace overlapping more or less, each one edged with black and marked with radiating black stripes ; neck unstriped ; a point at symphysis of upper jaw. La., N. to N. Ill. (*Rice & Davis.*)

FAMILY CXXII. **EMYDIDÆ.** (THE POND TURTLES.)

Carapace ovate, broadest behind, the margin having a tendency to flare outward, highest near the middle and usually not strongly convex ; plastron covering the whole under surface, its plates twelve in number; sometimes the anterior lobe (and rarely the posterior also) movable on a transverse hinge, enabling the animal to completely close the shell. Toes broadly webbed in the aquatic species; scarcely webbed in the others. The pond turtles feed largely upon animals, but they rarely catch active prey. Most of them will not bite except under much provocation. Species about 80, widely distributed, inhabiting marshes, ponds, and the shores of still streams; a few are strictly terrestrial.

a. Plastron without hinge, immovably joined to carapace.
  b. Alveolar surface of jaws broad; carapace depressed; toes short, broadly webbed.
    c. Alveolar surface of jaws smooth, a deep groove in front; upper jaw not notched in front; head covered with soft skin; carapace more or less keeled. . . . . MALACLEMMYS, 311.
    cc. Alveolar surface of upper jaw divided by a longitudinal ridge parallel to margin; upper jaw notched in front; head with thin hard skin; carapace scarcely keeled. . . . . PSEUDEMYS, 312.
  bb. Alveolar surface of jaws narrow.
    d. Carapace depressed (never keeled); toes strong, broadly webbed, the hind feet largest; alveolar groove of jaws well marked, except in front; upper jaw notched in front. . . . . CHRYSEMYS, 313.
    dd. Carapace considerably arched; feet subequal, the toes narrowly webbed. . . . . CHELOPUS, 314.
aa. Plastron with a movable transverse hinge across its middle; a movable cartilaginous lateral suture uniting plastron with carapace.
  e. Body depressed; plastron emarginate behind; toes well webbed.
                                   EMYS, 315.
  ee. Body short and high; plastron rounded or truncate behind; toes scarcely webbed; not aquatic. . . . . . . CISTUDO, 316.

311. MALACLEMMYS Gray. (μαλακός, soft ; κλεμμύς, tortoise.)

a. Lower jaw with a spoon-shaped dilatation at tip ; inland turtles. (*Graptemys* Ag.)
  b. Middle series of plates on carapace scarcely imbricated.

608. **M. geographicus** (Le Sueur). MAP TURTLE. Dark olive brown with greenish and yellow streaks and reticulations, especially distinct on neck, legs, and edges of carapace ; plastron yellowish ; carapace strongly notched behind and usually decidedly keeled. Miss. Valley, E. to N. Y., common W.

  bb. Middle series of dorsal plates distinctly imbricated.

609. **M. lesueuri** (Gray). Similar to the preceding but grayer, the markings on the shell paler, less distinct and in larger pattern;

keel of carapace stronger, each plate of the vertebral series with a blackish projection behind, which is more or less imbricated over the succeeding plate; plastron yellowish, marbled with blackish; head, neck, and legs with bright yellow stripes. Wis. to Ohio and S. W., not rare. (To Charles Albert Le Sueur, artist and naturalist.)

aa. Lower jaw without spoon-shaped dilatation; the cutting edges smooth. Salt-marsh turtles. (*Malaclemmys.*)

610. **M. palustris** (Gmelin). SALT-MARSH TURTLE. DIAMOND-BACK. Greenish or dark olive, rarely black; plates, both of carapace and plastron, usually with concentric dark stripes; shell smooth or with concentric grooves. N. Y. to Texas, along the coast; valued as food.

312 PSEUDEMYS Gray. (ψευδής, false; ἐμύς.)

a. Loose skin between legs without scales; ridge in alveolar surface tuberculate; young marked with confluent, lozenge-shaped figures.
  b. Jaws coarsely serrated; symphysis of upper with prominent hook.

611. **P. rugosa** (Shaw). RED-BELLIED TERRAPIN. Dusky, with irregular red markings above ; marginal plates with much red; plastron red or partly yellowish ; head and neck brown, with reddish lines: variable. N. J. to Va. (Lat., wrinkled.)

  bb. Jaws not serrated.

612. **P. hieroglyphica** (Holbrook). Shell smooth, depressed ; olive brown, variously marked with reticulated or concentric yellowish lines; plastron yellowish ; head and neck with yellow lines; head small. N. Y. to Wis. and S.

aa. Loose skin between legs not scaly; ridge in alveolar surface not tubercular; edge of marginal plates notched. Vertebral plates with lengthwise bands, other scales with transverse bands, these growing obscure with age.
  c. Carapace not keeled.

613. **P troosti** (Holbrook). YELLOW-BELLIED TERRAPIN. Greenish-black, lateral plates with horn-colored lines and spots; plastron dull yellow, with large, black blotches : throat with greenish stripes; shell never keeled. Miss. Valley, N. to Ill.

614. **P. elegans** (Wied). Brown, with yellowish wavy lines and blotches; a red or yellow band on each side of neck ; plastron yellow with a dusky blotch on each plate. Ill. to Idaho, and S.

  cc. Carapace strongly keeled.

615. **P. scabra** (L.). Dark brown, with irregular yellow stripes; plastron yellow with small black blotches in front : head and neck black, with yellow lines; carapace wrinkled. Va. to Fla. (Lat., rough.)

14

313. CHRYSEMYS Gray. (χρυσός, gold; ἐμύς.)

616. **C. picta** (Hermann). PAINTED TURTLE. MUD TURTLE. Greenish black ; plates margined with paler; marginal plates marked with bright red; plastron yellow, often blotched with brown. L. 8. E. U. S., one of the most common turtles. (Lat., painted) Westward it gives place to —

617. **C. marginata** (Agassiz). Plates of carapace alternating or in quincunx, the lateral rows out of line with the middle one, instead of forming sets of three as in the eastern form ; lateral plates with strong concentric striæ. W. N. Y. and W., common. Perhaps a variety of the preceding, but I have seen no intergradations.

618. **C. oregonensis** Holbrook. No red markings. Minn. to Ore.

314. CHELOPUS Rafinesque. (χέλυς, tortoise ; πούς, foot.)

a. Carapace usually more or less keeled; upper jaw deeply notched and arched downward. (*Chelopus.*)

b. Head not notably narrower below than above.

619. **C. muhlenbergi** (Schweigger). Brown with yellowish markings; plastron black with yellowish blotches; an orange spot on each side of neck; plates of back plain or concentrically grooved. L. 4½. E. Penn. and N. J.

bb. Head decidedly narrower below than above.

620. **C. insculptus** (Le Conte). WOOD TORTOISE. Shell carinated, its plates marked with concentric striæ and radiating black lines; reddish brown; plastron with a black blotch on each plate. L. 8. E. U. S., E. of Ohio, in woods and fields. (Lat., engraved.)

aa. Carapace not keeled; upper jaw slightly notched, its edge nearly straight. (*Nanemys* Agassiz.)

621. **C. guttatus** (Schneider). SPECKLED TORTOISE. Black, with round orange spots, these spots rarely obsolete ; plastron yellow, blotched with black. L. 4½. E. U. S., W. to N. Ind. (*Levette*), abundant E. (Lat., spotted.)

315. EMYS Brongniart. (ἐμύς, a mud turtle.)

622. **E. meleagris** (Shaw). Black ; usually with yellowish spots; plastron yellowish with black blotches; head with yellow spots ; young nearly circular, and black. L. 8. N. Y. to Wis., scarce. (μελεαγρίς, guinea-fowl.)

316. CISTUDO Fleming. (Lat., *cista*, box.)

623. **C. carolina** (L.). COMMON BOX TURTLE. Colors very variable, chiefly blackish and yellowish ; no two alike in pattern ; iris red in ♂ ; hind feet with 4 toes; young keeled, the keel grow-

ing obscure with age. N. Y. to Mo. and S., in dry woods. Represented S. by var. **triunguis** (Agassiz). Hind feet mostly 3-toed; color pale yellowish, with few spots. Southern, N. to Penn.

624. **C. ornata** Agassiz. "Shell round, broad, flat, without keel, even when young." Iowa and W.

FAMILY CXXIII. TESTUDINIDÆ. (THE LAND TORTOISES.)

Carapace strong, thick, ovate, generally very convex and falling off abruptly at both ends; caudal shields united into one; plastron very broad, covering the whole under surface, the anterior part sometimes movable on a transverse hinge. Legs and feet club-shaped; toes firmly bound together by the integument, only the blunt claws being exserted.

Herbivorous Turtles, entirely terrestrial, inhabiting the warmer parts of both continents; about 20 species are known.

317. **XEROBATES** Agassiz. (ξηρός, dry ; βάτης, walker.)

625. **X. polyphemus** (Daudin.) "GOPHER TURTLE." Brownish, head almost black; yellow below; fore limbs large and strong; hinder short, rounded; plastron projecting forward beyond carapace. L. 15. S. States, N. to N. C., in pine barrens; herbivorous and gregarious ; burrows in the ground like a wood-chuck. (πολύφημος, croaking.)

Passing over the order CROCODILIA, the highest in development among the recent reptiles, an order having no representatives within our limits, we take up next a group originally an offshoot from the Reptilian series, but now, if only living forms were taken into consideration, one of the most sharply defined of the classes of Vertebrata, the Birds.

Class II.—AVES. (The Birds.)

A Bird may be defined as an air-breathing vertebrate with a covering of feathers; warm blood; a complete double circulation; the two anterior limbs (wings) adapted for flying or swimming, the two posterior limbs (legs) adapted for walking or swimming; respiration never effected by gills or branchiæ, but, after leaving the egg, by lungs, which are connected with air cavities in various parts of the body. Reproduction by eggs, which are fertilized within the body and hatched externally, either by incubation or exposure to the heat of the sun; the shell calcareous, hard and brittle.

Much more might be added, but the obvious character is this: *All Birds have feathers, and no other animal has feathers*, or, as Stejneger puts it, " A bird is known by its feathers." There is probably no other character of importance which distinguishes birds living and extinct as a whole, from the *Reptilia*.

The classification of this group, as of most others, is still in an unsettled condition. Strictly speaking, the existing members of the class are so closely related that they might, with propriety, be combined into one order, which, by Professor Gill, has been named EURHIPIDURÆ. At present, however, the term " order " may be applied to the groups so designated below, without thereby implying any structural differences such as separate the " orders " of Reptiles or even of Fishes. The *Eurhipiduræ* are made a sub-class by Stejneger, while Coues divides them into two "sub-classes," the *Ratitæ* (Ostriches, etc.), and the *Carinatæ*. To the *Carinatæ*, characterized by the keeled sternum and more or less developed wings, all American birds belong. (Lat., *avis*, bird.)

The " orders " of the Carinate Birds, as now adopted, are rather temporary, pending investigation of certain groups. They are also in a degree conventional, some of them being admittedly unnatural in their composition, while none of them represent any such structural differences or differences of such long standing in time as those which characterize the orders of Mammals or Reptiles, or most of the orders of Fishes. For reasons which have been elsewhere given, I follow in this work without exception the classification, sequence, and nomenclature adopted by the American Ornithologists' Union. A system in some respects more in accord with modern investigations is outlined by Stejneger in the Standard Natural History. The following largely artificial key to the Orders recognized by the Am. Orn. Assoc. is partly arranged from the key given by Ridgway (Man. N. Am. Birds).

Orders of Aves.

A. Sternum keeled: (Carinat.e).

a. Feet totipalmate, the hind toe well developed and all four toes full-webbed; palate desmognathous. . . . . Steganopodes, XXXIV.
aa. Feet not totipalmate, the hind toe, if present, not connected with the others.

   b. Feet palmate, the anterior toes full-webbed or nearly so (or lobate, with the claws broad, nail-like) (tarsus not specially elongate, and the tibia little if any naked below, in our species).

      c. Bill not lamellate, its cutting edge entire; schizognathous. (*Cecomorphæ* Huxley, Stejneger.)

         d. Legs inserted far behind the middle of the body, which in standing position is more or less erect. . . . . . Pygopodes, XXXI.
         dd. Legs not inserted far behind the middle of the body, which in a standing position has its axis nearly horizontal; wings usually very long.

            e. Nostrils not tubular. . . . . Longipennes, XXXII.
            ee. Nostrils tubular. . . . . Tubinares, XXXIII.
      cc. Bill lamellate, its cutting edges serrated or fringed.

                            Anseres, XXXV.

  bb. Feet not palmate (with rare exceptions), the toes cleft, or webbed at base or on sides (full-webbed only in a few waders, with very long tarsus and the tibia partly naked).

     f. Waders, tibia usually more or less naked below; the tarsus more or less elongate.

        g. Hind toes well developed and usually inserted on same level as anterior toes, the claws never excessively lengthened; loral or orbital regions or both naked; desmognathous.

                      Herodiones, XXXVI.

        gg. Hind toe, if present, small and inserted above level of the the rest (or else size moderate). L. less than 36 inches; the loral and orbital regions feathered, the middle claw not pectinate); schizognathous. (*Grallæ.*)

           h. Hind toe short and elevated (or if the bird is less than 3 feet long, the hind toe almost on the level of anterior toes). . . . . Paludicolæ, XXXVII.
           hh. Hind toe (if present) short and distinctly elevated (length never more than 2 feet). . . Limicolæ, XXXVIII.

    ff. Not waders; tibia mostly entirely feathered; tarsus not greatly elongate.

        i. Bill strongly hooked, with a distinct cere at base; desmognathous.

           j. Toes 3 in front; 1 behind, the outer toe sometimes reversible. . . . . . . Raptores, XLI.
           jj. Toes 2 in front; 2 behind, the outer toe permanently reversed. . . . . . Psittaci, XLII.

ii. Bill not both strongly hooked and cered.

   k. Hind toe short, decidedly elevated; toes semipalmate; no soft membrane about nostrils; schizognathous. . . . . . GALLINÆ, XXXIX.

   kk. Hind toe little if at all above level of the rest (rarely absent).

     l. Nostrils opening beneath a soft swollen cere like membrane; hind claw short; doves. . . . . COLUMBÆ, XL.

     ll. Nostrils not opening beneath a swollen membrane or cere.

       m. Hind claw not longer than the others; mostly desmognathous. (*Picariæ.*)

         n. Wings not very long; gape not very wide nor deeply cleft. Feet zygodactyle or syndactyle. (Toes 2 in front, or if 3, then the outer and middle toes connected for at least half their length in our species.)

          o. Tail feathers soft; bill not chisel-like. . . COCCYGES, XLIII.

          oo. Tail feathers stiff and pointed; bill adapted for striking or boring. . . . . PICI, XLIV.

        nn. Wings very long, with 10 primaries (tail of 10 feathers and bill fissirostral, or else secondaries 6 and bill tenuirostral); toes 3 in front, 1 behind, the hinder a little elevated.

                            MACROCHIRES, XLV.

      mm. Hind claw at least as long as middle claw; toes always 3 in front, 1 behind, cleft to the base or with the basal joints only immovably coherent; palate ægithognathous. . . PASSERES, XLVI.

ORDER XXXI. **PYGOPODES.** (THE DIVING BIRDS.)

    Feet palmate or lobate; tibia feathered, included in the skin nearly to the heel-joint, hence the legs set far back, so that the birds are scarcely able to walk at all on land; hind toe small and elevated, or wanting; nostrils developed; bill horny, not lamellate or serrate; no gular pouch; palate schizognathous; wings very short; tail very short or rudimentary.

    This is apparently not a natural order. Stejneger (following Huxley) unites the *Pygopodes, Longipennes* and *Tubinares* in one order, *Cecomorphæ.* He remarks : " The fact is that not only are the gulls very nearly related to the auks, but their affinities with the *Grallæ* through the plovers are unmistakable. On the other hand, the grebes seem to be only distantly related to the other ' *Pygopodes* ' and the puffins and albatrosses similarly so to the *Longipennes* or gulls." The *Pygopodes* are water birds, expert divers, feeding chiefly on fishes. (πυγή, rump; πούς, foot.)

Families of Pygopodes.

a. Tail feathers wanting; anterior toes lobed, the claws very broad, flat, rounded at tip, resembling human nails. . . . . PODICIPIDÆ, 124.

aa. Tail feathers developed, but short.

   b. Toes 4; the hind toe present. . . . . URINATORIDÆ, 125.

   bb. Toes 3, the hind toe wanting. . . . . ALCIDÆ, 126.

Family CXXIV. PODICIPIDÆ. (The Grebes.)

Bill slender, or stout; lores naked; head often with crests, ruffs or ear tufts in the breeding season. Under plumage dense, lustrous, mostly white. Wings very short; tail rudimentary, without distinct quills. Feet four-toed, lobate, the toes webbed at base; toes flattened, provided with flat claws resembling human nails; tarsus scutellate, compressed. Genera 2; species about 20; in all parts of the world, chiefly about fresh waters; nest usually a floating mat of rushes.

a. Bill slender, straight, rather acute; its length more than twice its depth at base ; head in breeding season with conspicuous crests or ruff.
  b. Neck nearly as long as body. . . . . ÆCHMOPHORUS, 318.
  bb. Neck much shorter than body. . . . . COLYMBUS, 319.
aa. Bill stout, somewhat hooked; its length not quite twice its greatest depth ; no ruff nor crest. . . . . PODILYMBUS, 320.

318. ÆCHMOPHORUS Coues. (αἰχμή, spear; φόρος, bearing.)

626. Æ. occidentalis (Lawrence). WESTERN GREBE. Slate-color; satin-white below. L. 26. W. 8. B. 2¾. Minn. to Mexico and W.

319. COLYMBUS Linnæus. (κόλυμβος, diver.)

a. Bill about as long as head. (Colymbus.)

627. C. holbölli (Reinhardt). RED-NECKED GREBE. Upper parts brown; front and sides of neck brownish red; head ashy gray, its top blackish; crests and ruffs not large; below silvery, the feathers gray within. L. 18. W. 8. B. 2. N. Am., U. S. in winter. (To the Danish naturalist, C. Holböll.)

aa. Bill much shorter than head. (Dytes Kaup.)

628. C. auritus L. HORNED GREBE. Dark brown; head glossy black in ♂; a buffy patch above and behind eye; fore-neck, breast and sides brownish red; bill compressed, deeper than wide at base, black, tipped with yellow: ruffs very large, in dense tufts. L. 14. W. 6. B. 3¾. N. U. S. and N. (Eu.) (Lat., having large ears.)

629. C. nigricollis (Brehm). EARED GREBE. Crest of ♂ in the form of a fan-shaped patch; head, throat and breast black in ♂; bill depressed, wider than deep at base. L. 13. W. 5¼. Northern regions, the American var. californicus Heermann (W. U. S., E. to Ill.), with inner webs of inner quills dusky. (Eu.) (Lat., niger, black ; collum, neck.)

320. PODILYMBUS Lesson. (Podiceps-Colymbus.)

630. P. podiceps (L.). DIEDAPPER. DAB-CHICK. HELL-DIVER. WATER WITCH. PIED-BILLED GREBE. Chiefly brownish gray; silvery ash below, spotted with dusky; chin and throat black; bill

bluish, with dark band; young and winter plumage different, the
bill unmarked. but the bird resembles nothing else. L. 14. W. 5.
B. 1. Whole of America, abundant. (Lat., *podex*, rump; *pes*, foot.)

FAMILY CXXV. **URINATORIDÆ.** (THE LOONS.)

Bill long, strong, tapering, acute, wholly hard; nostrils linear.
Head densely and evenly feathered, without ruffs or naked spaces;
eye large. Feet 4-toed, palmate; tarsus reticulate, strongly compressed. Wings comparatively long and strong; tail short but
well developed. Precocial. Genus 1; species 5. Birds of large
size, with strong powers of flight, and pre-eminent in swimming and
diving, but scarcely able to walk; they are migratory, breeding northward, but coming S. in winter; the voice is singularly sharp and wild.

321. URINATOR Cuvier.

a. Tarsus shorter than middle toe, without claw.

\ 631. **U. imber** (Gunner). COMMON LOON. DIVER. Black;
breast and below chiefly white; head and neck iridescent, black in
summer; a patch of white streaks on each side of neck and on the
throat; back with many white spots; ♀ duller, brownish above,
without the head markings. L. 28 to 36. W. 14. Ts. 3. B. 3.
Northern Hemisphere; whole U. S. in winter. (*Eu.*) (Norwegian
name.)

632. **U. arcticus** (L.). BLACK-THROATED LOON. Similar, but
head and neck behind bluish or hoary gray; foreneck purplish
black, with a crescent of white streaks; ♀ duller. L. 28. W. 12¼.
B. 2¼. Northern hemisphere, not common in U. S. (*Eu.*)

aa. Tarsus longer than middle toe with claw.

633. **U. lumme** (Gunner.) RED-THROATED LOON. Blackish,
streaked on neck, chiefly white below; head and neck mostly bluish
gray; throat with a large chestnut patch in summer; ♀ duller.
L. 27. W. 11. B. 2. Northern regions, U. S. in winter. (*Eu.*)
(Norwegian name.)

FAMILY CXXVI. **ALCIDÆ.** (THE AUKS.)

Feet palmate, three-toed; tarsus reticulate or partly scutellate;
suffrago naked; claws ordinary; bill and nostrils various; tail perfect, of few feathers; lores feathered; legs variable, set far back;
color variable, the head often with curly crests; altricial; eggs few.
Genera 12; species about 35, living about rocks on rugged shores
in Northern regions. Most of them fly well and all swim on or
under water with equal ease. They feed chiefly on fishes.

a. Inner claw much larger and more curved than the others; corner of mouth
with a "rosette" of thick naked skin; bill greatly compressed, almost as
deep as long. (*Fraterculinæ.*)

b. Eyelids with deciduous appendages; no crests; culmen with one curve; covering of bill moulted in 7 to 9 pieces. . . . FRATERCULA, 322.

aa. Inner claw similar in size and form to the others; no rosette at corner of mouth.

  b. Bill not very short, the angle of chin much nearer nostril than tip of bill.

    c. Nostril exposed, overhung by a horny scale (*Phalerinæ*); culmen straight till near tip, then abruptly decurved. . CEPPHUS, 323.

    cc. Nostril more or less completely concealed by dense velvety feathers. (*Alcinæ*.)

      d. Bill narrow; neither mandible grooved : tail rounded, the feathers not pointed. . . . . URIA, 324.

      dd. Bill very deep, much compressed, one or both mandibles grooved in adult; tail graduated, its feathers pointed.

        e. Wings well developed; bill (less than 1½) much shorter than head. ALCA, 325.

        ee. Wings rudimentary, incapable of flight; bill as long as head (about 3). . . . . PLAUTUS, 326.

bb. Bill very short and broad, the angle of chin nearer to tip of bill than to nostril; culmen curved. (*Allinæ*). . . . . ALLE, 327.

322. FRATERCULA Brisson. (Lat., *fraterculus*, little brother.)

634. **F. arctica** (L.). PUFFIN. Grooves of bill oblique, broad and distinct. Blackish above ; a black band across fore-neck ; white below. L. 13. B. 1¾. Arctic, S. to N. J.

323. CEPPHUS Pallas. (κέπφος, a kind of petrel.)

a. Greater wing-coverts white to their extreme base.

635. **C. mandti** (Lichtenstein.) Blackish, with large white wing-patch; nearly white in winter. L. 13. W. 7. B. 1. Arctic, S. to N. J. (*Eu.*)

aa. Greater wing-coverts with at least their basal half black.

✗636. **C. grylle** (L.). BLACK GUILLEMOT. Bill larger than in C. *mandti* ; size and colors similar. Arctic, S. to N. J. (*Eu.*) (Scandinavian name.)

324. URIA Brisson. (οὐρία, a water-bird.)

a. Depth of bill at angle of mouth less than ⅓ culmen.

637. **U. troile** (L.). MURRE. GUILLEMOT. Dusky ; secondaries white-tipped; basal part of upper tomium dusky. L. 17. W. 8. B. 1¾. N. Atl., S. to Mass. (*Eu.*) ("Possibly to Troil, the Icelander.")

aa. Depth of bill at angle more than ⅓ culmen.

638. **U. lomvia** (L.). THICK-BILLED MURRE. Dusky ; secondaries white-tipped ; basal part of upper tomium thickened and light-colored in adult. L. 16. W. 8¼. B. 1½. Arctic, S. to N. J. (*Eu.*) (Swedish name.)

325. ALCA Linnæus. (Lat., from *alk* or *auk*.)

639. **A. torda** L. RAZOR-BILLED AUK. Black; lower parts and tips of secondaries white; snuffy-brown in summer. L. 16½.

W. 8¼. T. 3½. B. 1¼. N. Atl., S. to Conn. (*Eu.*) (An old name.)

326. PLAUTUS Brünnich. (πλώς, swimmer.)

640. **P. impennis** (L.). GREAT AUK. Black above; lower parts and tips of secondaries white. L. 29. W. 5¾. B. 3¼. N. Atl., formerly S. to Mass.; now wholly extinct. "His GRACE, the AUK. who lost the use of his wings, and perished off the earth in consequence." (*Coues.*) (Lat., wingless.)

327. ALLE Link. (Swedish name.)

641. **A. alle** (L.). DOVEKIE. Upper parts black, the second-aries tipped with white ; sooty-brown in summer. L. 8. W. 4¾. B. ½. Arctic, S. to N. J. (*Eu.*)

ORDER XXXII. **LONGIPENNES.** (THE LONG-WINGED SWIMMERS.)

Feet palmate; tibia feathered; legs inserted near the centre of equilibrium so that the bird stands with the axis of the body nearly horizontal; hind toe small and elevated, sometimes wanting. Bill usually long, horny, not serrate nor lamellate; nostrils developed, not tubular: no gular pouch. Wings very long and pointed; tail well developed. Palate schizognathous. Altricial. Water birds, of great powers of flight. feeding chiefly on fishes.

a. Bill with the lower mandible not produced nor specially compressed.
  b. Covering of upper mandible in three parts, a hook at tip, a sort of cere overhanging nostrils, and a lateral piece. . . STERCORARIIDÆ, 127.
  bb. Covering of upper mandible of a single piece pierced by the nostrils.'
                                         LARIDÆ, 128.
aa. Bill with the lower mandible much longer than upper, the terminal part of both mandibles much compressed, like a knife-blade.
                                      RHYNCHOPIDÆ, 129.

FAMILY CXXVII. **STERCORARIIDÆ.** (THE JÆGERS.)

Gull-like birds, with the bill hooked and "cered;" tail square, with the middle pair of feathers long-exserted; tibia naked below; tarsus scutellate in front, granular behind. Two genera with 4 species, "marine Raptores," large, vigorous, rapacious, living by robbing the terns and smaller gulls.

a. Depth of bill at base at least half length of upper mandible (measured along side); tarsus shorter than middle toe and claw; tail short, nearly even.
                          MEGALESTRIS, 328.
aa. Depth of bill at base not half length of upper mandible; tarsus longer than middle toe and claw; middle tail feathers (in adult) very long.
                          STERCORARIUS, 329.

328. MEGALESTRIS Bonaparte. (μέγας, large; λῃστρίς, pirate.)

642. **M. skua** (Brünnich). SKUA GULL. Grayish-brown, L. 22. W. 16. B. 2 N. Atl., rarely S. to Mass. (*Eu.*) (Færoëse name.)

329. STERCORARIUS Brisson. (Lat., scavenger.)

643. **S. pomarinus** (Temminck). POMARINE JÆGER. Chiefly blackish, colors varying with age; middle tail feathers broad to the tip, projecting about 4 inches. L. 20. W. 15. T. 9. B. 1½. Arctic, S. in winter to N. J. (*Eu.*) (πῶμα, flap; ῥίς, nose.)

644. **S. parasiticus** (L.). PARASITIC JÆGER. Dark brown; middle tail feathers acuminate, projecting 4 inches. L. 20. W. 13. T. 5½. B. 1¼. Arctic, S. in winter to N. Y. (*Eu.*)

645. **S. longicaudus** (Vieillot). LONG-TAILED JÆGER. Sooty black; tail feathers filamentous, projecting 8 or 10 inches. L. 22. W. 12. T. 13. B. 1¼. Arctic, S. in winter. (*Eu.*)

FAMILY CXXVIII. **LARIDÆ.** (THE GULLS.)

Long-winged swimmers, with the nostrils not tubular. Bill usually long, horny, not serrate nor lamellate; nostrils developed; no gular pouch. Feet palmate; tibia feathered; legs near centre of equilibrium; hind toe elevated, small, often wanting. Wings very long and pointed. Tail well developed. General color usually white, with a darker mantle of a pearly bluish tint, and commonly with some black markings. Sexes alike in color, but the plumage varying much with age and season. Genera about 12; species 90; abounding about all large bodies of water, and of remarkable power of flight. Altricial; food chiefly fishes.

a. Bill more or less hooked; (general color chiefly white, with a darker, bluish, grayish, or slaty mantle); gulls. (*Larinæ.*)
  b. Tarsus rough or serrate behind; tail even. . . . . . GAVIA, 330.
  bb. Tarsus nearly entire behind.
    c. Hind toe rudimentary or wanting, with minute claw or none; tail slightly emarginate. . . . . RISSA, 331.
    cc. Hind toe small, but with a perfect claw.
      d. Tail even. . . . . LARUS, 332.
      dd. Tail forked. . . . . XEMA, 333.
aa. Bill not hooked, the mandibles even; tail deeply forked (in our species). (*Sterninæ.*)
  e. Tail much more than ½ wing, its outer feathers narrow and pointed; toes well webbed. (Color chiefly white, with a black cap in full plumage, and the quills dusky with a long white stripe.)
    f. Bill stout, its depth at base equal to ½ culmen. GELOCHELIDON, 334.
    ff. Bill slender, its depth at base not ½ its length. . . . STERNA, 335.
  ee. Tail little more than ½ wing, its outer feathers broad and rounded; toes scant-webbed; colors dark. . . . . HYDROCHELIDON, 336.

330. GAVIA Boie.

646. **G. alba** (Gunner). IVORY GULL. Adults pure white; young spotted; feet black. L. 16 to 20. W. 13. B. 1½. Arctic, rarely to U. S., in winter. (*Eu.*)

331. RISSA Leach.   (Icelandic name.)

647. R. tridactyla (L.). KITTIWAKE GULL. Mantle bluish-
gray; head, etc., white; hind claw a minute knob, sometimes absent. L. 16 to 18. W. 12. Arctic, S. in winter to N. Y. (*Eu.*)
(Lat., *tres*, three; *dactylus*, digit.)

332. LARUS Linnæus.   (λάρος, gull.)

a. Head entirely white in adult in summer (young more or less dusky on
   head, etc.); mantle grayish blue, or dusky; lower parts white. (*Larus.*)
b. Primaries, without any black, pearly-gray, whitish at tip.

648. L. glaucus Brünnich. ICE GULL. BURGOMASTER. Bill
yellow with red spot on lower mandible; large. L. 30. W. 18.
T. 8. B. 2½. Arctic regions; S. in winter to N. Y. (*Eu.*)
(γλαυκός, bluish.)

649. L. leucopterus Faber. ICELAND GULL. Similar but
smaller. L. 25. W. 15½. T. 6½. B. 1⅝. Same region. (*Eu.*)
(λευκός, white; πτερόν, wing.)

bb. Primaries with white and dusky (sometimes all black in young).
c. Dark spaces on primaries gray.

650. L. kumlieni Brewster. Similar to *L. leucopterus.* L. 24.
W. 16. B. 1¾. Greenland to N. Y. (To Ludwig Kumlien.)

cc. Dark spaces on primaries black.
d. Shafts of primaries white throughout.

651. L. marinus L. GREAT BLACK-BACKED GULL. Mantle
blackish slate color; largest of our gulls. L. 30. W. 18. B. 2¼.
Feet flesh-colored. N. Atl., S. in winter to N. Y. (*Eu.*)

dd. Shafts of primaries black in the black markings.

652. L. argentatus Brünnich. HERRING GULL. COMMON GULL.
Mantle pearly-gray; bill plain. L. 25. W. 17. B. 2¼. Feet flesh-
colored. Northern regions, abundant on all bodies of water. The
American form (var. *smithsonianus* Coues) has the white of outer
quill separated from the rest by a band of black. (*Eu.*)

653. L. delawarensis Ord. RING-BILLED GULL. Mantle
pearly-gray; feet yellowish; bill yellowish, a black band at the
tip in adult; smaller. L. 20. W. 15. B. 1⅝. N. Am., abundant,
S. to Mex.

aa. Head black or dusky in adult in summer (more or less pale in young);
   mantle gray; lower parts, etc., white, rosy in breeding season. (*Chræ-
   cocephalus* Eyton.)
e. Tarsus much longer than middle toe and claw.

654. L. atricilla (L.). BLACK-HEADED OR LAUGHING GULL.
Bill and feet dusky; reddish in summer. L. 15 to 17. W. 13. T.
5. B. 1¾. E. U. S., coastwise. (Lat., *ater*, black; *cilla*, tail.)

ee. Tarsus not longer than middle toe and claw.

655. **L. franklini** Swainson & Richardson. FRANKLIN'S ROSY
GULL. Bill and feet carmine; bill usually with a dark band near
tip; medium. L. 14½. W. 11. B. 1⅛. U. S., chiefly W. of the
Miss. R. (To Sir John Franklin.)

656. **L. philadelphia** Ord. BONAPARTE'S GULL. Bill black,
slender, tern-like; small. L. 13. W. 10. B. 1⅛. N. Am., abundant.

333. XEMA Leach. (A coined word.)

657. **X. sabinei** (Sabine). FORKED-TAIL GULL. Largely white,
a black hood and collar. L. 14. W. 11. Arctic, S. in winter
to N. Y. (To Edward Sabine.)

334. GELOCHELIDON Brehm. (γελάω, to laugh ; χελιδών, swallow.)

658. **G. nilotica** (Hasselquist). GULL-BILLED TERN. Bill
black, very short and stout; head black ; mantle pearly-gray. L.
15. W. 12. Atlantic, N. to Mass. (Eu.)

335. STERNA Linnæus. (Eng. *tern,* or *sterne.*)

a. Wing more than 9.
 b. Wing more than 12.
  c. Tail much less than half wing, not deeply forked; occipital feathers
   short. (*Thalasseus* Boie.)

659. **S. tschegrava** Lepechin. CASPIAN TERN. Primaries without white band; bill red. L. 22. W. 17. T. 6. B. 3. Northern
regions; scarce in Amer., much the largest of the terns. (Eu.)

   cc. Tail more than one half wing, forked half its length ; occiput crested.
   (*Actochelidon* Kaup.)

660. **S. maxima** Boddaert. ROYAL TERN. Bill orange. L.
18 to 21. W. 15. T. 8. B. 2½. U. S.

661. **S. sandvicensis** Latham. SANDWICH TERN. Bill black,
yellow at tip. L. 16. W. 12½. T. 6. B. 2¼. Atlantic, N. to
Mass., rare. (Eu.) Ours is var. **acuflavidus** Cabot.

 bb. Wing less than 12 ; tail deeply forked; no crest; mantle bluish-gray,
   the tail chiefly white; inner webs of quills largely white. (*Sterna.*)
  d. Top of head black in summer.
   e. Outer tail-feather with the inner web dusky, the outer web white.

662. **S. forsteri** Nuttall. FORSTER'S TERN. Larger than next,
tail longer and wings shorter; bill and feet orange in adult. L. 15.
W. 10. T. 7. B. 1¼. N. Am., common. (To John Reinhold
Forster.)

   ee. Outer tail-feather with inner web white; outer web dusky.

663. **S. hirundo** L. COMMON TERN. Bill red, blackening towards tip; feet orange. L. 14½ (13 to 16). W. 10 (9½ to 11¾).
T. 6 (5 to 7). B. 1¼. Atlantic coasts, abundant. (Eu.) (Lat.,
hirundo, swallow.)

664. **S. paradisæa** Brünnich. ARCTIC TERN. Bill carmine
throughout; plumage as in *hirundo,* but darker below. L. 14 to
17. W. 10 to 11. T. 7 to 8. B. 1½. Smaller than *hirundo,* but
tail proportionally much longer. Arctic, S. to N. Y. (*Eu.*)

eee. Outer tail-feather with both webs white.

665. **S. dougalli** Montagu. ROSEATE TERN. Bill black, usually orange at base below; mantle very pale; rosy-tinted below in
breeding season. L. 14 to 17. W. 9½. T. 5 to 8. B. 1½. Atlantic coast. (*Eu.*) (To Dr. McDougall, of Scotland.)

aa. Wing less than 7; tail deeply forked, about half wing. (*Sternula* Boie.)

666. **S. antillarum** Lesson. LEAST TERN. Bill yellow, usually
tipped with black; a white frontal crescent between cap and bill;
shafts of two or more outer primaries black above; mantle pale
gray; very small. L. 8 or 9. W. 6½. T. 3½. B. 1½. E. U. S.,
chiefly abundant coastwise.

336. **HYDROCHELIDON** Boie. (ὕδωρ, water; χελιδών, swallow.)

667. **H. nigra** (L.). BLACK TERN. Head, neck and under parts
black (in full plumage); wings and tail above dark like the back;
crissum white. L. 10. W. 8. T. 3½. B. 1$\frac{7}{16}$. N. Am., chiefly
inland. (*Eu.*) The American var. *surinamensis* (Gmelin) is
darker than the European form.

FAMILY CXXIX. **RHYNCHOPIDÆ.** (THE SKIMMERS.)

Gulls with the lower mandible much longer than the upper, compressed like a knife-blade; its two sides completely soldered together; the upper edge as sharp as the lower, and fitting in a
groove in upper mandible; tip of bill obtuse; upper jaw compressed, movable at base; tongue very short, stumpy. Wings very
long. Otherwise similar to the terns. One genus, with 3 species.

337. **RHYNCHOPS** Linnæus. (ῥύγχος, beak; ὤψ, face.)

668. **R. nigra** L. BLACK SKIMMER. CUTWATER. Glossy
black; white below; lower mandible about an inch longer than
upper. L. 17 to 20. W. 15. T. 5, sharply forked. B. 2½.
Tropical Amer., N. to N. J., abundant southward.

ORDER XXXIII. **TUBINARES.** (THE TUBE-NOSED
SWIMMERS.)

Nostrils tubular; bill with the upper mandible hooked, its covering composed of several pieces separated by deep grooves. Otherwise essentially like the *Longipennes* so far as external characters
are concerned.

a. Nostrils united in a double tube, placed horizontally on the culmen.

PROCELLARIID.E, 130.

Family CXXX. PROCELLARIIDÆ. (The Petrels.)

Nostrils tubular, united together in a double tube placed horizontally. Bill hooked at tip, its covering not continuous, consisting of several horny pieces separated by deep grooves; hind toe minute or absent. Wings long and pointed; tail moderate; feet short, the front toes full-webbed. Plumage compact and oily, not varying much with sex, age or season. Gregarious sea-birds, mostly silent, with remarkable powers of flight, rarely landing except to lay their eggs. Genera about 12; species about 70. Closely allied to the Petrels are the Albatrosses (*Diomedeidæ*), huge sea-birds with the nostrils disconnected, not united in a horizontal " double-barrelled tube." These families together constitute the order or suborder of " *Tubinares.*"

a. Secondaries 13 or more in number.
  b. Wing long (more than 7).
    c. Partition between nostrils very thin (much narrower than nostril).
      d. Gonys very slightly if at all concave, shorter than nasal tubes.
                      Fulmarus, 338.
      dd. Gonys very strongly concave, longer than nasal tubes.
                      Estrelata, 340.
    cc. Partition between nostrils very thick, as wide as nostril: nostrils visible from above. . . . . Puffinus, 339.
  bb. Wings shorter (less than 7).
    e. Tail even or slightly rounded. . . . . . . Procellaria, 341.
    ee. Tail slightly forked. . . . . Oceanodroma, 342.
aa. Secondaries 10; tarsus not scutellate; legs long; claws narrow, pointed.
                      Oceanites, 343.

338. FULMARUS Leach. (Eng. *fulmar.*)

669. F. glacialis (L.). Fulmar. Bill stout, nearly half as deep as long; nasal tubes dusky. Color bluish gray or dusky. L. 18. W. 12. B. 1¼. The American bird (var. **minor** Kjærb.) considerably smaller. N. Atl., S. to Mass. (*Eu.*)

339. PUFFINUS Brisson. (Eng., *puffin.*)

a. Dusky above; white below.
  b. Wing more than 12.
670. P. borealis Cory. White of throat shading gradually into dusky of head and neck. L. 21. W. 14. B. 2¼. Off Mass.

671. P. major Faber. Greater Shearwater. White of throat separated rather abruptly from dusky of head and neck; rump with white. L. 20. W. 12. B. 1¼. Atlantic, abundant. (*Eu.*)

  bb. Wing less than 10.
672. P. auduboni Finsch. Crissum with dusky. L. 11. W. 8. T. 3½. B. 1¼. Tropics, N. to N. J. (To John James Audubon.)
aa. Dusky below as well as above.

673. P. **stricklandi** Ridgway. SOOTY SHEARWATER. Bill
dusky. L. 16. W. 12. B. 1¾. Atlantic, N. to Grand Banks.
(To Hugh Strickland.)

340. ÆSTRELATA Bonaparte. (οἰστρήλατος, goaded on by a
gad-fly.)

674. Æ. **hæsitata**(Kuhl). BLACK-CAPPED PETREL. Upper tail-
coverts and lower parts white; upper parts mostly blackish; tail
graduated. L. 16. W. 11. T. 5. B. 1½. Atlantic, N. to N. Y.,
scarce. (*Eu.*) (Lat.. *hæsitatus*, stuck-fast, the describer being in
doubt.)

341. PROCELLARIA Linnæus. (Lat., stormy.)

675. P. **pelagica** L. STORM PETREL. Dusky; upper tail coverts white, edged with black. L. 5¾. W. 4⅞. T. 2½. N. Atl.,
rarely S. (*Eu.*)

342. OCEANODROMA Reichenbach. ('Ωκεανός ocean ; δρόμος,
running.)

676. O. **leucorhoa** Vieillot. LEACH'S PETREL. Sooty, upper
tail coverts white; feet black. L. 8. W. 6½. T. 4. Northern
Seas. (*Eu.*) (λευκός, white ; ὄρρος, rump.)

343. OCEANITES Keyserling & Blasius. (ὠκεανίτης, a son of
the sea.)

677. O. **oceanicus** (Kuhl). WILSON'S PETREL. Sooty ; wings
and tail black ; upper tail coverts white. L. 7. W. 6. T. 3.
Tarsus 1⅜. Cosmopolitan, common. (*Eu.*)

ORDER XXXIV. **STEGANOPODES.** (THE TOTIPALMATE
BIRDS.)

Desmognathous swimmers with all four toes full-webbed ; hind
toe lengthened, scarcely elevated; tibia feathered; bill horny, not
lamellate ; nostrils very small or abortive; no basipterygoids ; a
prominent gular pouch; tarsus reticulate. Altricial. " Notwithstanding the shortness of the legs and the character of the toes,
. . . the birds of the present order are unquestionably nearly related to the *Herodii*" (*Stejneger*). Of this small order, most of the
species are sea-birds, active and voracious, about half of all being
cormorants. (στεγανός, covered ; πούς, foot.)

Families of Steganopodes.

a. Upper mandible not hooked at tip.
  b. Bill very thick through base, the tip slightly curved; tail moderate,
    graduated, the feathers rather pointed. . . . . . SULIDÆ, 131.
  bb. Bill slender, nearly straight; neck very long and slender; tail long,
    fan-shaped when spread, the feathers very broad. ANHINGIDÆ, 132.
aa. Upper mandible hooked at tip.

c. Tarsus moderate, much longer than hind toe with claw.
  d. Bill compressed; gular sac small. . . PHALACROCORACIDÆ, 133.
  dd. Bill much flattened; gular sac very large. . PELECANIDÆ, 134.
  cc. Tarsus extremely short, not longer than hind toe with claw; wings
    and tail excessively long, the latter deeply forked.
                                            FREGATIDÆ, 135.

FAMILY CXXXI. SULIDÆ. (THE GANNETS.)

Bill long, cleft to beyond eyes, very stout at base, the tip not hooked, the tomia irregularly serrate ; a nasal groove, but the nostril abortive ; gular sac small, naked ; wings long, pointed ; tail long and stiff, with pointed feathers ; feet stout. Body heavy, similar to that of a goose, the tissues under the skin with air-chambers as in the Pelicans. One genus with 5 or 6 species. Gregarious sea-birds, found in most regions.

344. SULA Brisson. (French. *Sule.*)

a. Lower jaw, chin and throat densely feathered. (*Dysporus.*)

678. **S. bassana** L. GANNET. White, black on wings ; yellowish on head ; young dark brown, spotted. L. 36. W. 20. T. 10. *mus*
B. 6. N. Atl., S. to Florida, common N. (*Eu.*) (From Bass Rock, Eng., where Gannets breed.)

FAMILY CXXXII. ANHINGIDÆ. (THE DARTERS.)

Bill very long, straight, slender, sharp, the tomia finely serrate ; gular sac small, naked ; nostrils minute, becoming obsolete ; tail long, stiff, fan-shaped, when spread, the feathers broad, the middle pair in the adult transversely corrugated. Neck long, very slender, the vertebræ (20 in number) of peculiar structure ; feet short, far back. A single genus, with 3 or 4 species ; swift, wary birds, their movements in the water resembling those of a snake.

345. ANHINGA Brisson. (Port., *anhina* ; Lat., *anguina*, snaky.)

679. **A. anhinga** (L.). DARTER. SNAKE-BIRD. WATER TUR-KEY. Chiefly black, with greenish lustre above ; neck with hair-*mus*like plumes ; ♀ largely buffy, back with pale streaks. L. 35. W. 14. T. 11. B. 3¼. Tropical Am., N. to S. Ill.   ·

FAMILY CXXXIII. PHALACROCORACIDÆ. (THE CORMORANTS.)

Bill slender, about as long as head, nearly terete, but compressed, strongly hooked, the cutting edges uneven ; gular pouch small. Wings short ; tail very large, almost scansorial, of very stiff feathers, often used as a support for the body ; legs set far back ; a nasal groove with abortive nostrils. Colors in both sexes lustrous, iridescent black ; in the breeding season usually with long, white,

15

filamentous plumes; many species crested. Genus one; species 25; of most regions, chiefly inhabiting rocky coasts, where they are gregarious and voracious.

346. PHALACROCORAX Brisson.   (φαλακρός, bald ; κόραξ, raven.)

a. Tail of 14 feathers.

680. **P. carbo** (*L.*).   COMMON CORMORANT.   SHAG.   Head, neck and belly blue-black ; back brownish, streaked with black ; young grayish ; sac flesh-color, heart-shaped behind. L. 36. W. 14. T. 7½. B. 2¼. Northern regions, S. to N. J. (*Eu.*) (Lat., coal.)

aa. Tail of 12 feathers.

681. **P. dilophus** (Swainson).   DOUBLE-CRESTED CORMORANT. Glossy greenish black ; back and wing coverts slaty brown ; adult with two curly black lateral crests ; sac convex or straight-edged behind, yellowish. L. 33. W. 13. T. 7. N. Am. ; our commonest species. (δίς, twice ; λόφος, crest.)

682. **P. mexicanus** (Brandt).   MEXICAN CORMORANT.   Brownish black ; back slaty ; gular sac orange, white-edged. L. 24. W. 10. B. 2. S. W., N. to S. Ill.

FAMILY CXXXIV. **PELECANIDÆ.**   (THE PELICANS.)

Bill very long, rather slender, straight, grooved throughout, with a claw-like hook at the end ; the broad space between the branches of the lower jaw occupied by a huge membranous sac ; nostrils abortive ; wings very long ; tail very short, of 20 or more feathers ; feet short, stout. Skin of breast and belly with large air-cells beneath it, so that the body is rendered better able to float. These air-cells occupy the usual position of the fat-cells. Sexes alike. Genus one ; species 6 ; found in most warm regions. Gregarious, greedy fish-eating birds, clumsy on the wing.

347. PELECANUS Linnæus.   (πελεκάν, pelican.)

a. Tail-feathers 24 ; lower jaw feathered. (*Cyrtopelicanus* Reich.)

683. **P. erythrorhynchos** Gmelin.   WHITE PELICAN.   White with black on wings and some yellowish ; pouch reddish or yellowish. L. 60. W. 24. B. 12. N. Am., abundant S. and W.,) often inland.   Farther S. occurs the Brown Pelican, *P. fuscus* L., chiefly dusky grayish in color. (ἐρυθρός, red ; ῥύγχος, beak.)

FAMILY CXXXV. **FREGATIDÆ.**   (THE MAN-OF-WAR BIRDS.)

Bill long, rather slender, straight, strongly hooked at tip. Gular sac moderate. Wings very long and pointed ; tail very long, deeply forked ; feet very small, the short, feathered tarsus very short ; the webbing narrow ; middle claw pectinate.

Sea birds of tropical regions, the immense wings giving them a power of flight surpassed by no other bird. They live mainly by robbing the terns and gulls, which they watch, often from great heights in the air. The two species range widely in the warm seas.

348. FREGATA Cuvier. (Ital., frigate.)

684. **F aquila** (L.). MAN O' WAR BIRD. Black, the shoulders lustrous in ♂. L. 40. W. 25. T. 17½. B. 4⅔. Tropical seas, occasional N. (Lat., eagle.)

ORDER XXXV. ANSERES. (THE DUCKS AND GEESE.)

Desmognathous swimmers with the basipterygoids more or less developed and the feet not totipalmate; bill lamellate; no gular pouch. Feet 4-toed, palmate; hind toe small, elevated. Legs short. This order (often called *Lamellirostres*, associated with the Flamingoes, etc., to form the *Chenomorpha* of Huxley and Stejneger) " opens the series of desmognathous birds, which are characterized by having the palatal bones united across the middle either directly or by the intermediation of ossifications in the nasal septum." (*Stejneger*.)

This familiar order contains nearly all the Water-fowl which are valued in domestication or as game birds. As here understood, the *Anseres* comprise but a single family, the *Phœnicopteridæ* or Flamingoes, wading birds with a duck's bill, being placed in a distinct order *Odontoglossæ*, by the American Ornithologists' Union.

FAMILY CXXXVI. ANATIDÆ. (THE DUCKS.)

Bill lamellate, *i. e.*, furnished along each cutting edge with a regular series of tooth-like processes, which correspond to certain laciniate processes of the fleshy tongue, which ends in a horny tip; bill large, thick, high at base, depressed towards the end, membranous except at the obtuse tip which is occupied by a horny nail. Body heavy, flattened beneath. Head high, compressed, with sloping forehead; eyes small. Tail various, usually short, of 14 to 16 feathers, the lower coverts being long and full. Legs and feet short; anterior toes full-webbed. Tibia feathered. Sexes usually quite unlike (excepting among the Swans and Geese). Species about 175, of all parts of the world; migratory; all are good swimmers.

a. Neck shorter than body; lores feathered.
  b. Tarsus scutellate in front, shorter than middle toe without claw. Sexes unlike. Ducks.
    c Lower mandible without trace of lamellæ along the side, but with a series of distinct, tooth-like serrations along the upper edge (inner

tomium); bill narrow, head more or less crested; hind toe lobate.
Fish ducks. (*Merginæ*.)

d. Serrations of both mandibles very conspicuous, tooth-like, strongly
   recurved at tip. . . . . MERGANSER, 349.

dd. Serrations of mandibles short, blunt, not distinctly recurved at tips.
                                      LOPHODYTES, 350.

cc. Lower mandible with a very distinct series of lamellæ along side besides the series along upper edge; bill rather broad. (*Anatinæ*.)

e. Hind toe without distinct membranous lobe; "river ducks."

x. Bill not spatulate, scarcely widened toward tip.

y. Tail feathers narrow, rather pointed, no crest.

z. Tail not very acute, the middle feathers not produced in ♂
   (speculum green, violet or white). . . . . ANAS, 351.

zz. Tail pointed, the middle feathers much produced in ♂; ♀
   with tail much shorter (speculum violet).   DAFILA, 353.

yy. Tail feathers broad, rounded at tip; ♂ with a high crest.
                                       AIX, 354.

xx. Bill spatulate, narrow at base and very broad toward the tip.
                                     SPATULA, 352.

ee. Hind toe with a broad, membranaceous lobe; "sea-ducks."

f. Tail feathers with their bases well hidden by the coverts.

g. Feathering on lores or forehead not reaching forward beyond
   posterior border of nostril.

h. Graduation of tail less than length of bill from nostril;
   width of nail not one-third width of bill at middle.
                                     AYTHYA, 355.

hh. Graduation of tail much more than length of bill from
   nostril.

i. Bill ordinary, not gibbous nor appendaged.

j. Nail of bill narrow, distinct; tail moderate.

k. Nostril anterior, its front much nearer tip of bill than
   loral feathers (eyes yellow). GLAUCIONETTA, 356.

kk. Nostril sub-basal, its front much nearer loral feathers
   than tip of bill; (eyes brown).
                                 CHARITONETTA, 357.

jj. Nail of bill large, fused; tail in ♂ with its middle
   feathers produced, about as long as wing; (no speculum). . . . . CLANGULA, 358.

ii. Bill variously gibbous or else appendaged on base or on
   side.

l. Bill not gibbous, but appendaged with a lobe at
   base of commissure; (speculum violet).
                                HISTRIONICUS, 359.

ll. Bill not gibbous, but with a leathery expansion on
   side of upper mandible; cheeks bristly; (speculum white). . . . . CAMPTOLAIMUS, 360.

lll. Bill gibbous at base, then broad, depressed, with
   a large fused nail and without frontal processes.
                                  OIDEMIA, 362.

gg. Feathering on forehead or lores reaching anteriorly to or beyond posterior end of nostril; bill gibbous at base and
   with large frontal processes; (no speculum).
                                 SOMATERIA, 361.

ff. Tail feathers with their bases scarcely concealed by the short coverts; tail more than half length of wing, much graduated, the feathers with narrow webs and very stiff shafts.

    m. Nail of bill very small, bent backward beneath tip of upper mandible; outer toe longer than middle. . . . . ERISMATURA, 363.

    mm. Nail of bill normal, not very narrow; outer toe shorter than middle toe.

                           NOMONYX, 364.

bb. Tarsus reticulate all around, the plates rather larger in front; tarsus not shorter than middle toe without claw. Sexes similar. Geese. (*Anserinæ.*)

q. Serræ on tomium of upper mandible visible from outside for most of its length; tomium decidedly sinuate or concave; (bill and feet pale).

    r. Bill very stout, its depth at base more than half its length above. (Color largely white.) . . . . CHEN, 365.

    rr. Bill smaller and more depressed, its depth at base not half its length. (Color not white.) . . . . ANSER, 366.

qq. Serræ on upper tomium scarcely visible except near angle of mouth, the tomium scarcely sinuate; nostril near middle of nasal fossa; (head, bill and feet mostly black). . . . . BRANTA, 367.

aa. Neck not shorter than body; lores partly naked. Color white. Swans. (*Cygninæ.*)

    s. Bill not tuberculate; tail rounded: outer primaries with sinuate webs. . . . . OLOR, 368.

349. MERGANSER Brisson. (Lat., *mergus*, diver; *anser*, goose.)

a. Nostril nearer middle of bill than base.

685. **M. americanus** (Cassin). MERGANSER. GOOSANDER. FISH DUCK. ♂ black and white above, lower parts creamy white; a black bar across white of wing coverts; head glossy green, scarcely crested; ♀ smaller, ashy gray; head brownish. L. 24. W. 11. B. 2. T. 5. N. Am., common.

aa. Nostrils near base of bill.

686. **M. serrator** (L.). RED-BREASTED MERGANSER. Similar: head crested; ♂ with breast reddish brown, black-streaked; wing with two black bars, instead of one as in preceding. L. 24. W. 9. T. 4. B. 2¼. N. Am., abundant. (*Eu.*) (Lat., one who saws.)

350. LOPHODYTES Reichenbach. (λόφος, crest; δύτης, diver.)

687. **L. cucullatus** (L.). HOODED MERGANSER. SHELDRAKE. Black and white; speculum white with 2 dark bars; sides chestnut in ♂; ♀ duller and grayish; crest high and compressed; nostrils sub-basal. L. 19. W. 8. T. 4. B. 1½. N. Am., common. (*Eu.*) (Lat., hooded.)

351. ANAS Linnæus. (Lat., duck.)

a. Culmen longer than middle toe without claw.
  b. Speculum violet, bordered with black; bill greenish-yellow; L. more
    than 20. (*Anas.*)

688. **A. boschas** L. MALLARD DUCK. TAME DUCK. ♂ head
and upper neck rich glossy green, a white ring below; breast
purplish chestnut; speculum violet, with black and white before
and behind it; ♀ duller, chiefly dull ochraceous, streaked with
dark brown. L. 24. W. 12. N. Am., abundant; commonest
westward. Original of the common domestic duck; various hy-
brids of this species with others are described. (*Eu.*) (βοσκάς,
mallard.)

689. **A. obscura** Gmelin. BLACK DUCK. Size of mallard and
resembling the ♀, but darker; both sexes entirely dusky, varied
with brown; no decided white except under the wings. E. U. S.,
common W. to Iowa.

  bb. Speculum green; bill dusky; L. less than 20.
    c. Wing-coverts in both sexes sky-blue, the greater white tipped; scapu-
      lars in ♂ striped with blue and buff; bill rather broad; head not
      crested. (*Querquedula* Stephens.)

690. **A. discors** L. BLUE-WINGED TEAL. ♂ head and neck
blackish plumbeous, darkest on the crown; a white crescent in
front of eye; under parts thickly spotted; ♀ dull streaky brownish and buffy, known by the wings. L. 16. W. 7. T. 3. E. U. S.,
W. to Rocky Mts., abundant. (Lat., discordant.)

691. **A. cyanoptera** Vieillot. CINNAMON TEAL. ♂ chiefly
chestnut; top of head blackish; ♀ dull and streaky. L. 17. W.
7½. S. W., straying E. to Ill. (κύανος, blue; πτερόν, wing.)

  cc. Wing coverts leaden gray, without blue; scapulars waved with black
    and white; bill very narrow; head slightly crested. (*Nettion*
    Kaup.)

692. **A. carolinensis** Gmelin. GREEN-WINGED TEAL. Head
and upper neck rich chestnut in ♂; a green patch behind eye;
upper parts with wavings of black and white; white below; buffy
on breast, with dark spots; ♀ different, known by the small size
and color of wing; white crescent on sides in front of wings.
L. 15. W. 7½. T. 3½. N. Am., common; one of the best of the
ducks as food.

aa. Culmen shorter than middle toe without claw.
  d. Lamellæ numerous, fine, more than 30 visible from outside; bill not
    shorter than head. (*Chaulelasmus* Bonaparte.)

693. **A. strepera** L. GADWALL. GRAY DUCK. ♂ barred, black
and white, middle wing coverts chestnut, greater coverts black,
speculum white; ♀ dusky and tawny with little chestnut, known

by the wings. L. 22. W. 11. T. 4½. N. Am., not rare. (*Eu.*) (Lat., obstreperous.)

dd. Lamellæ coarser, less than 15 visible externally; bill shorter than head. (*Mareca* Stephens.)

694. **A. penelope** (L.). EUROPEAN WIDGEON. Head and neck cinnamon; in ♂ top of head brownish white; sides of head with slight traces of green. Europe; rare in America. (*Eu.*)

695. **A. americana** (Gmelin). AMERICAN WIDGEON. BALD-PATE. Head and neck grayish in ♂, speckled with dusky; top of head white; sides of head with bright green patch; speculum glossy-green, preceded by black, white, and gray on wing-coverts; ♀ duller. L. 20. W. 11. T. 4½. N. Am., abundant. (*Eu.*)

352. SPATULA Boie. (Lat., spoon.)

696. **S. clypeata** (L.). SHOVELLER. SPOON-BILL DUCK. ♂ head and neck green; breast white; belly chestnut; wing coverts blue; speculum green, bordered by black and white; rump and tail coverts black; ♀ streaky brownish, known by the bill and wings. L. 20. W. 9¼. B. 2¾. N. Am., common. (*Eu.*) (Lat., *clypeum*, shield.)

353. DAFILA Stephens. (A coined word.)

697. **D. acuta** (L.). PIN-TAIL. SPRIG-TAIL. ♂ head dark brown with purplish gloss; side of neck with a long white stripe; back gray, finely waved with darker; lower parts white; crissum black; sides finely waved; speculum violet, with black, white, and buffy; tail cuneate when developed, central feathers black and much projecting; ♀ speckled and streaked; tail much shorter; bill dusky; feet grayish blue. L. 20 to 30. W. 11. T. 9 or less. N. Am. common, a slender, trim-built duck. (*Eu.*)

354. AIX Boie. (αἴξ, a water-bird; αἴσσω, to spring.)

698. **A. sponsa** (L.). WOOD DUCK. SUMMER DUCK. Crested; ♂ head iridescent green and purple, with white stripes and a forked white throat patch; back varied, black, green, etc.; breast rich chestnut; sides buffy, very finely waved with dark; speculum green; tips of primaries frosted; ♀ duller, head mostly gray; varied with white. L. 20. W. 9½. T. 5. U. S. frequent; nesting in trees, the most elegant of all ducks. (Lat., bride.)

355. AYTHYA Boie. (αἴθυια, a sea-bird.)

a. Bill not wider toward its end than at base; ♂ with head and neck reddish.
b. Bill much shorter than middle toe without claw, the nail hooked. (*Aythya.*)

699. **A. americana** (Eyton). RED HEAD. POCHARD. ♂ head and neck chestnut with red reflections; back mixed silvery

and black, the dark waved lines unbroken ; breast, rump, etc.,
black ; belly white ; speculum bluish gray, tipped with white ; ♀
duller; bill and feet dull bluish. L. 20. W. 10. T. 3. B. 2⅛.
N. Am., abundant.

   b. Bill as long as middle toe without claw, its tip flattened, the nail little
   hooked. (*Aristonetta* Baird.)

   700. **A. vallisneria** (Wilson). CANVAS-BACK DUCK. Head
and neck dark reddish brown ; black wavy lines on back broken,
the whitish predominating; bill dusky. L. 23. W. 9. T. 3
B. 2⅝. N. Am.; especially coastwise in winter ; a bird highly
valued by epicures, but ordinarily not superior to any of the river-
ducks. (*Vallisneria spiralis*, the " Water Celery," on which the
bird feeds.)

   aa. Bill wider toward end than at base; ♂ with head and neck black. (*Fuligula* Stephens.)
   c. Speculum white, tipped with **black.**

   701. **A. marila** (L.). BIG SCAUP DUCK. BLUE BILL. RAFT
DUCK. FLOCKING FOWL. Head, neck, and breast black ; no
ring about neck ; back and sides whitish, the back finely waved
with black ; ♀ with face white ; the head and neck snuffy brown.
L. 20. W. 9. Northern regions, the American bird is var.
nearctica Stejneger, its 6 inner quills without white on inner web.
(μαρίλη, charcoal.)

   702. **A. affinis** (Eyton). LESSER SCAUP DUCK. Similar, but
smaller ; the sides vermiculate with blackish. L. 16. W. 8. T.
2½. N. Am. (Lat., related.)

   cc. Speculum bluish gray.

   703. **A. collaris** (Donovan). RING-NECKED DUCK. ♂ with an
orange brown collar about neck ; blackish above ; crissum black ;
lower parts white ; wings brown ; ♀ chiefly brown, without collar.
L. 18. W. 8½. T. 2¾. N. Am. (Lat., collared.)

   356. GLAUCIONETTA Stejneger. (γλαύκιον, an old name ;
νῆττα, duck.)

   704. **G. clangula** (L.). GOLDEN-EYE. GARROT. ♂ head puffy,
glossy green with a round white spot before eye not touching bill;
upper parts black ; white continuous on outer surface of wing; ♀
head duller, snuff-colored and scarcely puffy, the body brownish.
L. 16 to 19. W. 8½. T. 3½. N. Am., common; "meat bad, rank
and fishy." (*Eu.*) The Amer. bird is var. **americana** Bonap.

   705. **G. islandica** (Gmelin). BARROW'S GOLDEN-EYE. Similar ;
head almost crested in ♂ ; gloss of head purplish ; a roundish white
space before eye touching base of bill ; white of wing divided by a
dark line ; ♀ head dark brown. L. 22. W. 10. T. 3¾. N. U. S.
and N.; rare. (*Eu.*) (From Iceland.)

357. CHARITONETTA Stejneger. (χάρις, grace; νῆττα, duck.)

706. **C. albeola** (L.). DIPPER. BUFFLE-HEAD. BUTTER-BALL. SPIRIT DUCK. ♂ with head very puffy and iridescent; no white before eye but a large white ear patch; wing coverts and secondaries mostly white. L. 16. W. 7. ♀ smaller, dark gray, the head scarcely puffy, with white behind eye. N. Am., abundant; an expert diver. (Lat., whitish.)

358. CLANGULA Leach. (Lat., clangor, a noise.)

707. **C. hyemalis** (L.). SOUTH-SOUTHERLY. OLD SQUAW. LONG-TAILED DUCK. Blackish and whitish; head, neck and lower parts mostly white in winter; a patch of gray on head; breast brownish black; bill black and orange; tail very long; ♀ quite different, mostly grayish brown, with short tail. L. 20 W. 9. T. 3 (♀) to 9 (♂). Northern, S. in winter; said to be melodious. (Eu.)

359. HISTRIONICUS Lesson. (From Lat., histrio, harlequin.)

708. **H. histrionicus** (L.). HARLEQUIN DUCK. "LORDS AND LADIES." ♂ leaden bluish, much varied with black, white, and chestnut; a white patch before eye; speculum violet purple; ♀ dark brown, with gray, etc.; a white spot before eye and one behind ear; bill very short. L. 17. W. 8. T. 4. B. 1¹⁄₁₀. Atlantic, S. to N. Y. (Eu.)

360. CAMPTOLAIMUS Gray. (καμπτός, flexible; λαιμός, throat.)

709. **C. labradorius** (Gmelin). LABRADOR DUCK. ♂ head, neck, chest, and wings white; rest of body with ring about neck and strip on crown black; ♀ chiefly grayish. L. 24. W. 9. N. Atl., very rare, or perhaps extinct.

361. SOMATERIA Leach. EIDER DUCKS. (σῶμα, body; ἔριον, wool.)

a. Frontal processes long, acute, clubbed, extending in line with culmen on each side of forehead; feathers on side of bill advancing to below nostril; ♂ with scapulars white; top of head black; no V-mark on chin. (Somateria.)

710. **S. dresseri** Sharpe. AMERICAN EIDER DUCK. ♂ in breeding dress white; under parts, rump, quills, and crown patch black; ♀ reddish brown, streaked; angle on side of forehead broad and rounded. L. 24. W. 12. T. 4. Arctic Am.; S. to Maine in winter. (Eu.) (To H. E. Dresser, an Eng. Orn.)

aa. Frontal processes broad, squarish, nearly vertical, out of line of culmen; feathers on side of bill not reaching nostrils. (Erionetta Coues.)

711. **S. spectabilis** (L.). KING EIDER. ♂ chiefly black; neck, breast, etc., white; a black V-shaped mark on chin; ♀ brownish,

known by the bill. L. 22. W. 11. T. 4. Northern regions;
S. to N. J. (*Eu.*) (Lat., conspicuous.)

362. OIDEMIA Fleming. (οἴδημα, swelling.)

a. Bill in ♂ scarcely encroached upon by frontal feathers; nostrils median; no white on wings. (*Oidemia.*)

712. **O. americana** Swainson & Richardson. AMERICAN BLACK SCOTER. ♂ entirely black; ♀ sooty brown, paler below and on throat. L. 17 to 20. W. 10. T. 4. N. Am., coastwise; S. to N. J.

aa. Bill in ♂ broadly encroached upon by frontal feathers; nostrils beyond middle of bill; a large white wing patch. (*Melanitta* Boie.)

713. **O. deglandi** Bonaparte. WHITE WINGED SCOTER. SURF DUCK. ♂ black; white spot under eye and white on wings; bill orange-tipped; feet orange; ♀ sooty brown. L. 21. W. 11. N. Am., S. to Md.

aaa. Bill in ♂ narrowly encroached upon by frontal feathers; no white on wings; nostrils beyond middle of bill. (*Pelionetta* Kaup.)

714. **O. perspicillata** (L.). SURF DUCK. SEA COOT. ♂ black, with white spot on forehead and nape; ♀ sooty brown; white patch on lores and cheeks; size of *O. americana.* N. Am., coastwise. (Lat., conspicuous.)

363. ERISMATURA Bonaparte (ἔρεισμα, prop; οὐρά, tail.)

715. **E. rubida** (Wilson). RUDDY DUCK. Chiefly brownish or tawny, glossy chestnut in full plumage; considerably waved and dotted, lower parts mottled silver-white; crissum white; ♀ brown, mottled with dusky. L. 17. W. 6. T. 3½. N. Am., frequent; an expert diver.

364. NOMONYX Ridgway. (νόμος, regularity; ὄνυξ, nail.)

716. **N. dominicus** (L.). Redder than the last; forehead and chin black. L. 13½. W. 6¼. Trop. Amer., straying N. to Wis., etc. (From San Domingo.)

365. CHEN Boie. (χήν, goose.)

717. **C. cœrulescens** (L.). BLUE GOOSE. Grayish brown, the rump and wing coverts gray; size and form of next. N. Am., rare. (Lat., bluish.)

718. **C. hyperborea** (Pallas.) SNOW GOOSE. Adult pure white, washed with reddish on head; the primaries black; young bluish; feet reddish, the claws dark; bill red; lamellæ very prominent. L. 30 to 38. W. 17. T. 6½. B. 2¼. N. Am., chiefly W. The form E. of the Rocky Mts. is var. **nivalis** Forster; larger than the Western bird. (Lat., far-northern.)

366. ANSER Brisson. (Lat., goose.)

719. **A. albifrons** Gmelin. WHITE-FRONTED GOOSE. SPECKLE-BILL. Grayish-brown, mottled, forehead and tail coverts white;

bill pink ; feet yellow ; nostrils basal. L. 27. W. 17. T. 6. Ts. 3.
N. Am., common W. of Mts. (*Eu.*) The American form is
var. **gambeli** (Hartlaub), distinguished by longer bill. B. 1¾ to 2.
The tame goose, *A. anser* L., is a European relative. (Lat., white-
fronted.)

367. BRANTA Scopoli. (Eng., Brant.)

a. Forehead black; cheeks and chin white; no white stripes or collar on
neck.

720. **B. canadensis** (L.). WILD GOOSE. CANADA GOOSE.
Grayish brown, more or less barred with whitish, paler below ;
head and neck black ; tail black ; upper coverts and crissum white.
L. 36. W. 20. T. 7½. B. 2. N. Am., abundant, U. S. in winter ;
the commonest of our geese ; runs into varieties W.

aa. Forehead, cheeks, and chin black; white stripes on neck.

721. **B. bernicla** (L.). BRANT GOOSE. BARNACLE GOOSE.
Head, neck, front, quills, and tail black ; white patch on neck ;
white on rump, crissum, etc ; back brownish gray. L. 24. W. 13.
T. 5. B. 1½. N. Atl., rarely S. in winter. (*Eu.*) (Eng. barnacle ;
these geese once supposed to hatch from barnacles.)

368. OLOR Wagler. (Lat., Swan.)

722. **O. columbianus** (Ord). WHISTLING SWAN. Tail 20
feathered ; bill with a yellow spot, not longer than head ; nostrils
median. L. 50. W. 20. T. 8. B. 4. N. Am., mostly coastwise.
(From Columbia R.)

723. **O. buccinator** (Richardson). TRUMPETER SWAN. Plumage white, sometimes washed with rusty ; young grayish ; tail (normally) 24 feathered ; bill black without yellow spot, longer than
head ; nostrils sub-basal. L. 60. W. 27. T. 9. B. 4½. N. Am.
E. to Ill. (Lat., trumpeter.)

ORDER XXXVI. HERODIONES. (THE HERONS AND
STORKS.)

Desmognathous waders, without basipterygoid processes; the
feet not palmate. Birds mostly, but not always, of large size, with
compressed bodies, long legs, and a very long S-bent neck of 15 to
17 vertebræ; tibia naked below ; toes long and slender, cleft or
slightly webbed, the hind toe long and usually not elevated, provided with a large claw. Wings broad, rounded; tail short.
Head contracted to the stout base of the bill which is long and
usually hard and acute, with sharp cutting edges ; nostrils small,
elevated part of head often naked. Altricial. The species live
about water, feeding on fishes, reptiles, etc., which are speared by
a thrust of the bill. The leading families are represented in our
fauna.

Families of Herodiones.

a. Sides of upper mandible with a deep narrow groove, extending from the nostrils to the tip; skull schizorhinal. Ibises.

    b. Bill very broad, much flattened, and greatly widened toward tip, only the point decurved. . . . . PLATALEIDÆ, 137.

    bb. Bill slender, subterete, gradually decurved for its whole length.

                                       IBIDIDÆ, 138.

aa. Sides of upper mandible without long groove; skull holorhinal. Storks and Herons.

    c. Hind toe inserted more or less above the level of the others; its claw short; claws broad and flat, resting on a horny pad or shoe; middle claw not pectinate. . . . . CICONIIDÆ, 139.

    cc. Hind toe inserted on the level of the rest; claws narrow, arched; the middle one pectinate on its inner edge; bill straight, pointed.

                                      ARDEIDÆ, 140.

FAMILY CXXXVII. **PLATALEIDÆ.** (THE SPOONBILLS.)

Bill long, flat, broad and spoonshaped at the end, otherwise essentially as in the *Ibididœ.* Genera 2; species 6, in most regions.

a. Trachea simple (not convoluted within the thorax). . . . AJAJA, 369.

369. **AJAJA** Reichenbach. (Brazilian name.)

724. **A. ajaja** (L.). ROSEATE SPOONBILL. Chiefly white, back and wings rose-pink; tail buffy; skin of the bald head variegated. L. 34. W. 15. B. 7; its width 2. T. 5. Tropical America, N. to S. Ill.

FAMILY CXXXVIII. **IBIDIDÆ.** (THE IBISES.)

Bill very long and slender, compressed, cylindric, curved throughout, the upper mandible with a deep groove reaching nearly or quite to tip; legs rather long, the toes slightly webbed at base. Head more or less naked; plumage stork-like, without powder-down tracts; wings broad; tail short; tarsus scutellate in front, in our species. Skull schizorhinal. Genera 10 or more; species 24; of the lakes and swamps of warm regions. Sexes alike.

a. Head of adult wholly naked anteriorly; no crest; claws curved.

                                      GUARA, 370.

aa. Head of adult feathered except on lores; crown with a short crest; claws nearly straight. . . . . PLEGADIS, 371.

370. **GUARA** Reichenbach. (From *guarauna,* a Brazilian name.)

725. **G. alba** (L.). WHITE IBIS. SPANISH CURLEW. Pure white, tips of longer quills glossy black. L. 24. W. 11. T. 4. B. 7. Southern States, N. to S. Ind.

371. **PLEGADIS** Kaup. (πληγάς, scythe.)

726. **P. autumnalis** (Hasselquist). GLOSSY IBIS. Rich dark purplish-chestnut; head, back, wings and tail metallic purplish-

green. L. 24. W. 11. T. 4. B. 4½. Tropics, rarely N. to N. E. (*Eu.*)

FAMILY CXXXIX. **CICONIIDÆ.** (THE STORKS.)

Bill longer than head, very stout at base, not grooved, tapering to the tip; nostrils high, close to base of bill; tarsus reticulate; hind toe more or less elevated; claws short, not acute; skull holorhinal. Genera 7; species about 20, in damp places in warm regions. The famous migratory Stork of Europe (*Ciconia ciconia* L.) is the best known member of the group. (Lat., *ciconia*, stork.)

a. Bill decurved at tip; hind toe scarcely elevated; trachea simple, not convoluted within thorax. . . . . TANTALUS, 372.

372. TANTALUS Linnæus. (Τάνταλος, a mythological character.)

727. **T. loculator** L. WOOD "IBIS." White, wings and tail mostly glossy black; the bald head livid and yellowish. L. 46. W. 18. B. 7. Southern States, N. to N. Y. (Lat., one who places.)

FAMILY CXL. **ARDEIDÆ** (THE HERONS.)

Bill straight, longer than the head, compressed, acute, with sharp cutting edges; upper mandible grooved; nostrils linear; lores naked, the bill appearing to run directly to the eyes; rest of head feathered; parts of the body with "powder-down tracts," — strips of short, dusty, or greasy down-like feathers, usually three pairs of these strips, *i. e.*, on the back above the hips, on the belly under the hips, and on the breast; usually long plumes from the back of head in the breeding season. Wings broad. Tail very short. Tibiæ largely naked below; toes long and slender, hind toe on a level with the rest, its claw longer than middle claw; middle claw pectinate. Skull holorhinal. Sexes usually colored alike, but the changes due to age and season often considerable. Species about 75; in most parts of the world, abundant in the warmer regions, wading in shallow water and feeding chiefly on fishes.

a. Tail feathers 10, very short, scarcely stiffer than the coverts; outer toe shorter than inner; no conspicuous crest or train in breeding season; bill slender. (*Botaurinæ*.) . . . . BOTAURUS, 373.
aa. Tail feathers 12, rather long, stiffer than the coverts; outer toe not shorter than inner; claws shorter, and more curved. (*Ardeinæ*).
   b. Bill long and slender, at least five times as long as deep at base.
ARDEA, 374.
   bb. Bill rather short and thick, and not more than 4 times as long as deep at base . . . . NYCTICORAX, 375.

373. BOTAURUS Stephens. (An imitation of the bird's note.)
a. Size large; sexes alike; young similar. (*Botaurus.*)

728. **B. lentiginosus** (Montagu). BITTERN. INDIAN HEN. STAKE DRIVER. Tawny brown of various shades, excessively

238 AVES : HERODIONES. — XXXVI.

variegated everywhere; foreneck striped with buffy; a dark patch on each side of neck. L. 23 to 28. W. 12. T. 4½. B. 3. N. Am., abundant. (Lat., freckled.)

aa. Size very small; sexes unlike, the young unlike adult. (*Ardetta* Gray).

729. **B. exilis** (Gmelin). LEAST BITTERN. ♂ chiefly glossy greenish black above, brownish yellow below, neck, shoulders and wings with chestnut; a buffy area on wing coverts; ♀ with brown instead of black. L. 14. W. 5. T. 1⅘. B. 1¾. N. Am., in reedy swamps. (Lat., slender).

374. ARDEA Linnæus. (Lat. heron.)

a. Bill shorter than tarsus.
  b. Length more than 35 (in adult); tarsus not twice middle toe without claw.
    c. Color chiefly bluish; head crested in breeding season; dorsal plumes short. (*Ardea.*)

730. **A. herodias** L. GREAT BLUE HERON. Grayish blue, marked with black and white; crown black with white centre; forehead white; lower parts dusky, striped with white; tibia and edge of wing cinnamon brown. L. 42 to 50. W. 19. B. 6½. Ts. 7. T. 7. ♀ smaller. N. Am., generally common. (ἐρωδιός, heron.)

  cc. Color white; no crest; back in breeding season with long plumes. (*Herodias* Boie.)

731. **A. egretta** Gmelin. GREAT WHITE EGRET. L. 40. W. 17. B. 5. Ts. 6. Amer., chiefly S. (Fr., *aigrette*, a top-knot.)

  bb. Length 20 to 32.
    d. Tarsus not nearly twice middle toe without claw.
      e. Color pure white at all times; plumes of breeding season very long, recurved, with loose webs. (*Garzetta* Kaup.)

732. **A. candidissima** Gmelin. SNOWY EGRET. WHITE "CRANE." L. 24. W. 12. B. 3. T. 4. Tropical America, N. to N. Y., abundant. (Lat., very white.)

    ee. Color slaty blue, the head and neck maroon; the young white, and sometimes the adult nearly or quite white; plumes slender, with compact webs. (*Florida* Baird.)

733. **A. cærulea** L. LITTLE BLUE HERON. L. 24. W. 12. T. 4. B. 3. Ts. 4. E. Am., N. to Ill., common.

    dd. Tarsus twice as long as middle toe without claw. (*Dichromanassa* Ridgway.)

734. **A. rufescens** Gmelin. REDDISH EGRET. Slate color; head and neck cinnamon; young grayish. L. 30. W. 13. B. 4. Ts. 5. Southern, N. to Ill. (Lat., reddish.)

aa. Bill not shorter than tarsus; L. less than 30.
  f. Scapular plumes in ♂ straight, hair-like, reaching beyond tail; wing more than 8. (*Hydranassa* Baird.)

735. **A. tricolor** Müller. LOUISIANA HERON. Variegated; leaden blue, chestnut and white. L. 27. W. 10. B. 4. Ts. 4. Tropical, N. to Ind. The U. S. bird is var. **ruficollis** (Gosse).

ff. Scapular plumes not very long, soft, with compact webs; wing not more than 8. (*Butorides* Blyth.)

736. **A. virescens** (L.) GREEN HERON. Crown, back and wings lustrous dark green; neck purplish cinnamon : throat and fore-neck striped with whitish; young similar. L. 18. W. 7. B. 2¼. Ts. 2. Amer., abundant; N. to Ont. (Lat., greenish.)

- **375. NYCTICORAX** Stephens. NIGHT HERONS. (νύξ, night; κόραξ, raven.)

a. Bill about as long as tarsus; gonys nearly straight. (*Nycticorax.*)

737. **N. nycticorax** (L.). BLACK CROWNED NIGHT-HERON. QUA BIRD. SQUAWK. Bluish gray, crown, back and shoulders glossy green; lower parts mostly white; no peculiar feathers save two or three long, white occipital plumes; young grayish brown, speckled and streaked with whitish, very different. L. 24. W. 14. B. 3. Ts. 3. T. 5. U. S., frequent. The American bird is var. **nævius** (Boddaert). (*Eu.*)

aa. Bill much shorter than tarsus; gonys convex. (*Nyctanassa* Stejneger.)

× 738. **N. violaceus** (L.). YELLOW-CROWNED NIGHT HERON. Grayish plumbeous, darker on back and streaked with black; head mostly black, the crown and crest tawny white; a white streak behind eye; back with long plumes; young grayish brown, streaked and spotted with brown. L. 24. W. 12. T. 5. B. 2¾. Ts. 3¾. S. U. S., scarce; N. to N. Y.

ORDER XXXVII. **PALUDICOLÆ.** (THE CRANES AND RAILS.)

This small order includes the allies of the Cranes and Rails, wading birds with schizognathous palate, allied to the *Limicolæ*, but with the head rather compressed than globose, the bill hard and not sensitive, not adapted for probing in the mud, and the hind toe little elevated. Precocial. Birds of moderate or large size, skulking about in the reeds and rushes, and feeding upon substances found on the surface. The position and boundaries of this group have been unsettled. It seems nearly related to the *Limicolæ.* (Lat , *palus*, swamp; *colo*, I inhabit.)

Families of Paludicolæ.

a. Nasal bones schizorhinal; head partly unfeathered or else with ornamental plumes; hind toe short, much elevated. Very large. Cranes.
GRUIDÆ, 141.
aa. Nasal bones holorhinal ; head feathered, except sometimes a frontal shield; hind toe rather long, little elevated; size moderate or small. Rails. . . . . RALLIDÆ, 142.

FAMILY CXLI. GRUIDÆ. (THE CRANES.)

Very large birds, with the legs and neck extremely long, the latter of 17 vertebræ. Wings large, rather short. Tail short, of 12 broad feathers. Head more or less naked, with scattered hair-like feathers. Plumage compact, without downy tracts. Bill as long or longer than head, straight and slender; tibia extensively naked; tarsus scutellate; toes rather short; hind toe highly elevated; nasal bones schizorhinal. Genera 3; species 15; of various parts of the world, resembling herons in external form, but similar to the rails in general structure.

376. GRUS Linnæus. (Lat., crane.)

739. **G. americana** (L.). WHOOPING CRANE. WHITE CRANE. Adult pure white with black on wings; bare part of head very hairy; young rusty, the head feathered. L. 50. W. 24. T. 9. Ts. 12. B. 6. N. Am.; rare E., a wild bird, avoiding civilization. "The windpipe is quite as long as the bird itself, 50 inches or more, and over 2 feet of it coiled away in the keel of the breastbone, which is entirely hollowed out to receive these extraordinary convolutions; the voice is singularly raucous and resonant." (*Coues.*)

740. **G. mexicana** (Müller). SAND-HILL CRANE. BROWN CRANE. Slaty gray or brownish, never white; head sparsely hairy. L. 46. W. 22. B. 5½. T. 9. Ts. 10. U. S., chiefly S. and W.

FAMILY CXLII. RALLIDÆ. (THE RAILS.)

Birds of medium or small size, with compressed bodies and large muscular legs. Wings short, rounded, concave; tail very short, of 10 or 12 soft feathers. Hind toe rather short, a little elevated; front toes very long. Bill various, rather short, not sensitive at tip. Plumage blended, changing little with age, sex, or season. Species about 150, of most parts of the world, skulking in swamps and marshes, gathering their food chiefly from the surface.

a. Forehead feathered; no frontal shield. (*Rallinæ.*)
  b. Bill slender, decurved, longer than head, with narrow nasal groove, and linear nostril. . . . . RALLUS, 377.
  bb. Bill stout, straight, not longer than head, with broad nasal groove and oblong nostril. . . . . PORZANA, 378.
aa. Forehead covered with a broad, bare, horny shield.
  c. Toes scarcely or not lobate. (*Gallinulinæ.*)
    d. Nostrils small, oval. . . . . IONORNIS, 379.
    dd. Nostrils slit-like. . . . . GALLINULA, 380.
  cc. Toes lobate, edged with broad flaps. (*Fulicinæ.*) . . FULICA, 381.

377. RALLUS Linnæus.   (Fr., *râle*, from its note.)

a. Large rails; wing more than 5.

741. **R. longirostris** Boddaert. CLAPPER RAIL. SALT-WATER MARSH HEN. Olive brown, variegated with ashy; dull reddish brown below; little or no distinct chestnut anywhere. L. 14 to 16. W. 6. T. 2¼. B. 2½. ♀ smaller. Salt marshes; common S., N. to Mass. Ours is var. **crepitans** Gmelin.

742. **R. elegans** Audubon. KING RAIL. FRESH-WATER ˑˑ MARSH-HEN. Brownish black, with bright chestnut below and on wing coverts; much brighter colored than the last, and larger; a red, rather than a gray bird. L. 18. W. 7. B. 2¼. U. S., in fresh-water marshes, N. to Conn.

aa. Small rails; wing less than 5.

743. **R. virginianus** L. VIRGINIA RAIL. Colors exactly as in *R. elegans;* much smaller. L. 10. W. 4. T. 1½. B. 1½. N. Am., common E.

378. PORZANA Vieillot.   (Italian name.)

a. Secondaries without white.

b. Wing more than 4: olive-brown above, striped with black. (*Porzana.*)

744. **P. carolina** (L.). CAROLINA RAIL. SORA. Olive-brown, streaked; adult with face and middle line of throat black; breast slaty gray; back streaked; belly barred. L. 9. W. 4½. T. 2. N. Am., common.

b. Wings less than 3½; dusky, usually speckled with white. (*Creciscus* Cabanis.)

745. **P. jamaicensis** (Gmelin). BLACK RAIL. Blackish, with white markings. L. 5½. W. 3. T. 1½. Tropical Amer. etc., rarely N. to Ill.

aa. Secondaries white. (*Coturnicops* Bonaparte.)

746. **P. noveboracensis** (Gmelin). YELLOW CRAKE. Buffy, blackish-streaked above with white marks, buffy below. L. 6. W. 3¼. T. 1½. E. N. Am., not common. (Lat., of New York.)

379. IONORNIS Reichenbach.   (ἴον, violet; ὄρνις, bird.)

747. **I. martinica** (L.). PURPLE GALLINULE. Olive green; ˑ head and lower parts purplish blue; wings and tail greenish-black; crissum white; bill mostly red; the shield blue. L. 12. W. 7. T. 3. Tropical Amer., N. to N. E. (From Martinique.)

380. GALLINULA Brisson.   (Dim. of Lat. *gallina*, hen.)

748. **G. galeata** (Lichtenstein). FLORIDA GALLINULE. ˑˑ Brownish olive above, grayish black on head and below; wings and tail dusky; bill, frontal shield, and ring around tibia red; feet

greenish. L. 14. W. 7½. T. 3½. Ts. 2. S. States, straying N.
to N. E. and Wis. (Lat., helmeted.)

381. FULICA Linnæus. (Lat., coot.)

749. **F. americana** (Gmelin). COOT. MUD HEN. Dark slate
color or sooty, with white on wings and crissum ; bill pale in adult,
with a brown spot near tip; frontal shield dark brown. L. 14.
W. 8. T. 2. N. Am., abundant in reedy swamps ; an excellent
swimmer.

ORDER XXXVIII. LIMICOLÆ. (THE SHORE-BIRDS.)

This division of the old order of *Grallæ* includes the allies of
the Plover and Snipe, as distinguished from the nearly related
Cranes and Rails on the one hand and the remotely related
Herons and Ibises on the other. Some of the external characters
of the group are the following. Tibia more or less naked below;
legs long; hind toe free and elevated, often wanting. Head globose, abruptly sloping to the base of the bill; completely feathered
(except in the male of *Pavoncella*); gape short; bill weak, flexible,
more or less soft-skinned and sensitive at tip in most cases, adapted
for probing in the mud; nostrils slit-like, surrounded by soft skin.
Schizognathous ; precocial.

The *Limicolæ* are all birds of small size, abundant on sandy
shores and in marshes. In spite of the difference in appearance
and habits, these birds have much in common with the gulls, in
their anatomy. (Lat., *limus*, mud; *colo*, I inhabit.)

Families of Limicolæ.

a. Toes lobate, with distinct lateral membranes ; tarsus extremely compressed.
PHALAROPODIDÆ, 143.

aa. Toes not lobate; webbed or not.
  b. Tarsus more than twice middle toe with claw; naked part of tibia much
  longer than middle toe with claw; feet palmate or not.
RECURVIROSTRIDÆ, 144.

  bb. Tarsus less than twice middle toe with claw; naked portion of tibia
  shorter than middle toe with claw; toes cleft or semipalmate.
    c. Tarsus scutellate in front.
      d. Bill slender, with a bluntish tip ; soft-skinned and sensitive through-
      out. . . . . SCOLOPACIDÆ, 145.
      dd. Bill stout, hard, pointed and wedge-shaped at tip (in our species).
APHRIZIDÆ, 146.
    cc. Tarsus reticulate in front.
      e. Bill not longer than tarsus, not compressed ; contracted behind the
      horny tip, shaped somewhat like a pigeon's bill.
CHARADRIIDÆ, 147.
      ee. Bill longer than tarsus, much compressed at tip.
HÆMATOPODIDÆ, 148.

Family CXLIII. PHALAROPODIDÆ. (The Phalaropes.)

Small sand-piper-like birds, with the toes lobed, as in the Coots and Grebes, but the lobes narrower. Body depressed, the lower plumage thick, as in the ducks, and capable of resisting water: wings long, tail short; tarsus much compressed. Species 3 in two genera. They inhabit northern regions, ranging S. in winter.

a. Bill stoutish, flattened, with lancet-shaped tip. . . Crymophilus, 382.
aa. Bill subulate, very slender. . . . . Phalaropus, 383.

382. CRYMOPHILUS Vieillot. (κρυμός, cold; φίλος, loving.)

750. **C. fulicarius** (L.). Red Phalarope. Back black, the feathers tawny edged; top of head blackish, its sides white; rump white; quills mostly black; feet yellowish; lower parts purplish chestnut; young white below; membrane of toes scalloped. L. 8. W. 5. T. 2¾. B. 1. Ts. ⅞. Northern regions. (*Eu.*) (Lat., Coot-like.)

383. PHALAROPUS Brisson. (φαλαρίς, the coot; πούς, foot.)

a. Membranes of toes scalloped; wing less than 5½. (*Phalaropus.*)

751. **P. lobatus** (L.). Northern Phalarope. Adult grayish black, variegated with tawny; rump and under parts white; neck largely rusty red; bill and feet black. L. 7. W. 4½. T. 2. B. ⅘. Ts. ⅘. Northern regions, chiefly along sea-shores. (*Eu.*)

aa. Membrane of toes plain; wing more than 4½. (*Steganopus* Vieillot.)

752. **P. tricolor** (Vieillot). Wilson's Phalarope. Ashy above, more or less variegated with chestnut; rump pale; lower parts white; sides of head and neck with a stripe of dark wine-red, which changes to black above; tail marbled; winter plumage with no red or black; bill and feet black. L. 9. W. 5. T. 2¼. B. 1¼. Ts. 1¼. N. Am., chiefly in interior; largest and handsomest of the Phalaropes, varying much with the season.

Family CXLIV. RECURVIROSTRIDÆ. (The Avocets.)

A little family allied to the snipe, with the legs excessively long and the bill very slender, long, acute, straight or curved upward. Genera 3, species 8; in most parts of the world. *Himantopus* is said to have relatively longer legs than any other bird.

a. Toes 4; the anterior full webbed; bill recurved, flattened, tapering to a fine point; plumage beneath thickened, as in ducks; swimmers.
  Recurvirostra, 384.
aa. Toes 3, semipalmate; bill nearly straight, not flattened.
  Himantopus, 385.

384. RECURVIROSTRA Linnæus.   (Lat. *recurvus*, bent
upward; *rostrum*, beak.)

753. **R. americana** Gmelin.   AVOCET.   BLUE STOCKING.
White, with cinnamon brown on head and neck, the wings mostly
black; legs blue. L. 18. W. 8½. T. 3½. B. 3¼. Ts. 3¾. N. Am.

385. HIMANTOPUS Brisson.   (ἱμαντόπους, strap-leg.)

754. **H. mexicanus** (Müller).   STILT.   LONG SHANKS.   LAW-
YER.   Glossy black above, white below; tail ashy; ♀ slaty; legs
pink. L. 15. W. 9. T. 3. Ts. 4. B. 2½. N. Am.

FAMILY CXLV. **SCOLOPACIDÆ.** (THE SNIPE.)

Bill elongated, usually longer than the head; if short, not plover-
like, being soft-skinned throughout (hard when dry); nasal grooves
in the form of narrow channels ranging from half to nearly the
whole length of the bill; sides of lower mandible usually also
grooved; nostrils narrow exposed slits; head feathered. Wings
usually thin and pointed; tail short and soft; tibia rarely entirely
feathered. Tarsus never entirely reticulate and usually scutellate
in front and behind; hind toe present (except in *Calidris*); front
toes cleft or slightly webbed; size medium or small. Sexes alike
or female slightly larger; seasonal changes in plumage often
strongly marked. Eggs usually four, placed with the small ends
together in a slight nest or depression in the ground; notes various; mostly migratory or gregarious. Genera about 20; species
100; chiefly of northern regions, but not wanting in most parts of
the world. (σκολόπαξ, snipe.)

a. Tarsus scutellate behind as well as in front; bill not strongly decurved.
  b. Eyes far back, directly above the ears; bill long; tip of upper mandible
    thickened; plumage unchanging. (*Scolopacinæ.*)
    c. Tibia entirely feathered; 3 outer primaries attenuate; toes not webbed.
                                                        PHILOHELA, 386.
    cc. Tibia naked below; no attenuate primaries. . . . GALLINAGO, 387.
  bb. Eyes not far back, considerably before the ears; tip of upper mandible
    thin; summer and winter plumage different. (*Tringinæ.*)
    d. Toes not webbed at all (or with a single minute web).
      e. Hind toe wanting. . . . . CALIDRIS, 392.
      ee. Hind toe present.
        f. Bill not shorter than middle toe with claw; (inner webs of quills
          not mottled). . . . . TRINGA, 390.
        ff. Bill shorter than middle toe with claw; (inner webs of quills
          mottled). . . . . TRYNGITES, 397.
    dd. Toes more or less webbed at base.
      g. Tail graduated, more than half wing. . . BARTRAMIA, 396.
      gg. Tail not more than half wing, little graduated.
        h. Tail longer than bill (from frontal feathers); gape reaching
          beyond base of culmen.

i. (Wing less than 4; toes well webbed; both mandibles grooved to the tip; tail not barred). . . . . EREUNETES, 391.

ii. (Wing not less than 4.)

  j. Bill narrower at tip, its upper surface hard and smooth, not grooved to the tip; (tail barred).

    k. Tarsus about as long as middle toe and claw; (wings less than 4½).
ACTITIS, 398.

    kk. Tarsus rather longer than middle toe and claw; (wings more than 4½).

      x. Bill slender; (legs dusky or yellow). . . . TOTANUS, 394.

      xx. Bill stout; (legs bluish). . . . . . SYMPHEMIA, 395.

  jj. Bill slightly broadened at tip, its upper surface slightly wrinkled or pitted. . . . . MICROPALAMA, 389.

hh. Tail shorter than bill; gape not reaching behind base of culmen: (tail barred or else chiefly black).

    l. Culmen with a median groove; tip of both mandibles wrinkled or pitted. . . . . MACRORHAMPHUS, 388.

    ll. Culmen smooth, not grooved. . . . . LIMOSA, 393.

aa. Tarsus scutellate in front, reticulate behind; bill very long, decurved. (*Numeniinæ*). . . . . NUMENIUS, 399.

386. PHILOHELA Gray. (φίλος, lover ; ἕλος, swamp.)

755. **P. minor** (Gmelin). AMERICAN WOODCOCK. Variegated, black, brown, gray, and russet; occiput banded with blackish and rusty, below warm brown. L. 11. W. 5. B. 3. T. 1¼. E. U. S., in swamps, W. to Nebr. (The European woodcock, *Scolopax rusticola* L., a similar but considerably larger bird, is an occasional straggler to E. U. S.)

387. GALLINAGO Leach. (Lat., *gallus*, cock.)

756. **G. delicata** (Ord). WILSON'S SNIPE. Back varied with black and bay; crown black, with a pale median stripe; breast mottled; sides barred; bill straight, very long. L. 11. W. 5. B. 2¼. T. 2¼. N. Am., abundant; a favorite game bird.

388. MACRORHAMPHUS Leach. (μακρός, long ; ῥάμφος, beak.)

757. **M. griseus** (Gmelin). GRAY SNIPE. DOWITCHER. Blackish and grayish; breast rusty-red in summer; bill long, nearly as in *Gallinago*. L. 11. W. 5½. T. 2¼. E. N. Am., abundant coastwise.

389. MICROPALAMA Baird. (μικρός, small ; παλάμη, palm.)

758. **M. himantopus** (Bonaparte). STILT SANDPIPER. Blackish, marked with chestnut, etc.; ashy gray in winter; bill nearly as in *Gallinago*. L. 9. W. 5. T. 2¼. B. 1¾. E. N. Am., not common. (*Himantopus*, the stilt.)

390. TRINGA Linnæus. (Low Lat., sandpiper.)

a. Wing 6 or more; middle pair of tail feathers not longer than the rest. (*Tringa.*)

759. **T. canutus** L. ROBIN SNIPE. KNOT. Brownish black, reddish brown below; bill straight; tarsus not shorter than middle toe and claw. L. 11. W. $6\frac{1}{2}$. T. $2\frac{1}{3}$. Atlantic coasts, common. (*Eu.*) (For King Canute.)

aa. Wing less than 6; middle pair of tail-feathers longer and more pointed than the rest.

b. Tarsus shorter than middle toe with claw, the latter shorter than bill. (*Arquatella* Baird.)

760. **T. maritima** Brünnich. PURPLE SANDPIPER. Ashy black with purplish reflections; feathers with pale edgings; lower parts mostly white; bill nearly straight. L. 9. W. 5. T. $2\frac{2}{3}$. B. $1\frac{1}{4}$. Atlantic coasts. (*Eu.*)

bb. Tarsus longer than middle toe with claw (or else toes very slender, without distinct lateral membrane).

c. Bill scarcely longer than tarsus, and not half length of tail. (*Actodromas* Kaup.)

d. Wing more than $4\frac{1}{2}$.

e. Rump and middle tail coverts plain black or dusky; throat with an ashy or brownish suffusion and dusky streaks.

761. **T. maculata** Vieillot. PECTORAL SNIPE. JACK SNIPE. Clay-color, striped with blackish above; belly white; breast ashy-shaded and sharply streaked. L. 9. W. $5\frac{1}{2}$. B. $1\frac{1}{8}$. N. Am., abundant. (*Eu.*)

ee. Rump dusky, the feathers bordered by pale.

f. Upper tail coverts white, with or without dusky marks; throat sharply streaked, with little if any ashy suffusion.

762. **T. fuscicollis** Vieillot. WHITE-RUMPED SANDPIPER. Top of head buffy, streaked with black; middle tail-feathers mostly black. L. $7\frac{1}{2}$. W. $4\frac{3}{4}$. T. $2\frac{1}{4}$. E. U. S., coastwise. (*Eu.*) (Lat., *fuscus*, tawny; *collum*, neck.)

ff. Upper (median) tail coverts plain dusky.

763. **T. bairdi** (Coues). BAIRD'S SANDPIPER. Colors of next but larger; throat but little streaked. L. 7 to $7\frac{1}{2}$. W. $4\frac{2}{3}$. T. $2\frac{1}{4}$. B. $\frac{7}{8}$. America, rare E. (To Spencer Fullerton Baird.)

dd. Wing less than 4.

764. **T. minutilla** Vieillot. LEAST SANDPIPER. PEEP. Blackish. rusty and white, much variegated; throat streaked. Smallest of the sandpipers, resembling *Ereunetes*, but the feet different, being without webs. L. 6. W. $3\frac{1}{2}$. T. 2. N. Am., abundant.

cc. Bill considerably longer than tarsus and more than $\frac{2}{3}$ tail.

g. Tarsus less than $1\frac{1}{2}$ times middle toe without claw; upper tail coverts mostly dusky. (*Pelidna* Cuvier.)

765. **T. alpina** (L.). Dunlin. Ox-bird. Red-backed Sand-piper. Chestnut brown above; feathers black centrally; belly, in summer, with a broad black area. L. 8$\frac{1}{4}$. W. 5. T. 2$\frac{1}{4}$. B. 1$\frac{3}{4}$. Northern regions, the American var. **pacifica** Coues, larger than the European.

    gg. Tarsus 1$\frac{1}{2}$ times length of middle toe without claw; upper tail coverts white; bill decurved. (*Ancylocheilus* Kaup.)

766. **T. ferruginea** Brünnich. Curlew Sandpiper. Chiefly chestnut in summer, the back black and rusty; in winter largely brownish and streaky. L. 8. W. 5. B. 1$\frac{1}{2}$. Europe, straggling to N. E. (*Eu.*)

    391. EREUNETES Illiger. (ἐρευνητής, searcher.)

767. **E. pusillus** (L.). Semipalmated Sandpiper. Sand-peep. Grayish brown, often shaded with cinnamon, white below; small. L. 6$\frac{1}{4}$. W. 3$\frac{3}{4}$. T. 2. B. $\frac{7}{8}$ to $\frac{9}{10}$. N. Am.; abundant along beaches. (Lat., puerile.)

768. **E. occidentalis** Lawrence. Bill longer, $\frac{4}{5}$ to 1$\frac{1}{4}$; color chiefly rusty red above; chest and breast streaked. Pacific, frequently E.

    392. CALIDRIS Cuvier. (καλίδρις, old name of some bird.)

769. **C. arenaria** (L.). Sanderling. Rusty above, marked and spotted with grayish and whitish; white on wing coverts. L. 8. W. 5. T. 2$\frac{1}{4}$. B. 1. Northern regions, abundant coastwise, known by its lack of the hind toe. (*Eu.*) (Lat., relating to sand.)

    393. LIMOSA Brisson. (Lat., muddy.)

a. Tail distinctly barred.

770. **L. fedoa** (L.). Marbled Godwit. Marlin. Cinnamon brown, variegated above, nearly uniform below; no pure white; upper tail coverts cinnamon barred with black. L. 16 to 22. Ts. 3. W. 9. T. 3$\frac{1}{4}$. B. 4$\frac{1}{4}$. N. Am., abundant along shores. (Perhaps *fœdus,* ugly.)

aa. Tail black, white at base and tip.

771. **L. hæmastica** (L.). Black-tailed Godwit. Brownish black and reddish, more or less variegated above and below; some white; upper tail coverts with a white band. L. 15. W. 8. Ts. 2$\frac{1}{2}$. B. 3$\frac{1}{2}$. E. N. Am., rather northerly. (αἱμαστικός, blood-red.)

    394. TOTANUS Bechstein. (Ital., *totano.*)

a. Tarsus more than 1$\frac{1}{2}$ times middle toe without claw; legs yellow. (*Totanus.*)

772. **T. melanoleucus** (Gmelin). Greater Tell-tale. Yellow Shanks. Stone Snipe. Ashy brown, variegated with white, etc.; bill very slender, the nasal groove not half its length; legs

long. L. 12½. W. 7½. T. 3¼. B. 2¼. N. Am., frequent. (μέλας, black ; λευκός. white.)

773. **T. flavipes** (Gmelin). YELLOW LEGS. Colors as in pre- ～～ ceding; nasal groove more than half bill; smaller; legs longer. L. 11. W. 6½. T. 2½. B. 1¾. U. S., abundant. (Lat., *flavus*, yellow : *pes*, foot.)

aa. Tarsus much less than 1½ times middle toe and claw; legs dusky. (*Helodromas* Kaup.)

774. **T. solitarius** (Wilson). SOLITARY TATTLER. Olive brown, sparsely speckled with whitish above; below white; breast dusky; bill straight and slender. L. 9. W. 5. T. 2½. B. 1¼. N. Am., abundant about secluded ponds.

395. SYMPHEMIA Rafinesque. (σύν, with ; φημί, I speak ; in allusion to their noisy discussions.)

775. **S. semipalmata** (Gmelin). WILLET. Brownish gray, ～～ varied with dusky, mostly whitish below. L. 15 or 16. W. 7½. T. 3. B. 2⅛. N. Am., common coastwise. The larger western bird (Ill. and W.) is var. **inornata** Brewster. W. 8. B. 2½.

396. BARTRAMIA Lesson. (To William Bartram, "grandfather of American ornithology.")

776. **B. longicauda** (Bechstein). UPLAND SANDPIPER. Light brownish, marked with ochraceous and blackish; throat whitish; tail feathers mostly marked with white. L. 12½. W. 6½. T. 4. B. 1¼. E. N. Am., abundant in fields, etc. Allied to this species is the European Ruff (*Pavoncella pugnax* L.), occasionally taken in E. U. S., the male with a very conspicuous ruff.

397. TRYNGITES Cabanis. (From Tringa.)

777. **T. subruficollis** (Vieillot). BUFF-BREASTED SANDPIPER. Grayish, mottled with darker; buffy below; under primary coverts and quills with white, and finely mottled with black. L. 8. W. 5½. T. 2¼. B. ⅘. N. Am., chiefly in interior; not common. (Lat., *sub*, under ; *rufus*, reddish; *collum*, neck.)

398. ACTITIS Boie. (Lat., *acta*, shore.)

778. **A. macularia** (L.). TIP-UP. TEETER-TAIL. SPOTTED ～～ SANDPIPER. Lustrous drab above in summer, varied with black; pure white below, with round blackish spots in adult. L. 8. W. 4. T. 2. B. 1. N. Am., everywhere, common. (Lat. spotty.)

399. NUMENIUS Linnæus. (νέος, new ; μήνη, moon.)

a. Secondaries, quills, etc., rusty cinnamon; lower parts pale cinnamon.

779. **N. longirostris** Wilson. LONG-BILLED CURLEW. SICKLE ～ BILL. Cinnamon, varied with gray and blackish. L. 24. W. 12. T. 4. B. 5 to 9. N. Am., frequent.

aa. Secondaries and quills chiefly dusky brownish; lower parts dull buffy.

780. **N. hudsonicus** Latham. Jack Curlew. Crown with two broad dusky stripes, with a narrower median stripe of buffy. L. 18. W. 9. T. 3½. B. 3 or 4. N. Am.

781. **N. borealis** (Forster). Esquimaux Curlew. Dough Bird. Crown narrowly streaked with dusky, without paler median stripe. L. 14. W. 8¼. T. 3. B. 2½. N. Am., northwards.

Family CXLVI. **CHARADRIIDÆ.** (The Plovers.)

Head rather large, nearly globose; bill of moderate length, shaped somewhat like a pigeon's bill, with a constriction behind the horny terminal portion; nasal fossæ lined with soft skin, through which the slit-like nostrils open. Wings long and pointed, usually reaching beyond the tip of the short tail, sometimes spurred. Toes usually three, with basal web; tarsus reticulate; tibia naked below. Sexes similar, but seasonal changes of plumage great. Species about 75, in most parts of the world.

a. Plumage above speckled; below, black in breeding season; tarsus much longer than middle toe and claw. . . . . CHARADRIUS, 400.

aa. Plumage of upper parts not speckled; neck with dark rings; tarsus not much longer than middle toe and claw; hind toe wanting.
ÆGIALITIS, 401.

400. CHARADRIUS Linnæus. (χαραδριός, old name.)

a. Hind toe present, but very small. (*Squatarola* Cuvier.)

782. **C. squatarola** L. Black-Bellied Plover. Ox-eye. Grayish, speckled; black below in breeding season, at other times white; axillars sooty-black. L. 11½. W. 7. T. 3. B. 1⅙. Ts. 2. Northern regions; rather rare in U. S. (*Eu.*) (Venetian name.)

aa. Hind toe wholly wanting.

783. **C. dominicus** Müller. Golden Plover. Frost Bird. Dark and grayish above, profusely speckled, some of the spots bright yellow; black below in breeding season, at other times greyish; wing coverts smoky-gray. L. 10½. W. 7. T. 3. B. 1. Ts. 1⅔. N. Am., a well known game bird.

401. ÆGIALITIS Boie. (αἰγιαλίτης, one who lives along shore.)

a. Tail half or more length of wing; rump orange brown; two black bands on breast. (*Oxyechus* Reich.)

784. **Æ. vocifera** (L.). Kildeer. Grayish brown; tail with black, white, and pale orange; a black band above the white forehead. L. 10. W. 6½. T. 4. B. ⅘, black. N. Am., abundant in the Miss. Valley.

aa. Tail not half length of wing; rump colored like back; breast with one band or none. (*Ochthodromus* Reich.)

b. Bill as long as middle toe and claw.

785. **Æ. wilsonia** (Ord). WILSON'S PLOVER. Brownish gray; forehead and lower parts white, a black band on breast and one on front of crown; ♀ duller and rusty. L. $7\frac{3}{4}$. W. $4\frac{1}{2}$. B. $\frac{4}{5}$. Tropical shores, N. to N. Y. (To Alex. Wilson.)

bb. Bill shorter than middle toe without claw.

c. All toes distinctly webbed at base.

786. **Æ. semipalmata** Bonaparte. RING-NECK PLOVER. Dark grayish brown; black bands broad. L. 7. W. 5. B. $\frac{1}{2}$. N. Am.

cc. Inner toe not webbed at base.

787. **Æ. meloda** (Ord). PIPING PLOVER. Very pale ashy brown, clear white below; dark bands narrow and faint; toes slightly webbed. L. $6\frac{3}{4}$. W. $4\frac{3}{4}$. B. $\frac{1}{2}$. E. N. Am., along the coast; represented in Miss. Valley by var. **circumcincta** Ridgway, with the black patches on sides of breast coalescent.

FAMILY CXLVII. APHRIZIDÆ. (THE SURF-BIRDS.)

Toes 4, not webbed, the hinder short, well-developed; tarsus scutellate in front; legs rather long; wings long and pointed; tail short; bill rather short. Two genera, each with a species, found on most northern shores. (*ἀφρός*, surf; *ζάω*, I live.)

a. Bill as long as tarsus, hard, sharp-pointed; tail rounded. ARENARIA, 402.

402. ARENARIA Brisson. (Lat., relating to sand.)

788. **A. interpres** (L.). TURNSTONE. Variegated; black, white, and chestnut above, mostly white below; young without reddish; feet orange; throat white. L. $8\frac{1}{2}$. W. 6. T. $2\frac{1}{2}$. B. $\frac{4}{5}$. Northern regions, generally common. (*Eu.*) (Lat., a go-between.)

FAMILY CXLVIII. HÆMATOPODIDÆ. (THE OYSTER-CATCHERS.)

Toes 3, webbed at base; tarsus reticulate; legs stout, coarse and rough; wings long and pointed; tail short. Bill hard, long, constricted near base, much compressed, truncate at tip, nearly straight, adapted for opening shells; nasal groove short; nostril linear. Size large; sexes similar. One genus, with 6 or 7 species; shore-birds found in most countries.

403. HÆMATOPUS Linnæus. (*αἷμα*, blood; *πούς*, foot.)

789. **H. palliatus** Temminck. OYSTER-CATCHER. Back dark slate; head and neck black; bill and legs red; tail coverts white. L. 18. W. 10. T. $4\frac{1}{2}$. B. 3. American coasts. (Lat., wearing a cloak.)

Order XXXIX. **GALLINÆ.** (The Gallinaceous Birds.)

Bill short, stout, convex, horny, not constricted; nostrils scaled or feathered; cutting edge of upper mandible overlapping the lower. Head often partly or wholly naked, sometimes with fleshy processes. Legs moderate, stout; hind toe elevated (excepting in *Cracidæ*), smaller than the other toes, sometimes wanting. Feet usually slightly webbed. Tarsus broadly scutellate (sometimes feathered), occasionally spurred in the males; claws blunt, not much curved. Wings short, strong, concave; tail various, sometimes wanting, often immensely developed. Palate schizognathous, nasal bones schizorhinal; basipterygoid processes present. Precocial, often polygamous, terrestrial in habit and hence sometimes called *Rasores* or Scratchers.

A large order including the chief game birds of most countries, as well as most kinds of domesticated fowl. The Hen (*Gallus gallus*), the Guinea Hen (*Numida pucherani*), and the Peacock (*Pavo cristatus*), are familiar examples of the order. All these are now placed with the common turkey in the Old World family, *Phasianidæ*. (Lat., *gallus*, cock.)

Families of Gallinæ.

a. Hind toe short, small, inserted above level of the others.
  b. Tarsus without spurs; head feathered (or nearly so) and tail not vaulted.   TETRAONIDÆ, 149.
  bb. Tarsus with spurs in ♂; head often largely naked, the tail often vaulted.   PHASIANIDÆ, 150.

Family CXLIX. **TETRAONIDÆ.** (The Grouse.)

Hind toe small, short; tarsus without spurs; head nearly or quite feathered; tail not vaulted. Genera 12; species about 25. Game birds abounding in northern regions; the grouse mostly N. American. (Lat., *tetrao*, grouse.)

a. Tarsus bare, scutellate; nostril unfeathered, with a naked scale; sides of toes not pectinate (*Perdicinæ*).
  b. Head not crested; lower mandible with its tomia serrate toward the tip.   COLINUS, 404.
aa. Tarsus and nostrils more or less feathered; sides of toes pectinate in winter (*Tetraoninæ*).
  c. Tarsus feathered about half way; tail fan-shaped, of 18 broad, soft feathers; neck with a ruff. . . . . BONASA, 406.
  cc. Tarsus feathered to the toes.
    d. Tail more than half wing, rounded or even; no ruff or peculiar feathers on neck.
      e. Toes naked; plumage not white. . . . . DENDRAGAPUS, 405.
      ee. Toes feathered; winter plumage chiefly snow-white.   LAGOPUS, 407.

dd. Tail about half as long as wing; toes naked.

  f. Neck with a ruff of straight stiff feathers, beneath which is a
      bare, inflatable air-sac; tail rounded. . TYMPANUCHUS, 408.

  ff. Neck without peculiar feathers ; tail graduated, the middle
      feathers exserted. . . . . PEDIOCÆTES, 409.

404. COLINUS Lesson.   (Mex. name, *Acolin.*)

790. **C. virginianus** (L.).   BOB-WHITE.   QUAIL (North).
PARTRIDGE (South). Forehead, line through eyes, chin and throat
white, brownish yellow in ♀ ; crown dark ; plumage generally
chestnut red, barred and streaked. L. 9½. W. 5. T. 3. B. ⅔.
E. U. S., W. to Great Plains, abundant. The smaller European
quail, *Coturnix coturnix* L., with very short tail and lower mandible
entire, has been introduced E.

405. DENDRAGAPUS Elliott.   (δένδρον, tree ; ἀγάπη, love.)

a. Tail of 16 feathers ; no evident air-sac on side of neck (*Canachites*
Stejneger).

791. **D. canadensis** (L.).   SPRUCE PARTRIDGE.   CANADA
GROUSE. Black above with grayish markings; mostly black below
with white spots ; the sides streaked; tail black, often tipped with
reddish ; ♀ smaller, black interrupted or streaky. L. 16. W. 6¾.
T. 5½. Spruce swamps, northward; S. to N. Y. and Mich.

406. BONASA Stephens.   (βόνασος, wild bull.)

792. **B. umbellus** (L.).   RUFFED GROUSE.   PARTRIDGE
(North). PHEASANT (South). Crested ; sides of neck with a
ruff of soft dark feathers; color reddish or grayish brown, much
streaked and variegated with blackish and pale. L. 18. W. 7¼.
T. 7. E. U. S., abundant in woodland. (Lat., umbel.)

407. LAGOPUS Brisson.   (λαγώς, hare; πούς, foot.)

793. **L. lagopus** (L.).   WHITE PTARMIGAN.   WILLOW GROUSE.
Fore parts cinnamon brown, variegated with blackish, rest of body
chiefly white: winter plumage pure white, the tail black; bill stout.
L. 16. W. 8. T. 5. Arctic, S. to N. N. Y. in winter. (*Eu.*)
Some other species occur N.

408. TYMPANUCHUS Gloger.   (Lat., *tympanum*, drum; *nucha*, nape.)

a. Scapulars without conspicuous terminal whitish spots; neck tufts in ♂ of
more than 10 parallel-edged, obtuse feathers.

794. **T. americanus** (Reichenbach).   PINNATED GROUSE.
PRAIRIE HEN. PRAIRIE CHICKEN. Sides of neck with a tuft of
long pointed feathers, beneath which is a patch of bare, red skin.
capable of great inflation; color black, tawny and white, much

barred and streaked. L. 17. W. 9. T. 4½. ♀ smaller. Prairies, etc., Indiana to La. and N.; nearly exterminated eastward.

aa. Scapulars with large, conspicuous spots of buffy whitish; neck tufts in ♂ of not more than 10 lanceolate feathers.

795. **T. cupido** (L.). HEATH HEN. Rather smaller. W. 8¼. E. U. S., once from Mass. to Va., now extinct except on Martha's Vineyard. (To Cupid, the ruff on the neck likened to Cupid's wings.)

409. PEDIOCÆTES Baird. (πεδίον, plain; οἰκητής, inhabitant.)

796. **P. phasianellus** (L.). SHARP-TAILED GROUSE. Streaked and spotted, yellowish brown, black, and white; sexes alike. L. 18. W. 8¾. T. 5. Arctic Amer., S. to N. Ill.; the S. E. form with rusty grayish predominating, is var. **campestris** Ridgway. (Lat. *phasianus*, pheasant.)

FAMILY CL. **PHASIANIDÆ.** (THE PHEASANTS.)

The chief family of the *Gallinæ*, differing as a whole from the *Tetraonidæ* in having the tarsus in the ♂ armed with a spur. In many species the head is naked, in others the tail is long and vaulted, or otherwise peculiar. Genera 18; species 90; nearly all of the Old World, some of them among the most remarkable of birds in form and coloration. The two species of *Meleagrinæ* are American.

a. Head and neck unfeathered, with scattered hairs, and with caruncles ; forehead with a fleshy process; tail long, broad, truncate ; plumage metallic ; breast in ♂ with a tuft of bristles. (*Meleagrinæ*.)
MELEAGRIS, 410.

410. MELEAGRIS Linnæus. (μελεαγρίς, guinea-hen.)

797. **M. gallopavo** L. WILD TURKEY.[1] Glossy, coppery black. L. 48. W. 21. T. 18½. ♀ smaller, duller. Ontario to Rocky Mountains, S. to Mexico, becoming extinct eastward. The domestic Turkey is descended from a Mexican variety (var. **mexicana** Gould). (Lat., *gallus*, cock ; *pavo*, pea-fowl.)

ORDER XL. **COLUMBÆ.** (THE DOVES.)

Bill straight, compressed, the horny tip separated by a constriction from the soft part. Nostrils opening beneath a soft, tumid membrane or cere, at base of bill. Frontal feathers sweeping in a strongly convex outline across base of upper mandible ; tomiæ meeting. Hind toe on a level with the rest (except in *Starnœnas*,

---

[1] The account of the habits of the Turkey given by Linnæus is worth quoting : "Mas exæstuat inflato pectore, expansa cauda, sanguinea facie, relaxata frontis caruncula ; iræ tenax ; sapida caro."

etc.), the others usually not webbed. Tarsus mostly scutellate in front, elsewhere reticulate, the plates soft. Head small, skull schizognathous, the nasal bones schizorhinous ; basipterygoids present. Plumage soft, compact, the feathers very loosely inserted. Altricial: monogamous.

A small order, including some extinct forms, closely related to the *Gallinæ*. The principal family is the *Columbidæ*.

Families of Columbæ.

a. Wings and tail well developed. . . . . COLUMBIDÆ, 151.

FAMILY CLI. COLUMBIDÆ. (THE PIGEONS.)

Wings long, pointed; tail never forked, of 12 or 14 feathers; plumage compact, the feathers loosely inserted. Species about 300, found in most regions, but most abundant in the East Indies. Besides the following, quite a number of pigeons occur in the Southern States. The common tame dove (*Columba œnas* L.) is a fair type of the family.

a. Tarsus feathered at the suffrago, shorter than the lateral toes. (*Columbinæ*.)
  b. Tail very long, wedge-shaped, of 12 pointed feathers. ECTOPISTES, 411.
aa. Tarsus entirely bare, scutellate, longer than the lateral toes. (*Zenaidinæ*.)
  c. Tail long, pointed, of 14 pointed feathers, its length more than ⅔ wing.
                    ZENAIDURA, 412.
  cc. Tail short, rounded, of 12 broad feathers; less than ⅔ wing.
                   COLUMBIGALLINA, 413.

411. ECTOPISTES Swainson. (ἐκτοπιστής, wanderer.)

798. E. migratorius (L.). WILD PIGEON. PASSENGER PIGEON. Bluish drab, with reddish and violet tinges, reddish below; ♂ more reddish. L. 17. W. 7½. T. 8. E. N. A., abundant; gregarious.

412. ZENAIDURA Bonaparte. (*Zenaida*, a related genus; οὐρά, tail; *Zenaida* was named for Madame Zenaida Bonaparte.)

799. Z. macroura (L.). MOURNING DOVE. TURTLE DOVE. CAROLINA DOVE. Brownish olive, glossed with blue and wine color; plumage with metallic lustre; a dark ear spot; outer tail feathers with white; ♀ duller. L. 12. W. 5¾. T. 6¾. N. Am., N. to Canada, very abundant, feeding on the ground, its mournful note not an index to its merry disposition. (μακρός, large ; οὐρά, tail.)

413. COLUMBIGALLINA Boie. (Lat., *columba*, pigeon ; *gallina*, hen.)

800. C. passerina (L.). GROUND DOVE. Grayish olive, with bluish gloss; the head, breast, etc., wine-color in ♂. L. 6¼. W.

3¼. T. 2⅓. Tropical America, N. to Va.; common S. (Lat., like a sparrow.)

ORDER XLI. RAPTORES. (THE BIRDS OF PREY.)

Bill powerful, cered at base, strongly hooked at the end. Feet never zygodactyle; fourth toe sometimes versatile; hind toe developed, elevated or not; claws very strong in typical forms, weak in the vultures; tibia, and often tarsus, feathered. Primaries 10; tail feathers usually 12. Altricial, but young downy at birth. Carnivorous birds, often of large size and great strength, found in every part of the world. Some of them feed upon carrion, some of the smaller on insects, some on reptiles or fishes, the most of them on mammals and birds which are captured in open warfare. (Lat., *raptor*, robber.)

Families of Raptores.

a. Head entirely naked (downy in young); hind toe short, elevated; claws small; inner toe somewhat webbed; nostril longitudinal.
CATHARTIDÆ, 152.

aa. Head nearly or quite fully feathered; hind toe not elevated, its claw large and strong, like the others; inner toe not webbed; nostrils vertical or roundish.

b. Eyes lateral, not surrounded by a disk of radiating feathers; cere exposed; outer toe not reversible (except in *Pandion*).
FALCONIDÆ, 153.

bb. Eyes directed forward, surrounded by disks of radiating feathers; cere concealed by bristly feathers; outer toe reversible.

c. Facial disk sub-triangular; middle claw pectinate. . STRIGIDÆ, 154.

cc. Facial disk sub-circular, middle claw not serrate. . BUBONIDÆ, 155.

FAMILY CLII. CATHARTIDÆ. (THE NEW WORLD VULTURES.)

Head and part of neck bare. Eyes lateral; ears small. Bill lengthened, weak and but little hooked; nostrils perforate. Wings very long and strong, giving a strength and grace of flight which few birds possess. Hind toe short, and elevated; front toes long, somewhat webbed, with rather weak and straightish claws. Large turkey-like raptores, without the strength and spirit of the hawks and owls; "voracious and indiscriminate gormandizers of carrion and animal refuse of all sorts, hence efficient and almost indispensable scavengers in the warm countries where they abound." (*Coues.*) The vultures are voiceless. On the ground they walk rather clumsily. When disturbed they eject the fetid contents of their capacious crops. Two species, the Condor and the California Vulture, are among the largest birds of flight in the world. All are American, the Old World Vultures (*Vulturinæ*) being vulture-like hawks. Genera 5; species 6 or 8.

a. Wings very long, primaries reaching to end of tail or farther; tail rounded; nostrils large and broad. . . . . CATHARTES, 414.

aa. Wings short, scarcely reaching middle of tail; tail truncate; nostrils small and narrow. . . . . CATHARISTA, 415.

414. CATHARTES Illiger. (καθαρτής, purifier.)

801. C. aura (L.). TURKEY BUZZARD. Black, lustrous above and somewhat mottled with brown; skin of head and neck red. L. 30. W. 22. T. 12. Am., abundant, especially S. and S. W. (A South American name.)

415. CATHARISTA Vieillot. (καθαρίζω, to cleanse.)

802. C. atrata (Bartram). CARRION CROW. Uniform dull black. L. 24. W. 17. T. 8. Trop. Amer., straying N. to Ohio; a heavier bird than the Turkey Buzzard, although shorter. (Lat., blackened.)

FAMILY CLIII. FALCONIDÆ. (THE FALCONS.)

Head fully feathered (except in the Old World *Vulturinæ*); no ear tufts. Eyes lateral; eyelids provided with lashes; usually a projecting bony eyebrow; no complete facial disk. Toes always naked, and usually tarsus also; hind toe not elevated. Bill stout, strongly hooked, its base not hidden by feathers. Claws very strong and sharp, the hind claw not shorter than the others. Plumage usually of blended colors, barred or streaked; changes considerable; ♀ usually the larger. Genera 50; species 300; abounding everywhere.

a. Outer toe not reversible; claws graduated from the largest (hind-toe) to the smallest (outer).

  b. Nostril not circular, nor with an inner bony tubercle. (*Accipitrinæ.*)

    c. Tarsus naked, reticulate all around, much shorter than tibia.

      d. Tail very deeply forked. . . . . ELANOIDES, 416.

      dd. Tail merely emarginate; claws not grooved beneath.

                                    ELANUS, 417.

    cc. Tarsus not reticulate all around; claws grooved beneath.

      e. Tarsus decidedly shorter than tibia.

        f. Tarsus scutellate in front only; not fully feathered.

          g. Toes somewhat webbed at base; cutting edge of upper mandible notched. . . . . ICTINIA, 418.

          gg. Toes not webbed at all; neck feathers lanceolate.

                                    HALLÆETUS, 425.

      ff. Tarsus almost or quite entirely feathered.

        h. Tarsus densely feathered all around down to base of toes.

                                    AQUILA, 424.

        hh. Tarsus densely feathered to base of toes except a bare strip behind. . . . . ARCHIBUTEO, 423.

    fff. Tarsus scutellate in front and behind.

i. Wing rather pointed, more than 4 times length of tarsus.
BUTEO, 421.
ii. Wing rounded, less than 4 times length of tarsus.
ASTURINA, 422.
  ee. Tarsus about as long as tibia.
    j. Face without ruff; wings rounded, little longer than
tail; tarsus scutellate in front only, rarely booted.
ACCIPITER, 420.
    jj. Face with a slight ruff, somewhat as in owls; wings
very long, longer than the long tail, tarsus scutellate
in front and behind. . . . . . . CIRCUS, 419.
  bb. Nostrils small, circular, with a conspicuous central bony tubercle; cutting edge of upper mandible with a strong tooth, separated from
hooked tip of bill by a distinct notch; tarsus reticulate all around.
(*Falconinæ.*) . . . . FALCO, 426.
aa. Outer toe reversible; claws all of the same length, narrowed and rounded
on lower side; tarsus reticulate; plumage compact. (*Pandioninæ.*)
PANDION, 427.

416. ELANOIDES Gray. (Elanus; εἶδος. form.)

803. **E. forficatus** (L.). SWALLOW-TAILED KITE. Lustrous
black; head, neck, lower parts, and band on rump white; young
streaky. L. 25. W. 17. T. 14. Southern, N. to Penn. and Minn.
(Lat., *forfex*, shears; tail deeply forked, like shears.)

417. ELANUS Savigny. (Lat., kite.)

804. **E. leucurus** (Vieillot). WHITE-TAILED KITE. Bluish
gray, with white on head and tail, and black on shoulder. L. 17.
W. 13. T. 7. Tropical Am., N. to S. Ill. (λευκός, white; οὐρά, tail.)

418. ICTINIA Vieillot. (ἰκτῖνος, kite.)

805. **I. mississippiensis** (Wilson). MISSISSIPPI KITE. Chiefly
lead blue, wings with chestnut. L. 15. W. 12. T. 6½. S. E. U. S.,
N. to Penn. and Wis.

419. CIRCUS Lacépède. (κίρκος, a kind of hawk.)

806. **C. hudsonius** (L.). MARSH HARRIER. Chiefly pale
bluish gray; rump and under parts whitish; tail bluish, mottled
and tipped with white, and with dark bands; ♀ dusky brown.
L. 18. W. 15. T. 9. N. Am., abundant; readily known by the
white rump.

420. ACCIPITER Brisson. (Lat., hawk.)

a. Tarsus feathered less than ⅓ of the way down in front, the feathers well
separated behind. (*Accipiter.*)

807. **A. velox** (Wilson). SHARP-SHINNED HAWK. "PIGEON
HAWK." Tail truncate; tarsus sometimes "booted"; general
color bluish gray, breast, sides, etc., whitish, streaked with reddish

brown. L. 12. W. 7. T. 6. N. Am.; abundant: a small but courageous hawk. The species of this genus are more destructive among poultry than any other hawks. (Lat., swift.)

808. **A. cooperi** Bonaparte. CHICKEN HAWK. Tail rounded; tarsus never booted; colors similar, more blue, the top of head darker, the tail more plainly white tipped. L. 18. W. 10. T. 8. N. Am., common. (To Wm. Cooper, of New York.)

aa. Tarsus feathered about half way down in front, the feathers scarcely separated behind. (*Astur* Lacépède.)

809. **A. atricapillus** (Wilson). GOSHAWK. Chiefly slate blue with white superciliary stripe; lower parts white, finely barred with brown; tail with four dark bars. L. 24. W. 14. T. 11. Northern, S. to U. S. in winter. (Lat., black-haired.)

421. BUTEO Cuvier.   (Lat. buzzard, as these hawks are called in England.)

a. Outer web of primaries without white, buffy, or ochraceous spots.
   b. Four outer primaries emarginate on inner web.
      c. Head and neck uniform dark sooty brown, or streaked with white, never with buffy or reddish.

810. **B. harlani** (Audubon). BLACK HAWK. Tail irregularly mottled with grayish, rusty, white, or blackish and with a dark band near tip; general color usually very dark but variable. L. 21. W. 16. T. 10. S. W., E. to Ill., scarce.

   cc. Head and neck more or less streaked with ochraceous or rusty red.

811. **B. borealis** (Gmelin). HEN HAWK. RED-TAILED BUZZARD. Dark brown; much barred and streaked; tail in adult bright chestnut red above, with a narrow black bar near its tip. L. 23. W. 15½. T. 8½. N. Am. common, replaced W. by var. **calurus** Cassin, dark brown, sometimes uniform.

   bb. Three outer primaries emarginate on inner web.
      d. Wing more than 13.

812. **B. swainsoni** Bonaparte. SWAINSON'S BUZZARD. Gray, variously streaked, usually a bright chestnut or brownish area on breast; wings dusky; tail with nine or ten narrow dark bars; variable. L. 20. W. 16. T. 8½. W. U. S., E. to Ind. and Mass. (To Wm. Swainson.)

      dd. Wing less than 12.

813. **B. latissimus** (Wilson). BROAD-WINGED HAWK. Brown above, whitish or fulvous below, variously streaked and barred; conspicuous dark cheek patches; tail with broad dark bands alternating with narrower pale ones, white-tipped; lower parts brownish with whitish spots; in young whitish with darker streaks. L. 18.

W. 11. T. 7. E. N. Am., a handsome but small hawk. (Lat. broadest.)

aa. Outer webs of primaries spotted with white, buffy or ochraceous : 4 primaries emarginate.

814. **B. lineatus** (Gmelin). CHICKEN HAWK. RED-SHOULDERED BUZZARD. Dark reddish brown ; head and neck more or less rusty ; bend of wing orange brown in adult; tail with several white bars; young much streaked below and with little reddish. L. 22. W. 14. T. 9. Considerably lighter in weight than the red-tailed hawk, although nearly as long. N. Am., abundant.

422. ASTURINA Vieillot. (Lat., dim. of *Astur*, a hawk).

815. **A. plagiata** Schlegel. GRAY HAWK. GOSHAWK. Chiefly dark ashy gray, white below : wings and tail black, with white markings; upper tail coverts white. L. 18. W. 10. T. 7½. Mexican, straying to S. Ill. (Lat., striped.)

423. ARCHIBUTEO Brehm. (Lat. *archi*, chief ; *Buteo*.)

a. Bill small and weak, its gape, from corner to corner, 1⅞ inches.

816. **A. lagopus** (Brünnich). ROUGH-LEGGED HAWK. BLACK HAWK. Chiefly whitish, rusty streaked ; but sometimes entirely black. L. 24. W. 18. T. 10. Northern regions. The American form, var. **sancti-johannis** (Gmelin) is darker and more rusty than European. (*Eu.*) (λαγώς, hare ; πούς, foot.)

aa. Bill strong, the gape 1⅞ inches wide from corner to corner of mouth.

817. **A. ferrugineus** (Lichtenstein). Rusty brown, marked with gray, white, and black ; sometimes plain dark chocolate-brown. L. 23. W. 17. T. 10. W. N. Am., E. to Ill.

424. AQUILA Brisson. (Lat., eagle.)

818. **A. chrysaetos** (L.). GOLDEN EAGLE. Glossy dark brown ; head and neck paler tawny brown ; quills blackish ; tail clouded with whitish at base. L. 36. W. 25. T. 16. Northern regions, less common than the Bald Eagle, in the U. S. (*Eu.*) (χρυσός, gold ; ἀετός, eagle.)

425. HALIÆETUS Savigny. (ἅλς, sea ; ἀετός, eagle.)

819. **H. leucocephalus** (L.). BALD EAGLE. Dark brown ; head, neck, and tail white (after the third year); bill and feet yellow. L. 36. W. 25. T. 14. N. Am., everywhere. "Common, for an eagle; a piratical parasite of the Osprey, otherwise notorious as the emblem of the Republic." (*Coues.*) (λευκός, white ; κεφαλή, head.)

426. FALCO Linnæus. FALCONS. (Lat., falcon.)

. First primary only emarginate on inner web; tarsal plates small; sexes colored alike.

  b. Tarsus longer than middle toe and claw.

    c. Tarsus feathered in front more than half way down. (*Hierofalco* Cuvier.)

820. **F. rusticolus** (L.). GRAY GYRFALCON. Bluish gray above with dark bands; lower tail coverts always with ashy; young plain above, streaky below. L. 24. W. 16. T. 10. Northern regions; var. **obsoletus**, S. to U. S. in winter. This form is darker, the lower parts chiefly dusky. (*Eu.*) (Lat. rural.)

    cc. Tarsus not feathered more than half way down in front. (*Gennaia* Kaup.)

821. **F. mexicanus** Schlegel. PRAIRIE FALCON. Grayish brown, more or less barred and streaked. L. 18. W. 14. T. 8. S. W., E. to Ill., allied to the Lanier of Europe.

  bb. Tarsus not longer than middle toe, scarcely feathered below heel joint. (*Rhynchodon* Nitsch.)

822. **F. peregrinus** Tunstall. PEREGRINE FALCON. DUCK HAWK. Blackish ash with paler waves; top of head black ; below whitish ; black cheek patches. L. 16. W. 13. T. 7. Northern regions, not very common; the American bird, var. **anatum** Bonaparte, has the breast unstreaked. (*Eu.*)

  aa. Two primaries emarginate; tarsal plates enlarged in front, appearing like scutella.

    d. Tarsus about equal to middle toe; basal joints of toes with small hexagonal scales. (*Æsalon* Kaup.)

823. **F. columbarius** L. PIGEON HAWK. AMERICAN MERLIN. Ashy blue or brownish above with darker streaks ; lower parts whitish or buffy, streaked with brown ; middle tail feathers in ♂ with about 4 black bands; in ♀ with about 6 pale bands. L. 13. W. 8. T. 5. U. S. (Lat., pertaining to a pigeon.)

    dd. Tarsus longer than middle toe ; basal joints of toes with transverse scutella. (*Tinnunculus* Vieillot.)

824. **F. sparverius** L. SPARROW HAWK. RUSTY CROWNED FALCON. Back tawny; wings bluish and black in ♂ ; seven black blotches about head : tail chestnut, with a broad black band in ♂, and a narrow terminal one of white ; below white or tawny. ♀ different, more streaky, the tail tawny with numerous narrow darker bars: back and wing coverts rusty barred with black. One of the most active and courageous of the hawks; a genuine falcon, notwithstanding its small size. L. 11. W. 7. T. 5. U. S., abundant. (Lat., relating to a sparrow.)

427. PANDION Savigny. (πανδίων, a name in mythology.)

825. **P. haliaëtus (L.).** Osprey. Fish Hawk. Dark brown; tail grayish with narrow dark bars; head neck and lower parts mostly white; ♀ with the breast more spotted; feet very large. L. 24. W. 20. T. 10. In most parts of the world, about water, an expert fisher. The American bird is var. **carolinensis** Gmelin. (Gr., sea-eagle.)

Family CLIV. STRIGIDÆ. (The Barn Owls.)

A small family including those owls which have long faces, the facial disk being complete and subtriangular. All these have the sternum entire behind, with a central emargination, the furculum grown fast; the middle and inner toes are about equal in length, and the middle claw is pectinate below. Genera 2, species 6 or 8, chiefly of Asia and Europe.

a. Wings long, pointed, reaching beyond tail when folded; no ear tufts; tarsus scant-feathered; bill pale ; eyes black . . . . Strix, 428.

428. STRIX Linnæus. (Lat. screech-owl.)

826. **S. pratincola** Bonaparte. Barn Owl. Tawny of various shades, very finely mottled, streaked, and dotted with darker; below pale, with some spots. L. 17. W. 13. T. 5½. N. Am., chiefly S., a handsome, solemn-looking, and fierce little owl. (Lat., inhabitant of fields.)

Family CLV. BUBONIDÆ. (The Owls.)

Head very large, shortened lengthwise and greatly expanded laterally, the eyes directed forwards and partly surrounded by a more or less complete circular disk of radiating feathers of peculiar texture ; loral feathers antrorse, long and dense ; feathers on the sides of forehead often elongated into ear-like tufts. Plumage very soft and lax, rendering the flight almost noiseless ; its colors blended and mottled so as to baffle description. External ear very large, often provided with a movable flap. Outer toe reversible ; claws very sharp, long, and strong ; inner toe shorter than middle ; middle claw not pectinate. Sternum double notched or fenestrate ; furculum free. Eggs nearly spherical, pure white. Chiefly nocturnal. Sexes colored alike, ♀ usually the larger. Owls are found in every part of the globe, and most of the species have a wide range. Their habits are too well known to need description here.

a. Tarsus fully feathered.
  b. Eye in the centre of a nearly complete circular disk; external ear larger than eye, with a well developed opercle.
    c. Ear-tufts present, sometimes very short ; cere longer than rest of culmen; iris yellow. . . . . Asio, 429.

  cc. Ear-tufts not evident; cere short.

    d. Tail about ⅔ wing.

      c. Eyes not small, the iris dusky; 5 outer primaries emarginate.

                                SYRNIUM, 430.

      cc. Eyes rather small, the iris yellow; six outer primaries sinuate; bill small. . . . . SCOTIAPTEX, 431.

    dd. Tail short, nearly even, about half wing ; iris yellow.

                                NYCTALA, 432.

  bb. Eye nearer top than bottom of a more or less incomplete disk; external ear not larger than eye, without developed opercle; iris yellow.

    f. Head with very conspicuous ear-tufts.

      g. Tail about ½ wing ; bill pale. . . . . MEGASCOPS, 433.

      gg. Tail about ⅔ wing; bill blackish. . . . . BUBO, 434.

    ff. Head without evident "ear-tufts;" (these rudimentary in *Nyctea*).

      h. Tail rounded; plumage chiefly white. . NYCTEA, 435.

      hh. Tail graduated; plumage not white. . . SURNIA, 436.

aa. Tarsus nearly naked, its length more than twice middle toe; facial disk imperfect; no ear-tufts. . . . . SPEOTYTO, 437.

429. ASIO Brisson.

a. Ear-tufts well developed. of 8 to 12 feathers. (*Asio.*)

  827. A. wilsonianus (Lesson). LONG-EARED OWL. One primary emarginate. Dusky, more or less mottled and streaked with buffy and grayish, much variegated below. L. 15. W. 12. T. 6. N. Am., often using deserted crow's nests.

aa. Ear-tufts inconspicuous, few feathered. (*Brachyotus* Gould.)

  828. A. accipitrinus (Pallas). SHORT-EARED OWL. Two outer primaries usually emarginate. Buffy whitish, striped with dark brown, the dark streaks narrower below. L. 15. W. 12. T. 6. Nearly cosmopolitan, not rare in U. S. (*Eu.*)

430. SYRNIUM Savigny.

  829. S. nebulosum (Forster). BARRED OWL. Toes not concealed. Olive brown, barred with white above; breast similarly barred; belly streaked. L. 18. W. 14. T. 9. E. N. Am., common; the most noisy of our owls, but rather mild in temper.

431. SCOTIAPTEX Swainson. (σκότιος, dark ; πτέρυξ, wing.)

  830. S. cinereum (Gmelin). GREAT GRAY OWL. SPECTRAL OWL. Toes concealed by long feathers. Cinereous brown above, waved with white ; breast streaked, belly barred. L. 30. W. 18. T. 12. Northern, occasionally S. in winter to N. J. and Ill.; a huge bird, one of the largest of owls. (*Eu.*) (Lat., ashy.)

432. NYCTALA Brehm. (νυκταλός, drowsy.)

  831. N. tengmalmii (Gmelin). SPARROW OWL. Bill yellow; cere not tumid ; nostrils obliquely oval, opening laterally. Choco-

late brown, striped with white; below white, striped with brown.
L. 10. W. 7¼. T. 4½. Northern regions, S. to N. U. S.; the
American var. **richardsoni** Bonaparte, is larger and darker. (*Eu.*)
(To P. G. Tengmalm, a Swedish naturalist.)

832. **N. acadica** (Gmelin). SAW-WHET OWL. Bill black; cere
tumid; nostrils nearly circular, opening anteriorly. Color similar,
less white above, more reddish below. L. 8. W. 5¾. T. 2¾. N.
Amer., rather northerly.

433. MEGASCOPS Kaup. (μέγας, great; σκώψ, screech-owl.)

833. **M. asio** (L.). SCREECH OWL. RED OWL. Grayish,
streaked and barred, or else with the grayish replaced by bright
reddish; these two different styles of plumage about equally common and bearing no relation to age, sex, or season. L. 10. W. 7.
T. 3½. N. Am., abundant; our commonest owl.

434. BUBO Duméril. (Lat., horned owl.)

834. **B. virginianus** (Gmelin). GREAT HORNED OWL. Black,
gray, and buffy, variously mottled and barred; usually a whitish
half-collar; ear tufts large, their feathers mostly black. L. 22.
W. 16. T. 10. N. Am., abundant; one of the strongest and most
courageous of the owls.

435. NYCTEA Stephens. (νυκτεύς, nocturnal.)

835. **N. nyctea** (L.). SNOWY OWL. Pure white, more or less
barred with blackish. L. 23. W. 17. T. 10. Northern regions,
not rare; whole U. S. in winter; the handsomest of owls. (*Eu.*)

436. SURNIA Duméril. (Meaning unknown.)

836. **S. ulula** (L.). HAWK OWL. DAY OWL. Brown, much
mottled and barred; head with white spots; a dark collar about
neck; lower parts barred, brown and white; tail barred. L. 16.
W. 9. T. 7. Northern regions, S. to Wis. and Mass. (*Eu.*)
The American bird, darker in color, is var. **caparoch** (Müller).
(Lat., owl; an imitation of the bird's note.)

437. SPEOTYTO Gloger. (σπέος, cave: τυτώ. a kind of owl.)

837. **S. cunicularia** (Mol.). BURROWING OWL. Brownish,
much spotted, barred, and variegated with whitish. L. 10. W. 7½.
T. 4. W. America, very abundant W., living in the holes of
prairie dogs, accidental E. Our form is var. **hypogæa** (Bonaparte),
its lower parts buffy. (Lat., burrower.)

ORDER XLII. **PSITTACI.** (THE PARROTS.)

" *Frugivorous Raptores*," bill enormously thick, cered at base and
strongly hooked; tongue thick and fleshy. Feet zygodactyle by

reversion of outer toe; tarsus reticulate. Tongue short, fleshy; upper jaw unusually movable. Altricial. Plumage often brilliant. In all warm regions. Genera 26; species 354, nearly half of which are American. All of the latter, and many of the Old World forms, belong to the principal family, *Psittacidæ*.

Families of Psittaci.

a. Carotids two, the left superficial. . . . . PSITTACID.Æ, 156.

FAMILY CLVI. **PSITTACIDÆ.** (THE PARROTS.)

Parrots with two carotid arteries, the left superficial. This great group includes the great majority of the parrots, — all of the American species.

a. Ambiens muscle present: a tufted oil-gland; furculum complete. (*Arinæ.*)

b. Face entirely feathered except a curve about the eye; tail graduated, the feathers narrowed. . . . . CONURUS, 438.

438. **CONURUS** Kuhl. (κῶνος, cone; οὐρά, tail.)

838. **C. carolinensis** (L.). CAROLINA PAROQUET. Green; head and neck yellow; face orange red; wings with blue and yellow; bill white; cere feathered. L. 13. W. 7½. T. 6. Southwestern, formerly N. to the Great Lakes; now nearly exterminated, except in Fla.

PICARIÆ.

NOTE. — Between the Parrots and the Singing Birds comes the series or so-called order of *Picariæ*, a highly diversified group including all the non-passerine land birds, except the pheasants, doves and birds with cered and hooked bill. In all, the hind toe is small (if present), and sometimes elevated ; its claw is usually shorter than that of middle toe. The wing coverts are larger and in more numerous series than in the *Passeres*. The primaries are 10 in number, the first rarely short; tail usually of 10 feathers. *Sternum non-passerine ;* musical apparatus imperfect ; tarsus never presenting an undivided ridge behind. Nature altricial.

Recent writers usually subdivide the *Picariæ* into three groups, which are recognized as distinct orders by the American Ornithologists' Union, under the names of *Coccyges, Pici* and *Macrochires.* The last two are natural groups and well defined by anatomical characters. The *Coccyges*, however, are scarcely less varied than the *Picariæ*, of which they form the greater part. Dr. Coues says : "I have no faith whatever in the integrity of any such grouping as *Picariæ* implies, but if I should break up this conventional assemblage, I should not know what to do with the fragments." The so-called order *Acanthopteri* among fishes is a case somewhat parallel.

ORDER XLIII. **COCCYGES.** (THE CUCKOO-LIKE BIRDS.)

This order includes the majority of the Picarian birds, some 15 families, not having very much in common, except that they lack the special peculiarities of the *Pici* and the *Macrochires*. "The sternum is usually notched behind; the syringeal muscles are two pairs at most." Feet generally short, the toes variously arranged. Palate desmognathous. The group is "a mixed lot requiring to be reconstructed by exclusion of some of the families entering into its composition." (*Coues*). (κόκκυξ, cuckoo.)

Families of Coccyges.

a. Toes 2 in front, 2 behind; bill as long as head, compressed, the tomia entire; nostrils exposed; no rictal bristles; toes cleft to base.
CUCULIDÆ, 157.

aa. Toes 3 in front, 1 behind; bill straight, longer than head; feet syndactyle, the outer and middle toes grown together for half their length; tarsus very short. . . . . ALCEDINIDÆ, 158.

FAMILY CLVII. **CUCULIDÆ.** (THE CUCKOOS.)

Bill compressed, lengthened, without rictal bristles or nasal tufts. Tongue not extensible. Tarsus long, nearly naked; toes not webbed. Feet zygodactyle, by reversion of fourth toe. Species about 200, in various parts of the world. (Lat., *cuculus*, cuckoo.)

a. Tail feathers 10; bill gently curved; plumage blended; arboreal.
COCCYZUS, 439.

439. **COCCYZUS** Vieillot.

839. **C. americanus** (L.). YELLOW-BILLED CUCKOO. "RAIN CROW." Color lustrous drab; bill yellow below; wings with much cinnamon red; middle tail feathers like the back; outer ones black, with broad white tips. L. 12. W. 5½. T. 6. N. Am.

840. **C. erythrophthalmus** (Wilson). BLACK-BILLED CUCKOO. Lustrous drab; bill chiefly black; wings with little or no reddish; tail feathers all brownish, obscurely whitish at tips. L. 11½. W. 5. T. 6¼. E. N. Am., more common E. (ἐρυθρός, red; ὀφθαλμός, eye.)

FAMILY CLVIII. **ALCEDINIDÆ.** (THE KINGFISHERS.)

Head large; bill long, straight and strong, usually longer than head; gape deep, tomia not serrate. Wings long; tail short. Legs quite small; feet syndactyle, — the outer and middle toes united half their length, with a continuous sole beneath; tibia naked below. Tail feathers 12. Species about 100, chiefly of the tropical parts of the Old World and Australia. Many of them feed upon fishes, and nearly all are remarkable for their brilliant metallic coloration. In many the bill is disproportionately large. (Lat., *alcedo*, kingfisher.)

a. Bill compressed, the culmen carinate; head crested; aquatic, feeding on fishes. . . . . CERYLE, 440.

440. CERYLE Boie. (κηρύλος, kingfisher.)

841. C. alcyon (L.). BELTED KINGFISHER. Ashy blue above, a bluish band across breast; white below; tail black, speckled and barred with white; ♀ with sides and band across belly chestnut. L. 13. W. 6. T. 3½. B. 2. N. Am., everywhere common. (Lat., kingfisher.)

ORDER XLIV. PICI. (THE WOODPECKERS AND WRYNECKS.)

A small order composed of the Woodpeckers, and two closely related families. Feet zygodactyle, the outer toe permanently reversed, the hind toe wanting in one genus; metatarsus modified in connection with the reversed toe; wing with 10 primaries and short secondary coverts; tail with 10 quills, besides which, in the woodpeckers, is an outer pair of partly concealed spurious quills; bill straight, hard and strong; palate saurognathous; sternum double-notched behind; salivary glands large. (Lat., *picus*, woodpecker.)

a. Nostrils covered by feathers; tail feathers rigid and acute. PICIDÆ, 159.

FAMILY CLIX. PICIDÆ. (THE WOODPECKERS.)

Bill stout, usually straight, with the tip truncate or acute, fitted for hammering or boring into wood. Tongue long, flattish, barbed, capable of great protrusion, adapted for securing insects (except in *Sphyrapicus*): hyoid apparatus peculiar, its horns generally quite long, curving around the skull behind and over forward again to the ear or beyond. Feet zygodactyle, the hind toe sometimes wanting; claws compressed, sharp and strong. Tail feathers 12, rigid and acuminate, the outer pair short, concealed; tail never forked; nasal tufts present.

Chiefly arboreal; all (except *Sphyrapicus*, which is truly a "Sap-Sucker,") are pre-eminently insectivorous. For this reason these birds are of the greatest service to the farmer. Voice loud and harsh. Colors generally bright, the male at least having almost always red on the head; sexes usually slightly different. Species 250, abundant almost everywhere.

a. Outer hind toe longer than outer anterior (middle) toe.
 b. Head with a conspicuous crest; (size very large; bill pale).
CAMPEPHILUS, 441.
 bb. Head not crested.
  c. Tongue pointed, highly extensible, as usual among woodpeckers; (no yellowish on belly).

d. Hind toes 2; nasal groove running nearly to tip of bill, which is not
   much compressed toward tip. . . . . . . DRYOBATES, 442.
dd. Hind toe single (the real hind toe wanting); bill broad, much
   compressed. . . . . PICOIDES, 443.
cc. Tongue obtuse, brushy, scarcely extensible; nasal groove running
   into tomium near middle of bill; bill evidently compressed towards its tip; (belly more or less yellowish). SPHYRAPICUS, 444.
aa. Outer toe not longer than outer anterior toe.
   e. Head conspicuously crested; (size large; bill dark).
   CEOPHLŒUS, 445.
   ee. Head not crested.
      f. Upper mandible with a distinct low lateral ridge, the tip more
         or less truncate. . . . . MELANERPES, 446.
      ff. Upper mandible without distinct lateral ridge or nasal groove,
         the tip scarcely truncate. . . . . . . COLAPTES, 447.

441. CAMPEPHILUS Gray. (κάμπη, caterpillar; φίλος, loving.)

842. **C. principalis** (L.). IVORY-BILLED WOODPECKER. Black
with white on shoulders and wings; crest scarlet in ♂, black in ♀.
L. 21. W. 11. T. 8. Southern, formerly N. to S. Ill.

442. DRYOBATES Boie. (δρύς, oak; βάτης, walker.)

a. Back black, with a long white stripe; sides usually white.

843. **D. villosus** (L.). HAIRY WOODPECKER. BIG SAP-
SUCKER. Spotted and lengthwise streaked, but not banded; outer
tail feathers wholly white. L. 9. W. 5. T. 3½. A scarlet nuchal
band in ♂ only. N. Am., common.

844. **D. pubescens** (L.). DOWNY WOODPECKER. LITTLE
SAP-SUCKER. Outer tail feathers white, barred with black; other-
wise precisely like the other, but much smaller. L. 6¼. W. 3¾.
T. 2¾. N. Am., common.

aa. Back black, barred with white; sides usually spotted or streaked with
   black.

845. **D. borealis** (Vieillot). RED-COCKADED WOODPECKER.
Black and white, spotted and crosswise banded, but not streaked;
a red line on each side of head in ♂. L. 8½. W. 4½. T. 3½.
S. E. U. S., in swamps, N. to Penn.

443. PICOIDES Lacépède. (*Picus*; εἶδος, resemblance.)

846. **P. arcticus** (Swainson). BLACK-BACKED WOODPECKER.
Black and white; no white on back or top of head; crown yellow
in ♂, plain in ♀. L. 9. W. 5. T. 3¾. N. Am., S. in winter,
to N. U. S.

847. **P. americanus** Brehm. Back with white bars or a white
stripe; usually more or less white on head; otherwise as in the preceding. L. 8. W. 4½. T. 3½. Arctic Amer., S. in winter to
N. E.

444. SPHYRAPICUS Baird. (σφῦρα, hammer; *Picus.*)

848. **S. varius** (L.). YELLOW-BELLIED WOODPECKER. Black and whitish above; black on breast; rump mixed black and white; belly more or less yellowish; sides streaked with dusky; a white wing patch; quills with white spots; crown red in adult ♂ and usually ♀ also; chin scarlet, throat black in ♂; both white in ♀, young dull brownish. L. 8¼. W. 4¾. T. 3¼. N. Am., not rare, the only woodpecker which ever injures trees.

445. CEOPHLŒUS Cabanis. (*Hylatomus* [1] Baird.)
(κέω, to split; φλοιός, bark.)

849. **C. pileatus** (L.). LOGCOCK. Black; white streak down neck; crest and cheek patch scarlet in ♂; cheeks and front of crest black in ♀. L. 18. W. 9½. T. 7. N. Am.; in heavy timber, a shy bird, now becoming rare. (Lat., capped.)

446. MELANERPES Swainson. (μέλας, black; ἕρπης, creeper.)
a. Back, scapulars and wing-coverts glossy blue-black (grayish in young). (*Melanerpes.*)

850. **M. erythrocephalus** (L.). RED-HEADED WOODPECKER. Whole head and neck crimson in both sexes, bordered below by black: belly, rump, secondaries, etc., pure white; rest of body glossy blue-black. L. 9½. W. 5¼. T. 3¾. E. U. S., rare in N. E., very abundant W. (ἐρυθρός. red; κεφαλή, head.)
aa. Back, scapulars and wings barred with white. (*Centurus* Swainson.)

851. **M. carolinus** (L.). RED-BELLIED WOODPECKER. Grayish, much barred above with black and white; belly pale ashy, more or less reddish-tinged; crown and nape crimson in ♂, ashy in ♀. L. 9¾. W. 5. T. 3½. E. U. S., rather S., common W.

447. COLAPTES Swainson. (κολαπτής, chisel.)

852. **C. auratus** (L.). YELLOW-HAMMER. FLICKER. GOLDEN-WINGED WOODPECKER. HIGH-HOLER. Head ashy, with red nuchal crescent; back drab-color, barred with black; rump white; below pinkish brown shading into yellowish; a black crescent on breast; belly with numerous round black spots; shafts and under surfaces of quills golden yellow; ♂ with a black maxillary patch. L. 12½. W. 6. T. 4½. E. N. Am., abundant.

853. **C. cafer** (Gmelin). RED-SHAFTED FLICKER. Quills with orange red instead of golden; maxillary patches in ♂ red instead

---

[1] The earlier name *Hylatomus* is set aside by the A. O. U. on account of the still earlier *Hylotoma*, a genus of Insects. I have elsewhere maintained that "A name is a name without necessary meaning," and therefore that generic names are different unless spelled alike, even though derived from the same Greek root. I prefer to use *Hylatomus, Eremophila, Lagochila, Lucania, Icteria, Cestreus, Heterodontus,* and similar names, notwithstanding their similarity to *Hylotoma, Eremophilus, Lagocheilus, Lucanus, Icterus, Cestrœus,* and *Heterodon.*

of black; no nuchal crescent; no yellowish on belly; the black spots fewer and smaller. L. 14. W. 6⅜. T. 5. Western, E. to Kan. Runs into the preceding, of which it is often considered a variety.[1]

ORDER XLV. **MACROCHIRES.** (THE SWIFTS AND HUMMING BIRDS.)

Fissirostral and tenuirostral *Picariæ*. Wing very long and pointed, the fingers and primaries especially elongate. Feet small, weak, with three toes in front, one behind, the hind toe usually somewhat elevated; tail-feathers 10; palate ægithognathous, as in the *Passeres.* There are three families, all represented within our limits. (μακρός, long; χείρ, hand.)

a. Bill fissirostral, swallow-like; secondaries more than 6.
   b. Middle toe much longer than lateral toes, its claw pectinate; rictus with bristles; plumage very soft. . . . . CAPRIMULGIDÆ, 160.
   bb. Middle toe scarcely longer than lateral toes, its claw not pectinate; no rictal bristles; plumage compact. . . . . . MICROPODIDÆ, 161.
aa. Bill tenuirostral, very long and slender; secondaries 6; plumage compact, with metallic lustre. . . . . TROCHILIDÆ, 162.

FAMILY CLX. **CAPRIMULGIDÆ.** (THE GOATSUCKERS.)

Bill very short, fissirostral, the gape exceedingly deep and wide, reaching to below the eyes, and usually with prominent rictal bristles. Wings long and pointed; secondaries lengthened. Plumage long and loose. Tail feathers 10. Feet very small; tarsus short; toes slightly webbed at base, the middle claw pectinate; hind toe somewhat elevated and lateral. Genera 11; species 100 or more, widely diffused; chiefly insectivorous, largely nocturnal, and of noiseless flight, like the owls. (Lat., *capra*, goat; *mulgeo*, to suck, from an old tradition.)

a. Rictal bristles very long; tail rounded; tarsus largely feathered.
                                    ANTROSTOMUS, 448.
aa. Rictal bristles inconspicuous; tail emarginate. . . . . CHORDEILES, 449.

448. ANTROSTOMUS Gould. (ἄντρον, cave; στόμα, mouth.)
a. Rictal bristles with lateral branches.

854. **A. carolinensis** (Gmelin). CHUCKWILL'S WIDOW. More reddish than *A. vociferus.* L. 12. W. 9. T. 6½. U. S., N. to S. Ill.
aa. Rictal bristles simple.

855. **A. vociferus** (Wilson). WHIPPOORWILL. NIGHT JAR. Grayish, very much variegated with blackish and buffy; pectoral bar and ends of outer tail feathers white (♂) or tawny (♀). L. 10. W. 6. T. 5. E. U. S., abundant in damp woods; nocturnal; noted for its "solemn and prophetic cry," continually repeated in the night.

[1] In the west you will find specimens *auratus* on one side of body, *cafer* on the other, tail gilded on some feathers, rubricated on others. (*Coues.*)

449. CHORDEILES Swainson.   (χορδή, a musical instrument; δείλη, evening.)

856. **C. virginianus** (Gmelin).   NIGHT HAWK.   BULL BAT.
Blackish, barred and mottled with grayish and buffy; a large wing
spot, bar across tail, and V-shaped blotch on throat — white in ♂,
tawny or obscure in ♀; the wing spot placed in front of tip of 7th
quill. L. 9¼. W. 8. T. 5. N. Am., very abundant, flying high
in evening or cloudy weather.

FAMILY CLXI.   **MICROPODIDÆ.**   (THE SWIFTS.)

Bill fissirostral, as in the Goatsuckers and Swallows. Wings
very long, thin and pointed; secondaries very short. Feet small,
weak; hind toe often elevated or otherwise turned; toes completely cleft; middle claw not pectinate; no rictal bristles; tail
feathers 10; plumage compact. In most species the salivary
glands are highly developed, and their secretion is used as a glue
in the construction of the nest; species of *Collocalia* in China thus
form the edible bird's nest. Small birds of the warmer parts of
the world, bearing a superficial resemblance to Swallows, but struc-
turally very different, being closely related to the Humming Birds,
nearer to them even than to the Goatsuckers. Genera 6 or 8;
species 50. (μικρός, small; πούς, foot.)

a. Tarsus bare, longer than middle toe; tail rounded, its feathers with the
shafts spinous, projecting beyond the plumage . . . CHÆTURA, 450.

450. CHÆTURA Stephens.   (χαίτη, bristle; ούρά, tail.)

857. **C. pelagica** (L.).   CHIMNEY SWIFT.   CHIMNEY SWAL-
LOW. Sooty brown; throat paler. L. 5¼. W. 5. T. 2. E. N. Am.
abundant; now nesting in chimneys, as formerly in hollow-trees.

FAMILY CLXII.   **TROCHILIDÆ.**   (THE HUMMING BIRDS.)

Bill subulate, usually longer than the head, straight or curved;
tongue capable of great protrusion. Wings long and pointed, the
secondaries short, only 6 in number; tail of 10 feathers. Feet
very small, with long sharp claws. Smallest of all birds and among
the most brilliantly colored. Genera 75; species 300 or more, one
of the largest families in Ornithology. All are American, and most
of them tropical, but our common species ranges far into Canada.
Chiefly insectivorous; not musical.

a. First primary not attenuate, bowed or curved inwards; bill straight; frontal
feathers covering nasal scale. . . . . TROCHILUS, 451.

451. TROCHILUS Linnæus.   (τροχίλος, plover.)

858. **T. colubris** L.   RUBY-THROATED HUMMING BIRD.   ♂
metallic green above; a ruby-red gorget; tail deeply forked,

uniform purplish, its feathers narrow ; ♀ without red, the tail variegated; no scales on crown. L. 3¼. W. 1⅔. T. 1¼. B. ⅔. E. N. Am.; abundant in summer, hovering about flowers. (S. Am. name, *Colibri*.)

ORDER XLVI. PASSERES. (THE PASSERINE BIRDS.)

Toes always 4 ; feet fitted for perching ; the hind toe always on the level of the rest, its claw at least as long as that of the middle toe ; joints of toes 2, 3, 4, 5, respectively, from first to fourth ; none of the toes versatile, and none webbed ; wing coverts few, chiefly in two series; tail feathers 12 ; primaries 10, but in most of the families the first one is reduced in size, and often rudimentary and displaced ; musical apparatus more or less developed ; sternum of a uniform passerine pattern ; palate ægithognathous. Nature altricial.

This order includes about 6000 known species, or more than half of all the kinds of birds. They represent the "highest grade of development and the most complex organization of the class ; their high physical irritability is co-ordinate with the rapidity of their respiration and circulation ; they consume the most oxygen and live the fastest of all birds." (*Coues*.)

A considerable number of anatomical characters (for which see Stejneger, "Standard Natural History," p. 458, *et seq*.), are more or less perfectly distinctive of the *Passeres*. These cannot, however, be discussed here. The group is divided, on anatomical characters, into about 5 suborders. Two of these groups, the *Clamatores* and the *Oscines*, are represented in our fauna. The latter, characterized especially by the perfect musical apparatus, comprises the vast majority of the *Passeres*. (Lat., *passer*, sparrow.)

Families of Passeres.

a. *Tarsus with its hinder edge rounded;* encircled by a single horny envelope divided into scutella anteriorly and on outer side, this sometimes extending all round (though separated by a seam along inner side), but often widely separated on inner side or behind or both, the intervening space occupied by granular scales, reticulations, or plain naked skin; musical apparatus imperfect; primaries 10, the first about as long as second. (*Clamatores.*)

   b. Inner toe free at base from middle toe; tarsus not reticulate behind; bill hooked at tip, with long rictal bristles. . . . TYRANNIDÆ, 163.

aa. *Tarsus with its hinder edge compressed,* forming a sharp, nearly undivided ridge (except in the Larks, which may be known by the long, nearly straight hind claw); musical apparatus highly developed; primaries properly ten, but the first short, or spurious, or sometimes rudimentary and misplaced, so that but nine are evident, in which case the first developed primary is about as long as second. (*Oscines.*)

 c. Hinder edge of tarsus not compressed, rounded and scutellate like
    anterior edge; hind claw very long, straightish; developed pri-
    maries 9. . . . . ALAUDIDÆ, 164.
cc. Hinder edge of tarsus compressed, forming a sharp ridge, for the
    most part undivided.
  d. Primaries apparently but 9 (the first minute and displaced); the first
     developed (i. e. second) primary about as long as the next; bill not
     hooked at tip.
   e. Bill not fissirostral, the gape little longer than the culmen; outer
      primary never twice as long as inner.
     f. Bill "conirostral," stout at base, with the commissure forming
        a more or less distinct angle at base of bill, "the corners of
        the mouth" drawn downward.
       g. Bill rather long, often longer than head, without notch at tip
          or bristles at the rictus. . . . . . . ICTERIDÆ, 166.
       gg. Bill shorter than head, often notched near tip, and usually
           with bristles at the rictus. . . . . FRINGILLIDÆ, 167.
     ff. Bill not truly conirostral (the corners of mouth not evidently
         drawn downward).
       h. Bill stout (conical in our species, the cutting edge with one
          or more lobes or nicks near its middle); nostrils placed
          high, exposed; (plumage chiefly red or yellow, in our
          species). . . . . TANAGRIDÆ, 168.
       hh. Bill rather slender, not conical; angle of gonys not be-
           fore nostril.
         i. Hind claw short and curved, mostly shorter than its toe;
            tertials not elongate, not nearly reaching tips of pri-
            maries. . . . . MNIOTILTIDÆ, 173.
         ii. Hind claw long and straightish, mostly longer than its toe;
             tertials much elongate, nearly reaching tips of primaries.
                                          MOTACILLIDÆ, 174.
  ee. Bill fissirostral, — the culmen very short, the gape very broad,
      its length more than twice the culmen; wings very acute, the
      outer primary more than twice length of innermost.
                                          HIRUNDINIDÆ, 169.
 dd. Primaries evidently ten, the first developed, but short, rarely half
     the length of the next; (first primary obsolete in some Vireos,
     known by the slightly hooked bill).
       j. Tarsus distinctly scutellate.
         k. Tarsus not longer than middle toe with claw; bill
            short, depressed; (head crested; tail tipped with yel-
            low, in our species). . . . . AMPELIDÆ, 170.
         kk. Tarsus longer than middle toe and claw (or if not,
             other characters not as above).
           l. Bill strongly hooked and toothed at tip, somewhat
              like a hawk's bill. . . . . . LANIDÆ, 171.
           ll. Bill slightly hooked at tip; (plumage more or less
               olivaceous). . . . . . . VIREONIDÆ, 172.
           lll. Bill not evidently hooked at tip.
             m. Tail feathers stiff, pointed; bill decurved.
                                          CERTHIIDÆ, 176.
             mm. Tail feathers more or less soft and rounded.
               n. Nasal feathers directed forwards, usually cov-
                  ering the nostrils.

o. Birds of large size; (wing more than 4). . .   CORVIDÆ, 165.
oo. Birds of small size; (wing less than 4).
  p. Bill not notched. . . . . PARIDÆ, 177.
  pp. Bill notched toward the tip, very slender. . . . SYLVIIDÆ, 178.
nn. Nasal feathers erect or directed backward, not covering nostrils; bill rather slender, the culmen convex; first primary not very short.
            TROGLODYTIDÆ, 175.
jj. Tarsus booted, without distinct scutella except near the base; rictal bristles present.
  q. Birds of small size; (wing less than 3); young unspotted.
            SYLVIIDÆ, 178.
  qq. Birds of moderate size (wing more than 3); young spotted.
            TURDIDÆ, 179.

FAMILY CLXIII. **TYRANNIDÆ.** (THE FLYCATCHERS.)

Primaries 10; the first more than ¾ length of second, longer than in any other of our passerine birds; bill typically broad, triangular, depressed, abruptly hooked and notched at tip, with long rictal bristles; commissure nearly straight; nostrils small, usually partly concealed. Tarsus with its back and sides as well as the front covered with scutella, so that there is no undivided ridge behind, as in most other *Passeres.* Feet small. Mouth capacious; vocal apparatus mesomyodian, *i. e.* the "syrinx with fewer than 4 distinct pairs of intrinsic muscles inserted at the middle of the upper bronchial half rings, constituting an uncomplicated and ineffective musical apparatus." (*Coues.*) Changes of plumage slight; ours mostly olivaceous.

A large family of 80 genera, and more than 300 species; all American and mostly tropical. All are insectivorous, most of them pre-eminently so; they are, therefore, in our latitude, migratory.

a. Bill of typical form, depressed, hooked at tip, with strong rictal bristles.
  b. Outer primaries, one or more of them, attenuate; crown in adult with a concealed red or yellow crest.
    c. Tail deeply forked, much longer than wings. . . . MILVULUS, 452.
    cc. Tail not forked, not longer than wings. . . . . TYRANNUS, 453.
  bb. Outer primaries not attenuate; crown without concealed bright-colored crest.
    d. [Wings and tail with chestnut; length 8 or more]; head slightly crested; wings little longer than tail. . . . MYIARCHUS, 454.
    dd. [Wings and tail without chestnut; general color olivaceous; length less than 8.]
      e. Wings at least 6 times as long as tarsus; (W. 3 to 4¼).
            CONTOPUS, 456.
      ee. Wings not more than 5 times as long as tarsus, little longer than tail.
        f. Bill rather narrow (black in our species; wing more than 3¼).
            SAYORNIS, 455.
        ff. Bill broad (usually pale below in our species; wing less than 3¼).
            EMPIDONAX, 457.

18

452. MILVULUS Swainson, (Lat., *milvus*, kite.)

859. **M. tyrannus** (L.). FORK-TAILED FLYCATCHER. Larger than next; no red; tail black, still more elongate. Tropical, straying N. to N. J. and Ky.

860. **M. forficatus** (Gmelin). SCISSOR-TAIL. Ashy; tail chiefly white; crissum, shoulders, sides, etc., with much red. L. 13. W. 5. T. 8. S. W., N. to Mo., straying E. (Lat., forked, like scissors.)

453. TYRANNUS Cuvier. (τύραννος, ruler.)

861. **T. tyrannus** (L.). KING BIRD. BEE MARTIN. Blackish, white below; crown-patch orange; tail black, white-tipped. L. 8½. W. 4⅜. T. 3½. N. Am., chiefly E.; very abundant. "Destroys a thousand noxious insects for every bee it eats!" (*Coues.*)

862. **T. verticalis** Say. ARKANSAS KING-BIRD. Ashy-gray; yellow below; tail black, white-edged. W. N. Am., straying E. (Lat., *vertex*, top of head, which is ornate.)

454. MYIARCHUS Cabanis. (μυῖα, fly; ἀρχός, ruler.)

863. **M. crinitus** (L.). GREAT CRESTED FLYCATCHER. Scarcely crested; olivaceous, with bright chestnut on wings and tail; breast ashy-gray; belly clear yellow. L. 8¾. W. 4. T. 4. E. U. S., chiefly S., N. to N. Wis. A handsome bird, "noted for the habitual use of cast-off snake skins in the structure of its nest." (Lat., crested.)

455. SAYORNIS Bonaparte. (Say; ὄρνις, bird.)

864. **S. phœbe** (Latham). PEWEE. PHŒBE. Olive brown, head and tail darker; yellow or whitish below. L. 7. W. 3½. T. 3¼. E. U. S., abundant; known by its black bill. (From the bird's note.)

865. **S. saya** Bonaparte. Ashy-brown, the belly pale cinnamon, the tail black. L. 8. W. 4. T. 3¾. W. U. S., E. to Iowa. (To Thomas Say.)

456. CONTOPUS Cabanis. (κοντός, pole; πούς, foot.)

a. Tarsus shorter than middle toe with claw; wing about half longer than tail; a white cottony patch on each side of rump. (*Nuttallornis* Ridgway.)

866. **C. borealis** (Swainson). OLIVE-SIDED FLYCATCHER. Rictal bristles short, one-fourth length of bill; slaty brown above with darker streaks; quills blackish; middle line of belly distinctly and abruptly white, otherwise grayish below. L. 7½. W. 4¼. T. 3. N. N. Am., S. to N. Y.; in mts. and pine forests.

aa. Tarsus longer than middle toe with claw; wing not ¼ longer than tail: no conspicuous cottony tuft. (*Contopus.*)

867. **C. virens** (L.). Wood Pewee. Rictal bristles half length of bill; wing bands whitish or rusty; olive brown above; pale or yellowish below; lower mandible usually pale. L. 6¼. W. 3¼. T. 2¾. B. ½. U. S., very abundant; known from the common Pewee by its drawling notes. (Lat., greenish.)

868. **C. richardsoni** (Swainson). Darker and less olivaceous, more gray below; bill dusky below. L. 6½. W. 3⅓. T. 2¾. N. W., E. to Wis.; nearly like the preceding, but the notes and nesting different. (To John Richardson.)

457. **EMPIDONAX** Cabanis. (ἐμπίς, gnat; ἄναξ, king.)

a. Lower parts distinctly yellow.

869. **E. flaviventris** Baird. Yellow-bellied Flycatcher. Clear olive green; yellow below, becoming bright yellow (not merely slightly yellowish as in the others) on the belly; first primary about equal to sixth; feet as in *acadicus*; bill yellow below. L. 5¼. W. 2¾. T. 2½. B. ½. Ts. ⅝. E. N. Am. (Lat., *flavus*, yellow; *venter*, belly.)

aa. Lower parts not distinctly yellow.

870. **E. acadicus** (Gmelin). Small Green-crested Flycatcher. Clear olive green, wing bands buffy; whitish or slightly yellowish below; yellowish ring about eyes; bill pale below; primaries nearly an inch longer than secondaries; 2d, 3d and 4th primaries nearly equal, and much longer than 1st and 5th; 1st much longer that 6th. L. 6. W. 3. T. 2¾. Ts. ⅔. Tcl. ½. B. ⅝. E. U. S., frequent.

871. **E. pusillus** (Swainson). Olive brown, duller than preceding; bill pale below; 5th primary about as long as 4th, 1st not much longer than 6th; middle toe ⅔ length of tarsus; longest primary ⅝ inch longer than secondaries. L. 5¾. W. 2¾. T. 2½. B. ⅗. Ts. ⅝. Tcl. ⅝. U. S., represented E. by the more olivaceous var. **trailli** Audubon. (Lat., petty.)

872. **E. minimus** Baird. Least Flycatcher. Olive gray; bill blackish below; wings like preceding, but longest primary but ½ inch longer than secondaries; middle toe half as long as tarsus; tail slightly emarginate. L. 5. W. 2½. T. 2¼. B. ⅓. Ts. ⅔. E. N Am., abundant; very similar to the last, known by the measurements. (Lat., least.)

Family CLXIV. ALAUDIDÆ. (The Larks.)

First primary very short or obsolete. Tarsus obtuse and scutellate behind as well as in front (a character singular among *Oscines*). Bill short, of various forms in different species; nostrils concealed by tufts of antrose feathers; hind claw very long and nearly straight; inner secondaries lengthened and flowing. About 100 species, chiefly Old World birds, a single genus in America; some of them are renowned as vocalists. Pre-eminent is the Skylark, *Alauda arvensis* L., a species which has been lately introduced into this country (Long Island, etc.).

a. Spurious primary obsolete; a little tuft of lengthened black feathers over each ear (sometimes obscure in ♀); tail not forked. . OTOCORIS, 458.

458. OTOCORIS [1] Bonaparte. (*Eremophila* Boie.) (οὖς, ear; κόρυς, helmet.)

873. **O. alpestris** (Forster). SHORE LARK. HORNED LARK. Pinkish brown, thickly streaked: a crescent on breast and strip under eye black; white below; chin, throat, and line over eye more or less yellow; ♀ with less black; winter birds grayish, with the markings more obscure. L. 7¼. W. 4½. T. 3. Northern Hemisphere, common. A pleasant singer. Runs into many varieties, the prairie form (var. **praticola** Henshaw) averaging smaller, W. 4 to 4⅓, etc. (*Eu.*) (Lat., alpine.)

FAMILY CLXV. CORVIDÆ. (THE CROWS AND JAYS.)

Primaries 10; first about half length of second; nostrils usually concealed by tufts of bristly feathers, which are branched to their tips. Bill long and strong, usually notched, its commissure not angulated. Tarsus sharp behind, its sides undivided and separated from the scutella in front by a groove, which is either naked or filled in with small scales. Voice usually harsh and unmusical.

Birds of large size, the largest of the *Oscines*, found almost everywhere. Genera about 40; species 175.

a. Tail not shorter than the short, rounded wings. (*Garrulinæ.*)
  b. Tail much longer than wing, graduated for half its length, its feathers narrowed to the tips; head not crested. . . . . . . PICA, 459.
  bb. Tail not much longer than wings, not graduated for half its length.
    c. Head with a conspicuous crest; (chiefly blue). . CYANOCITTA, 460.
    cc. Head without crest; plumage lax; (no blue). . PERISOREUS, 461.
aa. Tail much shorter than the long, pointed wings. (*Corvinæ.*)
  d. Bill compressed, higher than broad; plumage glossy.
              CORVUS, 462.

459. PICA Cuvier. (Lat., magpie.)

874. **P. pica** (L.). MAGPIE. Lustrous black; belly, shoulders, and wing-edgings white. L. 19. W. 8½. T. 13. Northern regions. The American bird (var. **hudsonica** Sabine) is larger, with the feathers of throat spotted with white below the surface. Its range is chiefly N. W. in America, E. to Wis. (*Eu.*)

[1] *Otocoris* is used for *Eremophila* by the A. O. U., on account of the prior *Eremophilus*, a genus of fishes.

460. CYANOCITTA Strickland. (κύανος, blue; κίττα, jay.)

875. **C. cristata (L.).** BLUE JAY. Blue; collar and frontlet black; grayish below; wings and tail clear blue, barred; outer tail feathers and secondaries tipped with white. L. 12. W. 5½. T. 5¾. E. N. Am., very abundant. (Lat., crested.)

461. PERISOREUS Bonaparte. (περισωρεύω, to accumulate.)

876. **P. canadensis (L.).** CANADA JAY. GRAY JAY. WHIS-KEY JACK. Ashy gray, with blackish and whitish markings. L. 10¾. W. 5¾. T. 6. N. N. Am., S. in winter, to Mich. and Me.

462. CORVUS Linnæus. (Lat., crow.)

a. Plumage entirely lustrous black.

877. **C. corax** L. RAVEN. Feathers of throat stiffened, elon-gated, narrow, and lanceolate, their outlines very distinct. L. 25. W. 17. T. 10. Northern regions; rare E. of Miss. R. The Amer-ican forms are var. **principalis** Ridgway, — New Brunswick, N. with larger bill; and var. **sinuatus** Wagler, — W. U. S., with slender bill and tarsus. The Eur. bird has bill shorter and deeper. (*Eu.*) (κόραξ, raven.)

878. **C. americanus** Audubon. CROW. Feathers of throat short, broad, obtuse, with their webs blended ; gloss of plumage purplish violet; head and neck scarcely lustrous. L. 20. W. 13. T. 8. Ts. 2¼. B. 2. N. Am., abundant; variable.

879. **C. ossifragus** Wilson. FISH CROW. Gloss of plumage green and violet, evident on head and neck ; feathers of throat short, blended. L. 16. W. 11. T. 7. B. 1¾. Ts. 1¾. N. Y. to La., only along the coast. (Lat., bone-breaker.)

FAMILY CLXVI. ICTERIDÆ. (THE AMERICAN "ORI-OLES" AND "BLACKBIRDS.")

Primaries 9 ; bill with the commissure angulated, as in *Fringil-lidæ*, but usually lengthened, rarely shorter than head, straight or gently curved, without notch or rictal bristles; culmen usually ex-tending up on the forehead, dividing the frontal feathers. Legs stout, usually adapted for walking. Plumage usually brilliant or lustrous, the predominant color generally black, often with red or yellow ; females usually different, smaller in size, brown or streaky in the lustrous species, and yellowish or dusky in the brightly col-ored ones. Notes usually sharp, often richly melodious, in other cases harsh. Excepting the " Orioles," the species feed chiefly on seeds.

Genera about 20, species 100, all American, some of the short-billed forms forming a perfect transition to the *Fringillidæ ;* others

are as closely related to *Sturnidæ* (starlings), which in turn are
allied to the *Corvidæ*.

  a. Outlines of bill nearly or quite straight, the tip not evidently decurved;
  the commissure not sinuated. (*Icterinæ.*)
  b. Bill stout, conical; its depth at base at least ⅓ its length; sexes unlike;
  ♀ smaller.
    c. Tail feathers acute : middle toe with claw longer than tarsus; bill
    shorter than head, finch-like. . . . . . . DOLICHONYX, 463.
    cc. Tail feathers not acute; middle toe with claw not longer than tarsus.
      d. Bill much shorter than head, finch-like. . . . MOLOTHRUS, 464.
      dd. Bill about as long as head.
        e. Lateral claws elongate, reaching beyond base of middle claw.
        XANTHOCEPHALUS, 465.
        ee. Lateral claws shortish, scarcely reaching base of middle claw.
        AGELAIUS, 466.
  bb. Bill slender, its depth at base scarcely ⅓ its length.
    f. Tail not ⅔ length of wing, its feathers acute; tertials lengthened;
    bill longer than head; feathers of crown each tipped by the
    bristle-like shaft : sexes similar. . . . STURNELLA, 467.
    ff. Tail nearly as long as wing, its feathers not pointed ; bill
    shorter than head ; feathers of crown not bristle-tipped; sexes
    unlike. . . . . ICTERUS, 468.
aa. Outlines of bill distinctly curved, the tip decurved; the commissure evidently sinuated. (*Quiscalinæ.*)
  g. Tail much shorter than wing, nearly even; bill slender, shorter
  than head. . . . . SCOLECOPHAGUS, 469.
  gg. Tail longer than wing, graduated, the middle feathers lowermost when the tail is folded; bill stout, not shorter than head.
  QUISCALUS, 470.

463. DOLICHONYX Swainson. (δολιχός, long ; ὄνυξ, claw.)

880. **D. oryzivorus** (L.). BOBOLINK. REED BIRD. RICE
BIRD. ♂ in spring black. neck buffy, shoulders and rump ashy
white, back streaky : ♀. and fall ♂, yellowish brown, streaked
above, — dull yellowish birds, resembling sparrows, but known by
the acute tail feathers. L. 7½. W. 4. T. 3. E. N. Am., abundant in meadows northward, where, in the breeding season, it is
our merriest and most delightful songster. Retiring southward in
the fall, it fattens in the rice swamps and becomes a "game bird,"
slaughtered by the thousand for city markets. (Lat., *oryza*, rice;
voro, I devour.)

464. MOLOTHRUS Swainson. (μολοθρός, vagabond.)

881. **M. ater** (Boddaert). COW BIRD. ♂ iridescent black, head
and neck glossy brown ; ♀ much smaller, dusky brown. L. (♂) 8.
W. 4. T. 3. U. S., abundant: noted for its parasitic habits. It
builds no nests, but lays its eggs in the nests of warblers and other
small birds. (Lat., black.)

ICTERIDÆ. — CLXVL 279

465. XANTHOCEPHALUS Swainson.

882. X. xanthocephalus (Bonaparte). YELLOW-HEADED
BLACKBIRD. ♂ black with white wing patch; head and neck
deep yellow; ♀ smaller, browner, with less yellow. L. 10. W. 5½.
T. 4½. W. N. Am., E. to Ind., etc., in swamps. (ξανθός, yellow;
κεφαλή, head.)

466. AGELAIUS Vieillot. (ἀγελαῖος, gregarious.)

883. A. phœniceus (L.). RED-WINGED BLACKBIRD. SWAMP
BLACKBIRD. ♂ glossy (not iridescent) black, lesser wing covers
scarlet, with buffy and paler edgings; ♀ dusky, streaked; young
♂ streaked, with rusty on bend of wing. L. 9. W. 5. T. 4.
U. S., everywhere abundant. (φοινίκεος, phœnician-red.)

467. STURNELLA Vieillot. (Lat., dim. of *sturna*, starling.)

884. S. magna (L.). MEADOWLARK. Brownish and much
streaked above; chiefly yellow below, a black crescent on breast;
yellow of throat not encroaching on cheeks; sides and crissum
buffy. L. 10. W. 5. T. 3½. E. N. Am., very abundant. (Lat.,
large, as compared with the sky-lark.)

885. S. neglecta (Audubon). WESTERN MEADOWLARK. Very
similar, the colors duller and paler, the yellow of throat encroaching on sides of lower jaw; sides and crissum nearly white. W. N.
Am., E. to Ill.; almost exactly like the other, but the song quite
different, thrush-like.

468. ICTERUS Brisson. (ἴκτερος, yellow.)

a. Depth of bill at base not half its length above. (*Icterus.*)

886. I. spurius (L.). ORCHARD ORIOLE. ♂ black; rump,
bend of wing and lower parts deep chestnut; ♀ yellowish olive,
quite small; young yellow, with various black or chestnut traces;
young ♂ often yellowish, with black throat-patch. L. 7. W. 3½.
T. 3. E. U. S., common southerly; a fine singer and an artist in
nest-building.

aa. Depth of bill at base half its length. (*Yphantes* Vieillot.)

887. I. galbula (L.). BALTIMORE ORIOLE. GOLDEN ROBIN.
FIRE BIRD. Black; bend of wing, rump, most tail feathers, and
under parts from the breast orange of varying intensity; ♀ duller,
olivaceous and yellow. L. 7¾. W. 3¾. T. 3. E. N. Am., abundant; noted for its elaborate hanging nest as well as for its song.
(Lat., name of some bird.)

469. SCOLECOPHAGUS Swainson. (σκώληξ, worm;
φάγος, eater.)

888. S. carolinus (Müller). RUSTY GRACKLE. RUSTY BLACK-
BIRD. ♂ glossy black becoming rusty in autumn; ♀ dusky, lus-
treless. L. 9½. W. 4¾. T. 4. E. U. S.

889. **S. cyanocephalus** (Wagler). BREWER'S BLACKBIRD. ♂ black with green lustre, head glossed with violet ; ♀ dusky. L. 10. W. 5¼. T. 4½. W. N. Am., straying E. to Ill. (κύανος, blue; κεφαλή, head.)

470. **QUISCALUS** Vieillot.   (From the bird's note.)

890. **Q. quiscula** (L.).   CROW BLACKBIRD.   PURPLE GRACKLE. Iridescent black, lustre on head purplish, on body bronzy. L. 13. W. 5½. T. 5¼. E. U. S., abundant ; now divided into the typical variety, chiefly S. of N. Y. and E. of Alleghanies, and var. **æneus** Ridgway, the common form N. and W., the latter with the body with uniform bronze lustre, without mixed tints, this color abruptly defined against the iridescent violet of the neck. Var. **quiscula** is nearly uniform iridescent.

891. **Q. major** Vieillot.   BOAT-TAILED GRACKLE.   Iridescent green and blue. Larger. L. 17. W. 7¼. T. 7¼. Va. to Texas and S.

FAMILY CLXVII.   **FRINGILLIDÆ.**[1]   (THE FINCHES.)

Primaries 9, the first being obsolete.   Bill " conirostral," mostly shorter than head, robust, of a conical form, with the commissure more or less abruptly angulated near its base ; in other words, the " corners of the mouth drawn down."   This feature is usually strongly marked, and it is almost the only special character pertaining to all the members of the family.   Even this is also shared by the *Icteridæ*, which, however, may generally be distinguished by the greater length and slenderness of the bill.   Nostrils high up, exposed or (in northern species) partly covered by a ruff of small

---

[1] Sundevall and Stejneger have placed the *Fringillidæ* at the end or head of the series of birds, for reasons which seem to me sufficient ones.   " In order to find out the most specialized form of the *Passeres*, we must look for the bird which is most specialized in all directions, not only as to the coloration of its plumage, or the fusion of its tarsal covering.   The *ideally* highest form . . . would have booted tarsi, 9 primaries, long mandibular symphysis, powerful bill for grain crushing, a digestive system adapted for grain-feeding, and the coloration of young and adult unspotted and similar.   That this is the regular course and ultimate end of the evolution among the higher birds is evident from the fact that we can trace it in nearly all the groups, and in the individual development of the birds possessing these characters."   (*Stejneger*.)

Acting on this principle, Stejneger selects as the highest or most specialized bird the Evening Grosbeak.   " The number of its primaries is reduced to 9, the mandibular symphysis is well developed, the palatine and facial part of the skull is highly specialized, and so is the digestive canal.   Furthermore the plumage of the young is essentially like that of the adults."   It fails, then, in only one respect, — its tarsus is not booted.

In most recent American systems, however, the *Turdidæ* are placed at the head of the list ; and as the A. O. U. has adopted this arrangement it is retained here, the " post of honor " being given to the beautiful Arctic Bluebird rather than to the Evening Grosbeak, although the latter has certainly the better claim.

feathers. Tarsus scutellate in front, with an undivided ridge behind.

A very large family, the most extensive in Ornithology, comprising about 100 genera and 500 species, found in nearly every part of the world, except Australia. They are especially abundant in North America, where about one seventh of all the birds are *Fringillidæ*. "Any one United States locality of average attractiveness to birds, has a bird-fauna of over two hundred species, and if it be away from the sea-coast, and consequently uninhabited by marine birds, about one-fourth of the species are *Mniotiltidæ* and *Fringillidæ* together, the latter somewhat in excess of the former. It is not easy, therefore, to give undue prominence to these two families." (*Coues.*)

All the Finches are granivorous, feeding chiefly on seeds, but not rejecting either berries or insects; nearly all sing, and some most delightfully; most of them are plainly clad, a streaky brown being the prevailing tint, but others are among the most brilliantly colored birds. Among these latter only are the changes in plumage strongly marked. (Lat., *fringilla*, finch.)

A strictly natural analysis of the genera of *Fringillidæ* is practically impossible, as they do not fall naturally into definable groups. The characters drawn from the development of the palate are not available for the ordinary purposes of the student. The following semi-artificial key is largely adapted from Ridgway's Manual.

a. Mandibles falcate, crossed at tip; nostrils concealed by a small ruff.
LOXIA, 474.
aa. Mandibles not crossed at tip.
  b. Head with a conspicuous crest : bill very large; culmen strongly curved (bill, wings, and tail chiefly red). . . . . . . CARDINALIS, 491.
  bb. Head without crest.
    c. Bill very stout, its depth at base equal to length of hind toe with claw, and more than ⅔ tarsus; nostrils partly concealed.
COCCOTHRAUSTES, 471.
    cc. Bill less stout, its depth at base less than length of hind toe with claw.
      e. Nasal plumules long, covering the basal third of upper mandible; bill stout. . . . . PINICOLA, 472.
      ee. Nasal plumules, if present, covering much less than one-third of length of upper mandible.
        f. Introduced birds; gonys distinctly convex in profile; (plumage streaked above, not below ; no white, red, yellow, or blue).
PASSER, 473, note.
        ff. Native birds; gonys straight or nearly so.
          g. Primaries much longer than secondaries (exceeding them by length of tarsus).
            h. Wing at least 5 times as long as the short tarsus.
              i. Birds of moderate size, the wing more than 3½ inches.
                j. Base of gonys nearer base of bill than its tip (measuring

along side of bill); (tail feathers without white; tail coverts rosy);
hind claw moderate. . . . . LEUCOSTICTE, 475.

jj. Base of gonys as near tip of lower mandible as to its base on the
    side; (tail largely white; plumage with much white and no rosy);
    hind claw very long, nearly as long as bill.
                     PLECTROPHENAX, 478.

  ii. Birds of small size, the wing less than 3½; tail forked.

    k. Nasal tufts very long, nearly ½ length of bill; (tail feathers with-
      out white or yellow; adults with red). . . . ACANTHIS, 476.

    kk. Nasal tufts short or obsolete, not ⅓ length of bill; (tail feathers,
      blotched with white or yellow; adults with yellow but no red).
                         SPINUS, 477.

hh. Wing not five times as long as tarsus.

    l. First (developed) primary not shorter than fourth; (back
      streaked).

      m. Depth of bill at base about equal to length of (exposed) cul-
        men; nostrils with a small ruff; (plumage streaked above
        and below; ♂ with red; no white on tail).
                     CARPODACUS, 473.

      mm. Depth of bill at base less than length of culmen; (no red;
        tail with white).

      n. Tail emarginate, the middle feathers narrow and pointed
        at tip; hind claw very long and straightish, nearly as long
        as bill.

        o. Gonys shorter than hind toe without claw and not more
          than depth of bill. . . . . . . CALCARIUS, 479.

        oo. Gonys longer than hind toe and greater than depth of
          bill. . . . . RHYNCOPHANES, 480.

      nn. Tail rounded; the middle feathers broad and rounded at
        tip; hind toe short, curved. . . . CHONDESTES, 483.

    ll. First (developed) primary shorter than fourth; bill very stout;
      (plumage with red or yellow, tail with white). HABIA, 492.

gg. Primaries not much longer than secondaries (exceeding them by less than
    length of tarsus): (no red).

      p. Bill very stout ; its depth at base nearly equal to hind
        toe with claw; (♂ with blue). . . GUIRACA, 493.

      pp. Bill more slender, not as above.

      q. Tail-feathers narrow, at least the middle ones acu-
        minate; (back streaked).

        r. Middle toe with claw decidedly shorter than tarsus;
          (outer tail feathers with white; bend of wing
          chestnut). . . . . POOCÆTES, 481.

        rr. Middle toe with claw not shorter than tarsus;
          (outer tail feathers without white markings;
          edge of wing yellow).

        s. (Breast with yellow ; throat with more or less
          black; plumage not streaked below in ♂).
                    SPIZA, 495.

        ss. (Breast without yellow ; throat without black;
          plumage streaked below.)
                AMMODRAMUS, 482.

      qq. Tail feathers broader, not acuminate (except in
        worn plumage).

        t. Hind claw decidedly longer than its toe.

u. Bill tapering rapidly to the acute tip; nostrils concealed by small antrorse feathers; (plumage streaked above and below).

PASSERELLA, 489.

uu. Bill tapering gradually toward the rather obtuse tip; nostrils exposed; (plumage not streaked). . . . . PIPILO, 490.

tt. Hind claw scarcely longer than its toe.

   t. Tertials very long, longer than secondaries, not much shorter than longest primaries; (a white wing patch). . CALAMOSPIZA, 496.

   tt. Tertials scarcely or not longer than secondaries, not nearly reaching tips of longest primaries.

    w. (Outer tail feather largely white; plumage not streaked.)

JUNCO, 486.

    ww. (Outer tail feather not white.)

     x. Lower mandible much deeper than upper; (♂ with blue or green). . . . . PASSERINA, 494.

     xx. Lower mandible not deeper than upper; (plumage streaky above ; no blue); wings not much longer than tail.

      y. Tail more or less forked ; its middle feathers shortest ; (no yellow; plumage not streaked below). . SPIZELLA, 485.

      yy. Tail rounded (or slightly double-rounded).

       z. Primaries exceeding secondaries by more than length of bill; (head in adult striped; in young chestnut; plumage not streaked below). . . . . . ZONOTRICHIA, 484.

       zz. Primaries exceeding secondaries by not more than length of bill.

        a. (Edge of wing yellow; plumage not streaked below.)

PEUCÆA, 487.

        aa. (No yellow anywhere ; plumage streaked below, or else with the crown chestnut.) . . . MELOSPIZA, 488.

471. COCCOTHRAUSTES Brisson. (κόκκος, berry; θραύω, to crush.)

a. Tips of four inner primaries of normal form, not widened at end. (*Hesperiphona* Bonap.).

892. **C. vespertinus** (Cooper). EVENING GROSBEAK. Olivaceous; crown, wings, tail and tibia black; forehead, rump, and crissum yellow; inner secondaries and coverts white; bill very large, yellowish ; ♀ grayer, with little yellow. L. 8. W. 4¼. T. 2½. W. N. Am., irregularly E. to Ohio or beyond, one of the most striking of the finches. (Lat., of sunset.)

472. PINICOLA Vieillot. (Lat. *pinus*, pine; *colo*, I inhabit.)

893. **P. enucleator** (L.). PINE GROSBEAK. ♂ chiefly rose red; changing to ashy below and behind ; wings dusky, with two white wing bars; ♀ ashy gray, with brownish yellow on head and rump ; bill blackish. L. 8½. W. 4½. T. 4. Northern regions, S. in winter to Va., in pine woods, etc. (*Eu.*) The American bird (var. **canadensis** Cabanis) is larger and more brightly colored. (Lat., one who shells nuts.)

473. CARPODACUS [1] Kaup. (καρπός, fruit ; δάκος, biting.)

894. **C. purpureus** (Gmelin). PURPLE FINCH. Everywhere streaky; ♂ flushed with red, most intense on the crown, fading below and behind; ♀ olive brown and streaky, with no red; bill stout. L. 6. W. 3⅓. T. 2½. N. Am., a sweet singer. (Lat., purple, which the bird is not.)

474. LOXIA Linnæus. (λοξός, crooked.)

a. Wing with white.

895. **L. leucoptera** Gmelin. WHITE WINGED CROSSBILL. ♂ rose red ; two white wing bars; scapulars black ; ♀ brownish olive. speckled with dusky; rump yellow. L. 6¼. W. 3½. T. 2½. N. N. Am., S. in winter, with the next, less common; variable. (λευκός. white ; πτερόν, wing.)

aa. Wing with no white.

896. **L. curvirostra** L. RED CROSSBILL. ♂ brick-red; wings dusky, unmarked; ♀ brownish, washed with greenish yellow. L. 6. W. 3⅓. T. 2½. Northern regions, about pine woods; S. in winter, sometimes in large flocks, to Tenn. and Va. (Eu.) The rather small form in E. U. S. is var. **minor** Brehm. The singular bill is adapted for opening nuts. (Lat., curve-bill.)

475. LEUCOSTICTE Swainson. (λευκός, white ; στικτός, spotted.)

897. **L. tephrocotis** Swainson. Cinnamon-brown; head more or less ashy gray ; nasal tufts white; quills dusky; tail coverts edged with rose pink in adult. L. 6. W. 4. T. 3. Rocky Mts., E. to Iowa. (τεφρός, gray ; οὖς, ear.)

476. ACANTHIS Bechstein. (ἀκανθίς, thistle-bird.)

a. Crown red in both sexes, crimson in ♂, lustrous brownish-red in ♀ ; chin blackish ; no yellow.

898. **A. hornemanni** (Holböll). GREENLAND REDPOLL. Sides and rump scarcely streaked ; colors very pale. ♂ with breast merely pinkish. L. 5. W. 3. T. 2½. Greenland ; the small var. **exilipes** (Coues) S. to N. U. S. in winter. (To J. W. Hornemann.)

899. **A linarius** (L.). RED POLL LINNET. Throat, breast and rump rosy in ♂ ; much streaked above and on sides; rump

[1] In the vicinity of *Carpodacus* belongs the Old World genus : —
Passer Brisson. *P. domesticus* (L.). EUROPEAN HOUSE SPARROW. ♂ chestnut brown above, thickly streaked ; ashy below ; throat, lores and chin black ; ♀ duller, without black : feet small. L. 6. W. 2¾. T. 2¼. Introduced from Europe : abundant in all towns E., a nuisance unfortunately long past the possibility of abatement. (Eu.)
P. montanus (L.). EUROPEAN TREE SPARROW. Smaller ; ♂ with black of throat not continued over chest : top of head liver-brown. L. 5¼. W. 2¾. T. 2¼. Europe, naturalized about St Louis.

streaked. L. 5¾. W. 3. T. 2½. Northern regions, S. in winter in flocks to Ind. and Penn. (*Eu.*) Besides the common form a larger var. rostrata *Coues* (W. 3⅓, etc.), with shorter, less acute bill, sometimes ranges S. to Ill. and N. Y. (Lat., flaxen.)

aa. Crown without red; no dusky spot on chin; some yellow.

900. **A. brewsteri** Ridgway. No dusky on chin; rump yellow in ♀ ; the ♂ unknown. L. 5½. W. 3. T. 2½. Mass., one specimen known. (To Wm. Brewster.)

477. SPINUS[1] Boie. (Latin name : "thistle-bird," the thistle being a spinous plant.)

a. Sexes unlike; plumage scarcely or not streaked: adult ♂ with black on crown, wings and tail; bill not very acute, without distinct ruff at base. (*Astragalinus* Cabanis.)

901. **S. tristis** (L.) YELLOW BIRD. THISTLE BIRD. AMERICAN GOLDFINCH. ♂ rich yellow ; rump whitish ; wing bars white; a white spot on each tail feather ; ♀ more olivaceous; fall plumage pale yellow brown ; young variously buffy, with yellow or not. L. 5. W. 3. T. 2. N. Am., everywhere; notable for its lisping notes and undulating flight. (Lat., sad.)

aa. Sexes alike; plumage thickly streaked everywhere; no black on head; bill very sharp, with a distinct ruff at base. (*Spinus.*)

902. **S. pinus** (Wilson). PINE SISKIN. Plumage streaky brown, suffused with yellow in the breeding season ; bases of quills and tail feathers sulphur yellow. L. 4¾. W. 2¾. T. 2. N. Am., chiefly N., but liable to appear anywhere. (Lat., pine.)

478. PLECTROPHENAX Stejneger. (πλῆκτρον, spur; φέναξ, deceiver, the word made in imitation of the old name *Plectrophanes*, which is preoccupied.)

903. **P. nivalis** (L.). SNOW BUNTING. In breeding season, pure white, with black on back, wings and tail ; bill and feet black; only the winter plumage usually seen in U. S. ; bill pale, and white of body clouded with clear, warm brown. L. 7. W. 4½. T. 3. Northern regions, S. in winter to Ohio R. ; a most beautiful bird. (*Eu.*) (Lat., snowy.)

479. CALCARIUS Bechstein. (Lat., *calcar*, spur.)

904. **C. lapponicus** (L.). LAPLAND LONGSPUR. ♂ with head and throat mostly black ; a chestnut collar; back black and streaky ; whitish below; outer tail feathers with white ; inner web of outer feather dusky ; legs and feet black ; ♀ and winter birds

---

[1] Allied to *Spinus* is the Goldfinch of Europe (*Carduelis carduelis* L.), now naturalized in New York, Cambridge, etc. In both sexes, the head is black and white, crimson anteriorly, the wings and tail black and yellow : the rump white ; brownish below. Allied also is the Canary, *Serinus canarius*, a favorite cage-bird.

with less black. L. $6\frac{1}{4}$. W. 4. T. $2\frac{3}{4}$. Northern regions, S. in winter to N. Y. and Ky. (*Eu.*) (Lat., Lapp.)

905. **C. pictus** (Swainson). ♂ with head and upper parts mostly black; collar and under parts rich fawn color; legs pale; inner web of outer tail feather chiefly white. ♀ duller. L. $6\frac{1}{2}$. W. $2\frac{2}{3}$. N. N. Am., S. E., to Ill. and Kan. (Lat., painted.)

480. RHYNCHOPHANES Baird. (*ῥύγχος*, beak; *φαίνω*, I show.)

906. **R. maccowni** (Lawrence). Black-breasted Longspur. Crown and pectoral crescent black; the black often obscured by pale edgings; bend of wing chestnut; line over eye and under parts white; back and sides streaked. L. $6\frac{1}{2}$. W. $3\frac{3}{4}$. T. $2\frac{1}{2}$. B. nearly $\frac{1}{2}$. Great plains, rarely E. to Ill. (To Capt. J. P. McCown.)

481. POOCÆTES Baird. (*πόα*, blue grass; *οἰκητής*, inhabitant.)

907. **P. gramineus** (Gmelin). Bay-winged Bunting. Grass Sparrow. Ground Bird. "Vesper Sparrow." Thickly streaked everywhere; slightly buffy below. L. 6. W. 3. T. $2\frac{1}{2}$. N. Am., abundant in fields, etc., known at once by the chestnut bend of wing and white outer tail feathers; a good singer. (Lat., grassy.)

482. AMMODRAMUS Swainson. Shore Sparrows. (*ἄμμος*, sand; *δραμεῖν*, to run.)

a. Outer pair of tail feathers longer than middle pair; wing much longer than tail. (*Passerculus.*)

908. **A. princeps** (Maynard). Ipswich Sparrow. Grayish; streaks on back sandy brown, not sharply defined; superciliary line white in front; bill not longer than hind toe without claw. L. 6. W. $3\frac{1}{4}$. T. $2\frac{1}{2}$. Nova Scotia to Va. and Texas, coastwise. (Lat., chief.)

909. **A. sandwichensis** (Gmelin). Savanna Sparrow. Sharply streaked; streaks on back blackish; superciliary line and edge of wing yellowish. L. $5\frac{1}{2}$. W. $2\frac{3}{4}$. T. 2. N. Am., abundant on plains and shores. The form E. of Rocky Mts., smaller (W. $2\frac{3}{4}$, etc., instead of W. $3\frac{1}{8}$, etc.), is var. **savanna** Wilson. (From Sandwich Isl., Alaska.)

aa. Outer pair of tail feathers shorter than middle pair; wing not much, if any, longer than tail.

b. Bill stout; tail feathers acute but not rigid; crown with a median light stripe; inland species. (*Coturniculus* Bonaparte.)

c. Tail double-rounded, the lateral feathers not much shorter than middle ones.

910. **A. savannarum** (Gmelin). Grasshopper Sparrow. Much streaked above; feathers edged with bay; breast buffy,

unstreaked ; wings and tail short ; edge and bend of wing and line over eye yellow. L. 5. W. 2⅔. T. 2. N. Am., in fields ; notes sharp, grasshopper-like ; the bird of E. U. S. is var. **passerinus** Wilson. (Spanish, *savana*, meadow.)

cc. Tail graduated, the outer feathers much shorter than middle ones.

911. **A. henslowi** (Audubon). Smaller than preceding. more yellow above ; breast. etc., with some sharp black streaks. L. 5. W. 2¼. T. 2⅛. E. U. S., scarce : N. to Mass. (To Prof. J. S. Henslow.)

912. **A. lecontei** (Audubon). Intermediate between the preceding and the next ; bill small, blue-black ; back with rufous ; tail feathers very sharp and slender : breast unspotted ; a broad buffy superciliary stripe. L. 4⅔. W. 2⅛. T. 2⅛. Great Plains, E. to Ill. (To Major J. Le Conte.)

bb. Bill rather slender ; tail feathers sharp and rather stiff ; crown without distinct median stripe. Seashore sparrows. (*Ammodramus*.)

913. **A. caudacutus** (Gmelin). SHARP-TAILED FINCH. Ashy olive, the back streaked with ashy buff and whitish ; edge of wing pale yellowish ; no yellow spot about eye ; a bright buff superciliary stripe. L. 5. W. 2¼. T. 1⅞. Salt marshes, Nova Scotia to N. C. ; represented in fresh water swamps (Ill., S. and E.) by var. **nelsoni** Allen, which has colors of upper parts very sharply contrasted, especially the whitish streaks on umber-brown ground color ; breast less sharply streaked. (Lat., *cauda*, tail; *acutus*.)

914. **A. maritimus** (Wilson). SEA-SIDE FINCH. Olive gray ; back obscurely streaked ; a yellow spot before eye ; edge of wing yellow ; no superciliary stripe. L. 6. W. 2¼. T. 2. Salt marshes, Mass. to Texas.

483. **CHONDESTES** Swainson. (χόνδρος. grain ; ἐδεστής, eater.)

915. **C. grammacus** (Say). LARK SPARROW. Streaked above, ashy below ; ear coverts chestnut ; crown chestnut. black anteriorly, with whitish median and superciliary stripes ; a black line through and below eye ; a conspicuous black streak on each side of the white throat ; a black pectoral spot ; middle tail feathers like back, the rest blackish, white tipped ; a pale spot on primaries. L. 6¼. W. 3¼. T. 3. W. U. S., E. to Ohio ; abundant on prairies and river bluffs ; a fine songster, suggesting the Bobolink. (γραμμικός, streaked.)

484. **ZONOTRICHIA** Swainson. (ζώνη, band ; θρίξ, hair, *i. e.* head.)

a. No yellow markings anywhere.

916. **Z. querula** (Nuttall). BLACK-HOODED SPARROW. Crown, face and throat jet black ; no yellow ; ♀ with less black. L. 7½. W. 3¼. T. 3½. Missouri region, E. to W. Ill.

917. **Z. leucophrys** (Forster). WHITE-CROWNED SPARROW.
Streaked above, with but little chestnut ; crown black, with a broad
white median band ; lores blackish ; a white superciliary streak ;
throat like breast, but paler ; young with the crown chiefly rich
brown. L. 7. W. 3¼. T. 3¼. N. Am. ; not rare. (λευκός, white ;
ὀφρύς. eyebrow.)

aa. Head with yellow.

918. **Z. coronata** (Pallas). Similar to *Z. leucophrys*, but the
crown-stripe yellow anteriorly, ashy behind. L. 7½. W. 3¼. T. 3¼.
W. N. Am., rarely E. to Wis. (Lat., crowned.)

919. **Z. albicollis** (Gmelin). WHITE-THROATED SPARROW.
PEABODY BIRD. Much chestnut streaking above ; crown black,
with white median and superciliary stripes ; spot over eye and
edge of wing always yellow ashy below, whitening on throat ; ♀
duller. L. 7. W. 3. T. 3⅛. E. N. Am. ; an abundant and
handsome sparrow. (Lat.. *albus*, white ; *collum*, neck.)

485. **SPIZELLA** Bonaparte. (Dim. of σπίζα, a sparrow.)

920. **S. monticola** (Gmelin). TREE SPARROW. Streaked above ;
crown chestnut ; bill black above, pale below ; neck, line over eye
and under parts ashy gray ; a dark pectoral blotch ; white wing
bars distinct. L. 6¼. W. 3. T. 3. N. Am., chiefly northerly ;
U. S. in winter. (Lat., living on mountains.)

921. **S. socialis** (Wilson). CHIPPY. CHIPPING SPARROW.
Streaked above, with much dull bay ; crown chestnut ; forehead
and streak through eye black ; ashy white below ; bill blackish ;
wing bars faint, brownish. L. 5¼. W. 2⅔. T. 2½. N. Am.,
everywhere common.

922. **S. pusilla** (Wilson). FIELD SPARROW. General color of
S. monticola. but paler and duller ; bill pale ; no pectoral blotch ;
wing bands obscure, whitish. L. 5½. W. 2½. T. 2½. E. U. S.,
abundant. (Lat., petty.)

923. **S. pallida** (Swainson). CLAY-COLORED SPARROW. Pale
brownish yellow, streaked with black ; crown grayish, with median
stripe. L. 5½. W. 2½. T. 2⅜. Great Plains, rarely E. to Ill.

486. **JUNCO** Wagler. (Lat., *Juncus*, a rush.)

924. **J. hyemalis** (L.). SNOW BIRD. Slaty gray ; head darker ;
bill pale ; belly and outer tail feathers white ; ♀ more grayish ;
L. 6¼. W. 3. T. 3. N. Am., everywhere abundant, breeding
in cold regions, and moving S. as cold weather approaches, usually
in advance of the snow. Represented W. by numerous varieties ;
var. **oregonus** Townsend, with sides pinkish, ranging E. to Ill. (Lat.,
wintry.)

487. PEUCÆA Audubon.   (πεύκη, pine.)

925. **P. æstivalis** (Lichtenstein). Upper parts largely chestnut, with ashy edgings and dusky streaks; a broad pale superciliary line; ashy below, the breast buffy; yellow on bend and edge of wing, but none on head. L. 6. W. 2¼. T. 2½. Southern, N. to Central Ind., the form ranging N. rather paler, the back chiefly chestnut, is var. **bachmani** Audubon. (Lat., summer.)

488. MELOSPIZA Baird.   (μέλος, song; σπίζα, sparrow.)

926. **M. fasciata** (Gmelin). Song Sparrow. Much streaked above and on breast and sides; crown with an obscure ashy median stripe; below white, pectoral streaks often forming a dusky blotch. L. 6¼. W. 2½. T. 3. N. Am., everywhere; a hearty songster, beginning early in spring. (Lat., banded.)

927. **M. georgiana** (Latham). Swamp Sparrow. Crown bright dark chestnut, streaked with black; wings strongly tinged with chestnut; back sharply streaked; an ashy collar and superciliary line; breast and below ashy with few streaks or none; tail shorter than in the Song Sparrow, its quills edged with chestnut. L. 5¾. W. 2¼. T. 2½. E. U. S., in low thickets; a timid bird, seldom seen, although not rare.

928. **M. lincolni** (Audubon). Everywhere above and below thickly, sharply streaked with black, gray and buffy; breast with a broad band of pale buffy or yellowish brown; sides washed with buffy. L. 5¼. W. 2¼. T. 2¼. N. Am., rare E.; a shy species quite unlike the Song Sparrow. (To Robert Lincoln.)

489. PASSERELLA Swainson.   (Lat., passer, sparrow.)

929. **P. iliaca** (Merrem). Fox Sparrow. Ashy above, overlaid and much streaked with rusty red, which becomes bright bay on rump, tail and wings; white below with large arrow-shaped spots and streaks, numerous on breast; feet stout, with long claws. L. 7. W. 3¼. T. 3. E. N. Am., migrating early; one of the handsomest streaked sparrows and a good singer. (Lat., ilium, flank, which is streaked.)

490. PIPILO Vieillot.   (Lat., I peep or chirp.)

930. **P. erythrophthalmus** (L.). Chewink. Marsh Robin. Towhee. Black, belly white; sides chestnut; outer tail feathers, primaries and inner secondaries with white; ♀ with clear brown instead of black; iris red. L. 8¼. W. 3¼. T. 4. E. U. S., abundant everywhere. (Gr., red-eyed.)

491. CARDINALIS Bonaparte.

931. **C. cardinalis** (L.). Cardinal Grosbeak. Red Bird. Clear red, ashy on back; chin and forehead black; crest con-

19

spicuous; ♀ ashy brown, more or less washed with red. L. 8¼.
W. 4. T. 4½. E. U. S., southerly, N. to Mass. and N. Wis.;
abundant. A brilliant songster, much sought as a cage bird. (Lat.,
from color of cardinal's hat.)

492. HABIA Reichenbach. (A South American name.)

932. **H. ludoviciana** (L.). ROSE-BREASTED GROSBEAK. ♂
with head, neck and upper parts mostly black, with white on rump,
wings and tail; belly white; breast and under wing coverts of an
exquisite rose-red; bill very stout, pale; ♀ olive brown, much
streaked, with the under wing coverts saffron yellow; head with
whitish stripes. L. 8½. W. 4. T. 3¼. E. N. Am., abundant;
perhaps our handsomest bird, and one of the most brilliant songsters. (Lat., Louisianian.)

493. GUIRACA Swainson. (S. Am. name.)

933. **G. cœrulea** (L.). BLUE GROSBEAK. ♂ rich blue;
feathers about bill, wings and tail, black; wing bars chestnut; ♀
yellowish brown, with whitish wing bars. L. 7. W. 3½. T. 2¾.
Southern, N. to N. Y. and Wis., rare; a fine songster.

494. PASSERINA Vieillot. (From *passer*.)

934. **P. cyanea** (L.). INDIGO BIRD. ♂ indigo blue, clear on
head, greenish behind: ♀ plain warm brown, obscurely streaky,
known from other small sparrows by a dusky line along the gonys.
L. 5¾. W. 3. T. 2¾. E. U. S., abundant in summer; a tireless
songster. (Lat., blue.)

935. **P. ciris** (L.). NONPAREIL. PAINTED BUNTING. ♂ head
and neck blue; under parts, etc., vermilion; shoulders, etc., green;
rump and tail purplish-brown; ♀ green, yellowish below. L. 5¼.
W. 2⅔. T. 2¼. Southern, N. to S. Ill. (*Nelson.*) (κείρις, name
of some bird.)

495. SPIZA Bonaparte. (σπίζα, old name of some sparrow.)

936. **S. americana** (Gmelin). BLACK-THROATED BUNTING.
" DICK SISSEL." Grayish and streaked above; wing coverts chestnut; line over eye, maxillary stripe, edge of wing, breast and part
of belly yellow; throat patch black; otherwise white below; ♀
with little chestnut, and the black reduced to dark streaks. L. 6¾.
W. 3¼. T. 2¾. Fields, Conn. to Kansas, chiefly W.; a handsome
bird with sleek plumage, and a peculiar, but scarcely musical song,
incessantly repeated in hot weather.

937. **S. townsendi** (Audubon). Upper parts, head, neck, etc.,
slaty blue; no chestnut, and little yellow or black. A single specimen known from Penn., perhaps a hybrid. (To. J. K. Townsend.)

496. CALAMOSPIZA Bonaparte. (κάλαμος, reed ; σπίζα.)

938. **C. melanocorys** Stejneger. LARK BUNTING. WHITE WING BLACKBIRD. ♂ black, with a large white wing-patch and white on quills ; ♀ streaky, like the ♀ bobolink, known by the whitish wing-patch and long tertials. L. 6½. W. 3½. T. 2¾. Western plains, occasional E. (μέλας, black ; κόρυς, helmet.)

FAMILY CLXVIII. **TANAGRIDÆ.** (THE TANAGERS.)

Primaries 9 ; bill usually conical, sometimes depressed or attenuate, the culmen curved; cutting edges not much inflected, sometimes toothed, notched or serrated ; tarsus scutellate ; legs short ; claws long. Colors usually brilliant. A large family of more than 300 species, confined to the warmer parts of America, and embracing a wide diversity of forms. Some have slender bills and are scarcely distinguishable from the Warblers, and might well be referred to the same family. Others, like our *Piranga*, have stout conical bills, and are equally closely related to the Finches.

a. Bill stout, finch-like, considerably longer than broad, and more or less evidently teothed or lobed near middle of upper mandible. PIRANGA, 497.

497. PIRANGA Vieillot. (S. Am. name.)

939. **P. rubra** (L.). SUMMER RED BIRD. ♂ bright rose red throughout; wings a little dusky ; ♀ dull brownish olive, dull yellowish below ; no wing bars ; bill and feet paler than in the Scarlet Tanager; size the same. E. U. S., chiefly S.; N. to N. J. and Ill.; abundant. (Lat., red.)

940. **P. erythromelas** Vieillot. SCARLET TANAGER. ♂ brilliant scarlet; wings and tail black; no wing bars ; ♀ clear olive green ; clear greenish yellow below. L. 7⅓. W. 4. T. 3. E. N. Am., abundant in woodland ; a most beautiful bird and a respectable songster. (ἐρυθρός, red ; μέλας, black.)

FAMILY CLXIX. **HIRUNDINIDÆ.** (THE SWALLOWS.)

Primaries 9, the first being obsolete; bill "fissirostral," *i. e.*, short, broad, triangular, depressed, the gape wide and about twice as long as the culmen, reaching to about opposite the eyes, similar in its form to that of the Swifts and the Goatsuckers, with which birds the Swallows have no real affinity. Rictus without bristles; wings very long and pointed, the first primary usually longest, and twice as long as the last; secondaries very short. Tail more or less forked. Feet weak; tarsus scutellate, shorter than middle toe and claw. Plumage compact, and more or less lustrous.

A very natural family of about 100 species, found in all parts of the world. All are strong on the wing, insectivorous, and in our latitude migratory. (Lat., *hirundo*, swallow.)

  a. Nostrils opening directly upward, with very little membrane bordering inner edge.

    b. Tail forked for a distance more than half tarsus; bill very stout, curved, (plumage lustrous, ♂ all black). . . . . PROGNE, 498.

    bb. Tail even.

      c. Outer web of outer primary without recurved hooks; (plumage lustrous).
                PETROCHELIDON, 499.

      cc. Outer web of outer primary with stiff recurved hooks, obscure in ♀; (plumage plain brown.) . . . . . . STELGIDOPTERYX, 503.

  aa. Nostril opening laterally, and bordered above by a broad membrane or overhanging scale.

    d. Tail forked for more than half its length, the outer feathers very narrow toward tip; no tarsal tuft. . . . . CHELIDON, 500.

    dd. Tail forked for less than half its length.

      e. Tarsus without tuft of feathers on its lower part; (plumage lustrous). . . . . TACHYCINETA, 501.

      ee. Tarsus with a small tuft of feathers on its lower part; (plumage plain brownish). . . . . CLIVICOLA, 502.

498. PROGNE Boie. (Πρόκνη. a character in mythology, turned into a swallow.)

941. **P. subis** (L.). PURPLE MARTIN. Lustrous blue-black throughout; ♀ duller, whitish and streaky below. L. 7¼. W. 6. T. 3⅓. N. Am., abundant. (Lat., old name of some bird.)

499. PETROCHELIDON Cabanis. (πέτρα, rock; χελιδών.)

942. **P. lunifrons** (Say). CLIFF SWALLOW. Lustrous steel blue; forehead, sides of head, throat, rump, etc., of various shades of chestnut; a blue spot on breast; belly whitish. L. 5¼. W. 4½. T. 2⅛. N. Am., abundant, formerly nesting in cliffs, but now building under the eaves of barns. (Lat., *luna*, moon; *frons*, forehead.)

500. CHELIDON Forster. (χελιδών, swallow).

943. **C. erythrogaster** (Boddaert). BARN SWALLOW. Lustrous steel-blue, buffy below; forehead and throat deep chestnut; an imperfect steel-blue collar; tail feathers with white spots. L. 7. W. 5. T. 4½. N. Am., very abundant; breeding in colonies about barns, etc. (Gr., red-belly.)

501. TACHYCINETA Cabanis. (ταχυκίνητος, moving swiftly.)

  a. Ear coverts steel-blue; upper parts with metallic lustre. (*Iridoprocne* Coues.)

944. **T. bicolor** (Vieillot). WHITE-BELLIED SWALLOW. Lustrous blue-green, pure white below; ♀ duller. L. 6¼. W. 5. T. 2⅜. N. Am., abundant about water; very handsome.

502. CLIVICOLA Forster. (Lat., *clivus*, cliff; *colo*, I inhabit.)

945. **C. riparia** (L.). BANK SWALLOW. SAND MARTIN. Dark gray, not iridescent, pale below, a brown shade across the breast. L. 4¾. W. 4. T. 2. N. Am., abundant, breeding in holes in sandbanks. (*Eu.*). (Lat., of the bank of a stream.)

503. STELGIDOPTERYX Baird. (στελγίς, scraper; πτέρυξ, wing.)

946. **S. serripennis** (Audubon). ROUGH-WINGED SWALLOW. Brownish gray, pale below. L. 5⅓. W. 4½. T. 2¼. U. S., common W., breeding in banks, etc. (Lat., *serra*, saw; *penna*, feather.)

FAMILY CLXX. **AMPELIDÆ.** (THE CHATTERERS.)

Primaries 10, or apparently 9, the first in our species rudimentary and displaced; bill stout, triangular, depressed, decidedly notched and hooked, with the gape very wide. Tarsus short, with the lateral plates more or less subdivided, their covering often unlike that of the other *Oscines;* lateral toes nearly equal. As now recognized, a small family of 6 or 8 species, constituting two groups which bear little resemblance to each other.

The *Ampelinæ* includes the three species of *Ampelis.* They are crested birds with a soft plumage of a handsome cinnamon drab color; the ends of the secondaries, and sometimes of the tail feathers also, are tipped with horny appendages, looking like red sealing-wax; these often absent in ♀. The tail is tipped with yellow or red. The Wax Wings are migratory and gregarious, feeding on insects and soft fruits. Their voices are weak and wheezy, and they can scarcely be considered as songsters.

a. Wings pointed; tail short, truncate; primaries apparently 9; the first very minute; no rictal bristles; nostrils concealed by bristles. (*Ampelinæ.*)
AMPELIS, 504.

504. AMPELIS Linnæus. (Lat., name of some bird frequenting grape-vines.)

947. **A. garrulus** L. BOHEMIAN WAX WING. NORTHERN WAX WING. General color a soft silky, ashy brown; front and sides of head shaded with purplish cinnamon; a pale-edged black band across forehead through eye, around crest; throat black; crissum chestnut red; two broad white wing bars. L. 7⅓. W. 4½. T. 3. Northern regions, S. in winter in large flocks to the Great Lakes; an interesting and beautiful bird. (*Eu.*)

948. **A. cedrorum** (Vieillot). CEDAR BIRD. CHERRY BIRD. Similar but smaller and less cinnamon-tinged; chin black; strip across face black, bordered above by whitish; belly yellowish posteriorly; crissum white; no wing bars; ♀ with the wax-like ap-

pendages small or wanting. L. 6½. W. 3¾. T. 2½. N. Am.,
abundant. (Lat., of the cedars.)

FAMILY CLXXI. LANIIDÆ. (THE SHRIKES.)

Primaries 10, the first short (rarely wanting); bill hawklike,
very strong, the upper mandible toothed and abruptly hooked at
the tip; both mandibles distinctly notched. Wings short, rounded.
Tail long. Tarsus scutellate on the outside as well as in front.
Sexes alike.

Species about 100, found in most parts of the world, remarkable
for their vigor and pugnacity. Their habits, corresponding with
the form of the bill, are similar to those of birds of prey, for which
reason they were placed by Linnæus among the *Accipitres.* They
have a remarkable habit of impaling small animals on thorns and
leaving them there.

a. Rictus with bristles; nostrils concealed by bristly tufts; first primary not
very short. LANIUS, 505.

505. LANIUS Linnæus. (Lat., butcher.)

949. **L. borealis** Vieillot. GREAT NORTHERN SHRIKE. BUTCH-
ERBIRD. Ashy above, rump paler; black bars on side of head
narrow, *not* meeting in front, and interrupted by a white crescent
on under eyelid : rump and shoulders whitish; wings and tail black,
outer tail feathers with white; white below always waved with
blackish. L. 9½. W. 4¼. T. 4¾. N. N. Am., S. in winter to Ky.
and Va.

950. **L. ludovicianus** L. LOGGER-HEAD SHRIKE. Clear
ashy blue; a whitish superciliary line; black bars on sides of head
broad, meeting across forehead: no white on under eyelid; adults
white below, not dark-waved. L. 9. W. 3⅔. T. 4. S. U. S., the
typical variety, S. E., N. to Ohio and Vt.; a paler form, var. **ex-
cubitorides** Swainson (*White Rumped Shrike*), common W., E. to
N. Y. This has the tail coverts whitish. L. 9. W. 4. T. 4.
(Lat., Louisianian.)

FAMILY CLXXII. VIREONIDÆ. (THE VIREOS.)

Primaries 10. or apparently only 9, the first being often rudimentary and displaced. Bill shorter than head, stout, compressed,
decidedly notched and hooked. Rictus with bristles. Nostrils
exposed, overhung by a scale, reached by the bristly frontal feathers. Tarsus scutellate ; toes soldered at base for the whole length
of basal joint of middle one, which is united with the basal joint of
the inner and the two basal joints of the outer; lateral toes usually
unequal.

A rather small family, comprising 5 genera and 60 to 70 species
of small olivaceous birds, all American. They are allied to the

Laniidæ, being in fact small insectivorous Shrikes. The coloration is usually blended, and varies little with age or sex. Many of them are remarkable as songsters.

Concerning the "nine-primaried" species, Professor Baird remarks: —

"In *V. flavifrons*, in which the outer primary is supposed to be wanting, its presence may be easily appreciated. One of the peculiar characters of this species consists in a narrow edging of white to all the primary quills, while the primary coverts (the small feathers covering their bases, as distinguished from what are usually termed the wing coverts, which more properly belong to the forearm or secondaries) are without them. If these coverts are carefully pushed aside, two small feathers considerably shorter than the others will be disclosed, one overlying the other, which (the under one) springs from the base of the exposed portion of the long outermost primary, and lies immediately against its outer edge. This small feather is stiff, falcate, and edged with white like the other quills, and can be brought partly around on the inner edge of the large primary, when it will look like any spurious quill. The overlying feather is soft, and without light edge. In the other Vireos, with appreciable spurious or short outer primary, a similar examination will reveal only one small feather at the outer side of the base of the exterior large primary. In all the families of *Passeres*, where the existence of nine primaries is supposed to be characteristic, I have invariably found, as far as my observations have extended, that there were two of the small feathers referred to, while in those of ten primaries but one would be detected."

a. Wings not shorter than tail; outer toe longer than inner. . VIREO, 506.

506. VIREO Vieillot. (Lat., I grow green.)

a. Wings long and pointed, ¼ or more longer than tail; first primary very small or apparently wanting, not ¼ second.

   b. Slender species; bill slender, light horn color, pale below; commissure straight and culmen relatively so; no wing bars nor conspicuous orbital ring; feet weak. (*Vireosylva* Bonaparte.)

      c. Primaries apparently 9, the first obsolete.

951. **V. olivaceus** (L.). RED-EYED VIREO. GREENLET. Olive green, crown ashy, edged on each side with blackish; a white superciliary line, and below this a dusky streak; white below, somewhat olive shaded; iris red. L. 6. W. 3¼. T. 2½. E. N. Am., very abundant in woodland; an energetic songster.

952. **V. philadelphicus** (Cassin). Dull olive green, becoming ashy on crown; no black lines on head; a whitish superciliary line; below faintly yellowish, fading to white on throat. L. 4¾. W. 2⅔. T. 2¼. E. N. Am., scarce. (φιλέω, I love; ἀδελφός, brother.)

   cc. Primaries evidently 10, the first well developed.

953. **V. gilvus** (Vieillot). WARBLING VIREO. Colors exactly as in the preceding, but the spurious quill evident. L. 5⅛. W. 2⅞. T. 2¼. E. N. A., frequent; an exquisite songster, nesting in tall trees in cities. (Lat., yellowish.)

bb. Stout species, the bill short and stout, blue-black; a pale stripe running to and around eye; two white wing bars; quills blackish, mostly edged with pale; feet stout. (*Lanivireo* Baird.)

d. Primaries apparently 9, the first obsolete.

954. **V. flavifrons** Vieillot. YELLOW-THROATED VIREO. Rich olive green above, becoming ashy on rump; bright yellow below; belly white; superciliary line and orbital ring yellow. L. 5¾. W. 3. T. 2. E. U. S., abundant, the most brightly colored species. (Lat., yellow-fronted.)

dd. Primaries evidently 10, the first small but distinct.

955. **V. solitarius** (Wilson). BLUE-HEADED VIREO. Bright olive green; crown and sides of head bluish-ash; stripe to and around eye white; a dusky line below it; white below, washed with yellow. L. 5⅔. W. 3. T. 2⅓. U. S. in woodland; a handsome species. Var. **alticola** Brewster, is a larger form, darker in color, in the Great Smoky region and S.

aa. Wings relatively short and rounded, not ¼ longer than tail, first primary ⅔ or more length of second; bill stout. (*Vireo.*)

956. **V. noveboracensis** (Gmelin). WHITE-EYED VIREO. Bright olive green, white below; sides and crissum bright yellow; pale wing bars; stripe from bill to and around eye, yellow; iris white. L. 5. W. 2⅓. T. 2¼. E. U. S., in thickets; a sprightly bird, with a loud and varied song. (Lat., of New York.)

957. **V. belli** Audubon. BELL'S VIREO. Olive green, yellow below, chin and superciliary line whitish; wing bars whitish. L. 4¼. W. 2⅕. T. 2. Ill. to Dak. and W. (To J. G. Bell.)

FAMILY CLXXIII. MNIOTILTIDÆ. (THE NEW WORLD WARBLERS.)

Primaries 9; inner secondaries not enlarged, nor the hind toe long and straight, as in *Alaudidæ* and *Motacillidæ*. Bill usually rather slender, notched or not; the commissure not angulated at base, as in *Fringillidæ*, nor toothed in the middle, as in some *Tanagridæ*; the end not notched and abruptly hooked, as in *Vireonidæ* and *Laniidæ*; the gape not broad and reaching to the eyes, as in *Hirundinidæ*.

The Warblers are small birds; all, except *Icteria*, are less than 6½ inches in length, and very many are less than 5. The colors are usually brilliant and variegated, but the sexes are unlike, and the variations due to age and season are great, so that the identification of immature birds is often very difficult. Many of the Warblers are pleasing songsters, but none exhibit any remarkable powers in that line. The name " Warbler " comes from their resemblance to the warblers of Europe (*Sylviidæ*) and not from any distinguished musical quality of their own. All are insectivorous and migratory.

This family consists of more than 100 species, all American. The *Mniotiltidæ* grade perfectly into the *Cærebidæ* and *Tanagridæ*, and the last as perfectly into the *Fringillidæ*. Convenience is the only excuse for retaining any of these groups as distinct families.

a. Bill not depressed and fly-catcher-like; rictal bristles if present scarcely reaching beyond nostrils.
  b. Bill rather slender, little compressed; (small birds; length less than 6¼).
    d. Hind toe with claw very long, as long as tarsus in front; claw of middle toe in same line as axis of the toe; (color black and white, no yellow). . . . . MNIOTILTA, 507.
    dd. Hind toe with claw much shorter than naked portion of tarsus in front; claw of middle toe (seen from above) set obliquely to axis of the toe.
      e. Middle toe with claw not shorter than tarsus; (no white wing bars); bill rather long.
        f. (Tail feathers blotched with white.) . . PROTONOTARIA, 508.
        ff. (Tail feathers without white.)
          g. Bill very much compressed; culmen straight, with a ridge at base. . . . . HELINAIA, 509.
          gg. Bill slightly compressed; culmen gently curved, the basal portion not ridged. . . . . . . HELMITHERUS, 510.
      ee. Middle toe with claw decidedly shorter than naked portion of tarsus in front (except in *Dendroica dominica*, a species with white wing bars).
        h. Rictus without bristles; bill very acute, scarcely notched; (tail feathers with or without white). HELMINTHOPHILA, 511.
        hh. Rictus with bristles.
          i. Tail scarcely rounded, usually much shorter than wing; (tail blotched with white or with the inner web bright yellow; legs and feet moderate, usually dark colored).
            j. Hind toe evidently longer than its claw; bill acute, not notched. . . . . COMPSOTHLYPIS, 512.
            jj. Hind toe scarcely longer than its claw; bill usually not very acute, and with a slight notch toward its tip. . . . . DENDROICA, 513.
          ii. Tail usually more or less rounded, not very much shorter than wing; legs and feet strong, usually pale; (no white or bright yellow on tail feathers).
            k. (Lower parts much streaked.) . . SEIURUS, 514.
            kk. (Lower parts not streaked.). . GEOTHLYPIS, 515.
  bb. Bill stout, much compressed, its greatest depth half its length from nostril to tip; outer side of tarsus smooth on its upper half; tail longer than wings; bill without notch or bristles; (large, more than 7). Chats. ICTERIA, 516.
aa. Bill depressed, broader than deep at base, notched and slightly hooked, with strong rictal bristles about half the length of bill; length 5½ or less. Fly-catching Warblers.

l. Bill fully twice as long as wide at base ; tail a little shorter than wings.
                               SYLVANIA, 517.

ll. Bill scarcely twice as long as wide at base, formed much as in a Flycatcher ; tail about as long as wings. . . . . . SETOPHAGA, 518.

507. MNIOTILTA Vieillot. (μνίον, moss ; τίλλω, I pluck.)

958. **M. varia** (L.). BLACK AND WHITE CREEPER. Everywhere black and white, streaked; crown with a broad white stripe;
wing bars white ; ♀ similar, grayer. L. 5. W. 2¾. T. 2¼. E. N.
Am., not rare; a beautiful warbler, with the habits of a nut-hatch.

508. PROTONOTARIA Baird. (Lat., first notary.)

959. **P. citrea** (Boddaert). PROTHONOTARY WARBLER.
GOLDEN-HEADED WARBLER. Front and lower parts brilliant
yellow; back olivaceous ; wings and tail dusky; rump ashy ; bill
long. L. 5¼. W. 3. T. 2¼. S. U. S., N. to Wabash Valley, in
bushy swamps; rather rare, a most beautiful bird. (Lat., lemon-
yellow.)

509. HELINAIA Audubon. (ἕλος, swamp ; ναίω, to dwell.)

960. **H. swainsoni** Audubon. Chiefly olive-brown, reddish on
top of head; a dusky loral streak, bordered above by a brownish
white superciliary stripe ; head with a paler median streak ; yellowish white below. L. 6. W. 2. T. 2. S. C. to Texas, N. to
S. Ind.; rare. (To Wm. Swainson.)

510. HELMITHERUS Rafinesque. (ἕλμινς, bug ; θηράω,
                               to hunt.)

, 961. **H. vermivorus** (Gmelin). WORM-EATING SWAMP
WARBLER. Olive green; head buffy, with four black stripes;
buffy below : ♀ similar. L. 5½. W. 3. T. 2¼. E. U. S., N. to
L. Erie. (Lat., worm-eating.)

511. HELMINTHOPHILA Ridgway. (ἕλμινς, bug; φιλός,
                                 loving.)

a. Tail feathers with distinct white blotches; wings with bands or patches of
white or yellow.

b. Throat and ear-coverts black in ♂, dusky gray in ♀.

962. **H. chrysoptera** [1] (L.). GOLDEN-WINGED WARBLER.
Ashy blue; forehead, crown and wing patch bright yellow ; throat
and broad stripe through eye black ; a white streak above eye and
one below black of cheek; belly mostly white; ♀ duller, L. 5.
W. 2½. T. 2½. E. U. S., N. to N. Mich.; a beautiful bird. (χρυσός,
gold ; πτερόν, wing.)

[1] *H. lawrencei* (Herrick). Similar to *H. chrysoptera;* cheeks and lower parts pure
yellow ; wing bars white ; back, etc., olive-green. N. J., etc., rare. Either a hybrid
of *chrysoptera* and *pinus* or else a yellow dichromatic phase of the former. The
latter view is considered by Ridgway the most probable. (To Geo. N. Lawrence.)

bb. Throat yellow or white; ear coverts olive or ashy above, pale below.

963. **H. pinus** [1] (L.). BLUE-WINGED YELLOW WARBLER. Olive green; crown and all under parts bright yellow; wing bars whitish ; loral strip black ; ♀ similar. L. 4½. W. 2½. T. 2. E. U. S., N. to N. Y; a handsome bird, like a miniature *Protonotaria.* (Lat., pine.)

aa. Tail feathers without white; no wing bars.

964. **H. peregrina** (Wilson). TENNESSEE WARBLER. Olive green; head more or less ashy and without crown patch ; white or slightly yellowish below. L. 4½. W. 2¾. T. 1¾. N. Am., rare E. of Ohio ; closely resembles the young of the two following, but its wings are nearly half longer than the short tail; *celata* has no ashy on head, and *ruficapilla* is yellower below.

965. **H. celata** (Say). ORANGE-CROWNED WARBLER. Olive green, never ashy on head ; crown patch orange brown, more or less concealed ; greenish yellow below ; ♀ duller, sometimes without crown patch, known from the next by the more olive color of the head, which is similar to the back ; belly less yellow. L. 4¾. W. 2¼. T. 2. N. Am.; rare E. (Lat., concealed.)

966. **H. ruficapilla** (Wilson). NASHVILLE WARBLER. Olive green, ashy on head and neck, the color contrasting with back; crown patch bright chestnut, more or less concealed ; bright yellow below ; lores and orbital ring pale ; ♀ duller, crown patch obscure. L. 4¾. W. 2¼. T. 2. E. N. Am., common. (Lat., *rufus*, red ; *capillus*, hair.)

512. **COMPSOTHLYPIS** Cabanis. (*Parula* Bonaparte ; changed on account of the earlier *Parulus.*) (κομψός, comely ; θλυπίς, a little bird or warbler.)

967. **C. americana** (L.). BLUE YELLOW-BACKED WARBLER. Clear ashy blue ; back with a large golden green patch ; yellow below, belly white ; a brown band across breast ; white wing bars ; tail feathers with white ; ♀ obscurely marked. L. 4¾. W. 2¼. T. 2. E. N. Am., not rare ; very elegant.

513. **DENDROICA** Gray. (δένδρον, tree ; οἰκέω, I inhabit.)

A large genus comprising about 30 species of brightly colored little birds, very abundant in the United States during the migra-

---

[1] *H. leucobronchialis* (Brewster). Ashy gray ; throat and lower parts white ; wing bands yellow or white; variable. E. U. S., not common; now considered as probably a white phase of *H. pinus*, as *H. lawrencei* is a yellow phase of *chrysoptera*. It is further thought that the two species in both yellow and white condition hybridize. (λευκός, white ; βρόγχος, throat.)
H. cincinnatiensis Langdon. Olive green, lores and part of ear coverts black ; spot below eye and entire lower parts yellow. Cincinnati ; now regarded as a hybrid of *H. pinus* and *Geothlypis formosus.* (See Ridgway, N. Am. Birds, p. 486.)

tions. The adult males of the different species are readily distinguished, but ♀ and young offer difficulties. The tail feathers are always marked with white or yellow, and the bill is usually little pointed, notched, and with evident bristles at the rictus.

The following artificial analysis, partly taken from Coues's "Key," will generally enable the student to distinguish specimens, at least the males in full plumage : —

a. Tail feathers edged with yellow; plumage chiefly yellow. . *æstiva,* 969.
aa. Tail feathers blotched with white.
　　b. A white blotch on the primaries near their bases; no wing bars.
　　　　　　　　　　　　　　　　　　　　　　cærulescens, 970.
　　bb. No white blotch on primaries.
　　　c. Wing bars, if present, not white.
　　　　d. White below; crown and wing patch more or less yellow.
　　　　　　　　　　　　　　　　　　　　　pennsylvanica, 974.
　　　　dd. Yellow below; sides reddish-streaked; crown chestnut.
　　　　　　　　　　　　　　　　　　　　　　palmarum, 983.
　　　　ddd. Yellow below; sides black-streaked.
　　　　　e. Back olive with reddish spots. . . . . *discolor,* 982.
　　　　　ee. Back ashy. . . . . *kirtlandi,* 981.
　　　cc. Wing bars or wing patch white.
　　　　f. Rump yellow.
　　　　　g. Crown clear ash; yellow and streaked below. *maculosa,* 972.
　　　　　gg. Crown with yellow spot; white and streaked below.
　　　　　　　　　　　　　　　　　　　　　　coronata, 971.
　　　　　ggg. Crown black with a median stripe of orange brown; an orange brown ear-spot. . . . . . . *tigrina,* 968.
　　　　ff. Rump not yellow.
　　　　　h. Crown with orange or yellow spot; throat orange or yellow.
　　　　　　　　　　　　　　　　　　　　　　blackburniæ, 978.
　　　　　hh. Crown black; no distinct yellow anywhere; much streaked.
　　　　　　　　　　　　　　　　　　　　　　striata, 976.
　　　　　hhh. Crown blue or greenish, like the back; no definite yellow.
　　　　　　　　　　　　　　　　　　　　　　cærulea, 973.
　　　　　hhhh. Crown chestnut, like the throat; no definite yellow; buffy below. . . . . *castanea,* 975.
　　　　　hhhhh. Crown bluish or yellowish, not as above, — some yellow.
　　　　　　i. Throat black (sometimes obscured by yellow tips to feathers); outer tail feather white-edged. *virens,* 979.
　　　　　　ii. Throat yellow.
　　　　　　　j. Back ashy blue; cheeks black. . . . *dominica,* 977.
　　　　　　　jj. Back yellowish olive; cheeks same. . . *vigorsi,* 980.

We copy from Coues's Key the following valuable diagnostic marks of Warblers in any plumage : —

A white spot at base of primaries. . . . . *cærulescens,* 970.
Wings and tail dusky, edged with yellow. . . . . *æstiva,* 969.
Wing bars and belly yellow. . . . . *discolor,* 982.

Wing bars yellow and belly pure white. . . . . . . *pennsylvanica*, 974.
Wing bars white and tail spots oblique, at end of 2 (rarely 3) outer feathers
only.. . . . . *vigorsi*, 980.
Wing bars brownish; tail spots square at end of two outer feathers only.
palmarum, 983.
Wing bars not conspicuous; whole under parts yellow; back with no greenish.
kirtlandi, 981.
Tail spots at end of nearly all the feathers, and no definite yellow anywhere.
cærulea, 973.
Tail spots at middle of nearly all the feathers; rump and belly yellow.
maculosa, 972.
Rump, sides of breast (usually) and crown with yellow; throat white.
coronata, 971.
Throat definitely yellow; belly white; back with no greenish. *dominica*, 977.
Throat yellow or orange; crown with at least a trace of a central yellow or
orange spot, and outer tail feather white-edged externally.
blackburniæ, 978.
Throat, breast and sides black, or with black traces (seen on parting the
feathers); sides of head with diffuse yellow; outer tail feather white-
edged externally. . . . . *virens*, 979.
Bill acute, perceptibly curved; rump usually yellow. . . . . *tigrina*, 968.
With none of the foregoing special marks; crissum buffy. . *castanea*, 975.
Crissum white. *striata*, 976.

a. Bill very acute, the tip appreciably decurved, terminal half of tongue with
its edges folded over upon the upper surface, the tip deeply cleft and
fringed. (*Perissoglossa* Baird.)

968. **D. tigrina** (Gmelin). CAPE MAY WARBLER. Olivaceous
above with black streaks; rump and sides of neck bright yellow;
yellow below, much streaked with black; crown mostly black; ear
coverts orange brown; a white wing patch; ♀ duller, with no black
or reddish about head. L. 5¼. W. 2¾. T. 2. E. U. S. A fine species with a peculiar structure of the tongue, which is somewhat as
in the Honey Creepers (*Cœrebidæ*) of the Tropics.

aa. Bill not very acute nor distinctly decurved at tip; tongue gradually tapering to the slightly cleft and fringed tip. (*Dendroica*.)

b. Tail feathers without white, the inner web yellow.

969. **D. æstiva** (Gmelin). SUMMER WARBLER. GOLDEN
WARBLER. Chiefly golden yellow; breast and sides with orange
brown streaks; quills dusky, edged with yellow; ♀ similar, the
brown streaks obsolete. L. 5¼. W. 2½. T. 2¼. America; everywhere abundant. (Lat., summer.)

bb. Tail feathers blotched with white.

c. A white spot on some of the primary quills, near their bases.

970. **D. cærulescens** (L.). BLACK-THROATED BLUE WAR-
BLER. Rich gray blue, with a few black streaks on back; throat,
sides of head, neck and sides of body black, otherwise pure white
below; quills black, edged with blue; ♀ dull olive greenish, ob-

scurely marked, known by the blotch on the primaries. L. 5¼.
W. 2¾. T. 2¼. E. N. Am.; an elegant species, common. (Lat.,
bluish.)

cc. No white spot on primary quills.

971. **D. coronata** (L.). Yellow-rumped Warbler. Bluish
ash above, streaked with black ; white below with large black area
on breast ; crown patch, rump and sides of breast bright yellow,
there being four definite yellow places; ♀ and young brownish,
with less yellow on breast and head. L. 5¾. W. 3. T. 2½. U. S.,
very abundant. The earliest migrant: represented W. of Rocky
Mts. by **D. auduboni** Townsend, very similar but with the throat
yellow. (Lat., crowned.)

972. **D. maculosa** (Gmelin). Black and Yellow Warbler.
Back black, with olive skirtings ; rump yellow ; head clear ash ; a
white stripe behind eye ; sides of head black ; under parts rich yellow, with black streaks which are confluent on breast; crissum
white; ♀ similar, more olivaceous, with much less black. L. 5.
W. 2¼. T. 2¼. E. N. Am.; a brilliant little bird, common. (Lat.,
spotty.)

973. **D. cærulea** (Wilson). Cerulean Warbler. Bright
blue with black streaks; white below; breast and sides with streaks
of slaty blue : ♀ not streaked, greenish above, slightly yellowish below. L. 4¼. W. 2½. T. 2. E. U. S., N. to L. Erie, common S. W.;
a dainty species.

974. **D. pennsylvanica** (L.). Chestnut-sided Warbler.
Blackish above, much streaked with whitish and olive; crown clear
yellow ; black patch about eye ; pure white below; a line of bright
chestnut streaks along sides ; wing patch yellowish (never clear
white) ; ♀ similar but with less chestnut and black. L. 5. W.
2½. T. 2¼. E. N. Am., abundant, especially N. ; very pretty.

975. **D. castanea** (Wilson). Bay-breasted Warbler.
Autumn Warbler. Back ashy olive, streaked with black ;
forehead and sides of head black, enclosing a large deep chestnut
crown patch ; chin, throat and sides chestnut, otherwise pale buffy
below ; ♀ more olivaceous with less chestnut ; young scarcely distinguishable from *striata*, but the latter has crissum white instead
of buffy: *castanea* is less streaked on sides. L. 5. W. 3. T. 2¼.
E. N. Am., not rare. (Lat., chestnut.)

976. **D. striata** (Forster). Black-poll Warbler. Ashy
olive, white below; almost everywhere streaked with black, the
streaks below narrow ; whole top of head pure black ; ♀ more olivaceous, slightly yellowish below; rather large. L. 5¾. W. 3.
T. 2¼. E. N. Am.; the last to migrate, "bringing up the rear of
the warbler-hosts; when the Black-Polls appear in force, the collecting season is about over." (*Coues.*) (Lat., striped.)

977. **D. dominica** (L.). YELLOW-THROATED WARBLER. Ashy
blue; throat bright yellow; belly white; cheeks and top of head
black; superciliary line white or yellowish in front. L. 5. W. 2¾.
T. 2¼. Southern States; N. to Pa. and N. Ind., rare N. A neat,
plain species with the habits of a creeper; represented W. by var.
albilora Ridgway, smaller, with shorter bill, the superciliary streak
chiefly or entirely white, instead of yellowish as in var. *dominica*.
Miss. Valley and S. (From St. Domingo.)

978. **D. blackburniæ** (Gmelin). ORANGE-THROATED WAR-
BLER. Black above with buffy streaks; crown patch, superciliary
line, sides of neck and the whole throat brilliant orange, becoming
yellowish on the belly; ♀ similar, but olive and yellow instead of
black and orange. L. 5½. W. 2¾. T. 2¼. E. N. Am., abundant
in migration, among the tree-tops; the most brilliant species.
(To Mrs. Blackburn.)

979. **D. virens** (Gmelin). BLACK-THROATED GREEN WAR-
BLER. Clear yellow olive; rump ashy; sides of head rich yellow; whole throat and breast jet black, the color extending along
the sides; otherwise whitish below; ♀ and winter birds with
the black interrupted or veiled with yellowish. L. 5. W. 2½. T.
2¼. E. N. Am.; abundant. (Lat., greenish.)

980. **D. vigorsi** (Audubon). PINE-CREEPING WARBLER. Yellow olive above; under parts and superciliary line dark yellow; no
sharp markings anywhere; wing bands dull whitish, distinct only
in adult ♂; ♀ more grayish. L. 5¾. W. 3. T. 2¼. E. U. S., N.
to Me. and N. Mich.; abundant in pine forests, the dullest in color
of our species.

981. **D. kirtlandi** Baird. KIRTLAND'S WARBLER. Ashy blue
above, back and sides streaked with black; yellow below; chin
and crissum white; no distinct white wing bars; lores black; ♀
similar, duller. L. 5¼. W. 2¾. T. 2¾. E. U. S., quite rare. (To
Dr. J. P. Kirtland.)

982. **D. discolor** (Vieillot). PRAIRIE WARBLER. Olive yellow; back with a patch of red spots; forehead, superciliary line,
wing bars and under parts bright yellow; streaked below; sides of
head with black; ♀ similar. L. 4¾. W. 2¼. T. 2. E. U. S., N.
to Mass. and Mich.; chiefly in evergreen thickets. An elegant
species. (Lat., two-colored.)

983. **D. palmarum** (Gmelin). RED-POLL WARBLER. Brownish olive above, somewhat streaked, rump brighter; crown bright
chestnut; superciliary line and under parts yellow with brown
streaks; no wing bars; ♀ similar. L. 5. W. 2¾. T. 2¼. E. N.
Am.; abundant; terrestrial; represented along the Atlantic coast
by var. **hypochrysea** Ridgway, larger, and much more deeply colored,
entire lower parts bright yellow. (Lat., of the palms.)

514. SEIURUS Swainson. (σείω, I wag; οὐρά, tail.)

a. Crown orange brown with a black stripe on each side.

984. **S. aurocapillus** (L.). OVEN-BIRD. GOLDEN-CROWNED "THRUSH." Bright olive green, white below, sharply spotted on breast and sides, like a thrush. L. 6¼. W. 3. T. 2½. U. S.; abundant in woodland, spending most of its time on the ground, like the other species of this genus, and the next; remarkable for its ringing song and its curious oven-shaped nest; the largest of the true Warblers. (Lat., *aurum*, gold; *capillus*, hair.)

aa. Crown plain brownish.

985. **S. noveboracensis** (Gmelin). WATER WAGTAIL. WATER THRUSH. Dark olive brown above, pale yellowish beneath; thickly streaked everywhere with the color of the back; superciliary line buffy; bill about half inch long; feet dark. L. 6. W. 3. T. 2⅓. N. Am., in thickets; moves its tail like a Wagtail. The Western form, var. **notabilis** Grinnell is larger and darker; it ranges E. to Ind. (Lat., of New York.)

986. **S. motacilla** (Vieillot). LARGE-BILLED WATER THRUSH. Color of preceding, but paler below, the streaks below broader and less sharply defined; superciliary stripe white; bill larger, about ¾ inch; feet pale. L. 6¼. W. 3¼. T. 2⅓. E. U. S., scarce; N. to Mass. and N. Wis. (Lat., wagtail.)

515. GEOTHLYPIS Cabanis. (γέα, earth; θλυπίς, some small bird like a warbler.)

a. Tail evidently shorter than wing, more than half hidden by the coverts. (*Oporornis* Baird.)

987. **G. formosa** (Wilson). KENTUCKY WARBLER. Clear olive green, bright yellow below; crown and sides of head and neck black, with a rich yellow superciliary stripe, which bends around the eye behind; ♀ with the black replaced by dusky olive. L. 5¾. W. 3. T. 2¼. E. U. S., chiefly S. W., N. to Wis. and Conn.; in low thickets; a handsome and active species. (Lat., comely.)

988. **G. agilis** (Wilson). CONNECTICUT WARBLER. Olive green, ashy on head; throat and breast brownish ash, otherwise yellow below; no sharp markings; in fall almost uniform olivaceous. L. 5¾. W. 3. T. 2¼. E. N. Am.; a shy, quiet bird, rarely seen in spring.

989. **G. philadelphia** (Wilson). MOURNING WARBLER. Bright olive, clear yellow below; head ashy; throat and breast black, the feathers usually ashy-skirted (as though the bird wore crape, hence "Mourning Warbler"); ♀ and ♂ not in full plumage ashy anteriorly, almost exactly like *G. agilis*, but the tail more nearly length of wings; no white spot on eyelid. L. 5½. W. 2⅓. T. 2⅕. E. U. S., rather rare, in dense thickets.

aa. Tail not shorter than wing; its feathers not half concealed by coverts. (*Geothlypis.*)

990. **G. trichas** (L.). MARYLAND YELLOW THROAT. Olive green; forehead and broad mask extending down sides of head and neck jet black, bordered behind with clear ash; under parts yellow, clear on throat and breast; ♀ obscurely marked, without black mask and with less yellow. L. 4½. W. 2¼. T. 2⅜. U. S., abundant in thickets; a pretty bird with a lively song. Replaced W., by var. **occidentalis** Brewster, larger and brighter, the belly clear yellow instead of buffy whitish. Rocky Mts., E. to Ga. and Ill. (τριχάς, some small bird.)

516. **ICTERIA** Vieillot. (ἴκτερος, yellowness, as jaundice.)

991. **I. virens** (L.). YELLOW-BREASTED CHAT. Olive green; throat and breast bright yellow; belly abruptly white; lores black: a white superciliary line; wings and tail plain; tarsus almost booted. L. 7½. W. 3¼. T. 3¼. U. S., southerly, N. to Mass. and Wis.; a loud, quaint songster.

517. **SYLVANIA** Nuttall. (Lat., sylvan.)

a. Tail feathers blotched with white; no wing bars.

992. **S. mitrata** (Gmelin). HOODED WARBLER. Bright yellow olive; breast, crown, and neck all around jet black, enclosing a broad golden mask; under parts from the breast bright yellow; ♀ olive instead of black. L. 5. W. 2¾. T. 2¼. E. U. S., southerly, N. to L. Erie; a singular and beautiful species. (Lat., mitred.)

aa. Tail feathers plain dusky; no wing bars.

993. **S. pusilla** (Wilson). GREEN BLACK-CAPPED WARBLER. Clear yellow olive; crown glossy black; forehead, lores, sides of head and entire under parts bright yellow; ♀ with less black. L. 4¾. W. 2¼. T. 2¼. U. S. (Lat., weak.)

994. **S. canadensis** (L.). CANADA WARBLER. Bluish ash; crown streaked with black; under parts clear yellow; crissum white; lores black, continuous with black under the eye; this passing as a chain of black streaks down the side of the neck encircling the breast like a necklace; a yellow superciliary streak; ♀ similar, with less black. L. 5½. W. 2¾. T. 2¼. E. U. S., to the Missouri, frequent; one of the handsomest Warblers.

518. **SETOPHAGA** Swainson. (σής, moth; φαγός, eating.)

995. **S. ruticilla** (L.). AMERICAN REDSTART. Black; sides of breast and large blotches on wings and tail orange-red; belly white, reddish tinged; no wing bars; ♀ olive, marked with creamy yellow instead of red. L. 5¼. W. 2½. T. 2½. E. N. Am., very abundant; a handsome and active fly-catcher.

FAMILY CLXXIV. **MOTACILLIDÆ.** (THE WAGTAILS.)

Primaries 9, the first about as long as second; inner secondaries enlarged, the longest one about as long as the primaries in the closed wing. Bill shorter than the head, very slender, straight, acute, notched at tip. Feet large, fitted for walking; hind claw long, little curved, as in the Larks; inner toe cleft; basal joint of outer toe united with middle one; tarsus as in Oscines generally, ending in a sharp, undivided ridge behind. Rictal bristles not conspicuous; nostrils exposed.

A group of about 100 species, mostly of the Old World. Terrestrial birds, with the habit (shared by various others) of moving the tail up and down, as if "balancing themselves on unsteady footing;" hence the name "Wagtail." (Lat., *motacilla*, wag-tail.)

a. Tail shorter than wings, its feathers tapering; hind claw long and straightish. . . . . ANTHUS, 519.

519. ANTHUS Bechstein. (ἄνθος, some small bird.)

a. Tarsus longer than hind toe with claw. (*Anthus.*)

996. **A. pensilvanicus** (Latham). BROWN LARK. TITLARK. PIPIT. Dark brown above, slightly streaked; superciliary line and under parts buffy ; breast and sides streaked; outer tail feathers with white. L. 6½. W. 3½. T. 3. N. Am., not rare.

aa. Tarsus shorter than hind toe with claw. (*Neocorys* Sclater.)

997. **A. spraguei** (Audubon). MISSOURI SKYLARK. Buffy and dusky streaked. W. U. S., E. to Minn., abundant W.; its habits similar to those of the Skylark, its song not inferior. (To Isaac Sprague.)

FAMILY CLXXV. **TROGLODYTIDÆ.** (THE WRENS
AND MOCKING-BIRDS.)

Primaries 10, the first short, hardly spurious; wings moderate or long. Bill usually more or less slender, with or without a notch near the tip; nostrils not covered by bristles. Tarsus scutellate, the plates usually distinct.

The *Miminæ*, now associated with the wrens by the A. O. U., are in many respects intermediate between wrens and thrushes. Their reference to either group is chiefly a matter of convenience. The wrens "are sprightly, fearless and impudent little creatures, apt to show bad temper when they fancy themselves aggrieved by cats or people, or anything else that is big or unpleasant to them; they quarrel a good deal, and are particularly spiteful towards martins and swallows, whose homes they often invade and occupy. Their song is bright and hearty, and they are fond of their own music; when disturbed at it they make a great ado with noisy scolding. Part of them (*Cistothorus*) live in reedy swamps and marshes, where they hang astonishingly big globular nests, with a little hole on one side, on tufts of rushes, and lay six or eight dark-colored eggs; the others nest anywhere." (*Coues.*) To the *Miminæ* belongs the first of song-birds, the mocking-bird. All of the *Troglodytinæ* and *Miminæ* are plainly colored, being chiefly brown. All are insectivorous, and most of them migratory. Genera about 23, species 150, most abundant in tropical America.

a. Bill with bristles at the rictus; inner toe free to the base. Mockers. (*Miminæ.*)
  b. Tail longer than wing.
    c. Bill shorter than middle toe without claw; bill notched at tip.
      d. Tarsal scutella distinct; (tail with white). . . . . Mimus, 520.
      dd. Tarsal scutella indistinct; (tail without white).
                          Galeoscoptes, 521.
    cc. Bill not shorter than middle toe with claw, often decurved; bill scarcely notched at tip. . . . . Harporhynchus, 522.
aa. Bill not notched, without evident bristles at the rictus ; inner toe somewhat joined at base to middle; nostril with a small scale. Wrens. (*Troglodytinæ.*)
  f. Outer tail feathers reaching decidedly beyond tips of longest lower coverts; (back without lengthwise streaks).
    g. Bill rather stout, somewhat decurved at tip; (back without cross-bars : superciliary streak distinct). . . . . Thryothorus, 523.
    gg. Bill more slender, straight or slightly decurved ; (back with cross-bars more or less distinct; no distinct superciliary stripe).
                          Troglodytes, 524.
  ff. Outer tail-feathers reaching little beyond tips of lower coverts; (back streaked lengthwise). . . . . Cistothorus, 525.

520. MIMUS Boie. (Lat., mimic.)

998. **M. polyglottos** (L.). Mocking-bird. Ashy brown above, nearly white below ; wings blackish, with white wing bars; tail blackish, outer feathers white ; ♀ with less white. L. 9¼. W. 4½. T. 5. U. S., chiefly southerly ; N. to Mass., Iowa, etc. A famous singer, easily first among birds in the range and variety of its notes. (πολύς. many ; γλῶττα, tongue.)

521. GALEOSCOPTES Cabanis. (γαλῆ, weasel ; σκώπτης, mocker.)

999. **G. carolinensis** (L.). Cat-Bird. Dark slate color; crown and tail black; crissum chestnut. L. 8¾. W. 3¾. T. 4. N. Am., generally common; a fine singer.

522. HARPORHYNCHUS Cabanis. (ἅρπη, sickle ; ῥύγχος, bill, true of the typical species.)

a. Tarsus longer than bill; lower parts spotted and streaked. (*Methriopterus* Reichenbach.)

1000. **H. rufus** (L.). Brown Thrush. Thrasher. Cinnamon red above; lower parts thickly spotted; bill nearly straight, shorter and much less curved than in the other *Harporhynchi*, five species of which occur in the S. W. L. 11. W. 4. T. 5¼. B. 1. E. U. S., abundant. A brilliant songster, its notes similar to those of the mocking-bird, but softer and less varied.

523. THRYOTHORUS Vieillot.   (θρύον, reed ; θοῦρος, leaping.)

a. Tail not longer than wings, its feathers all brown with fine black bars. (*Thryothorus.*)

1001. **T. ludovicianus** (Gmelin). CAROLINA WREN. MOCK-ING WREN. Clear reddish brown, brightest on rump ; pale buffy below ; wings barred; a pale superciliary stripe. L. 6. W. 2¼. T. 2¼. E. U. S., southerly, N. to Penn; a remarkable singer. (Lat., of Louisiana.)

aa. Tail longer than wings, its feathers mostly blackish, the middle ones grayish, barred. (*Thryomanes* Sclater.)

1002. **T. bewickii** (Audubon). Umber brown above; brownish white below ; white streak above eye and on neck. L. 5½. W. 2¼. T. 2½. S. U. S., N. to Penn. and Minn.   (To Thos. Bewick.)

524. TROGLODYTES Vieillot.   (τρωγλοδύτης, cave-dweller.)

a. Tail more than ⅔ wing. (*Troglodytes.*)

1003. **T. aedon** Vieillot. HOUSE WREN. Brown, brightest behind ; rusty below : everywhere above and behind barred or waved with darker, distinctly so on wings, tail, and crissum. L. 5. W. 2. T. 2. E. U. S., abundant ; an active and familiar little bird. (ἀηδών, a singer.)

aa. Tail very short, less than ⅔ wing. (*Anorthura* Rennie.)

1004. **T. hiemalis** Vieillot. WINTER WREN. Deep reddish-brown, waved with dusky ; wings, tail, and belly posteriorly sharply barred. L. 4. W. 1⅔. T. 1¼. N. Am., U. S. in winter, common N.; a fine singer. (Lat., wintry.)

525. CISTOTHORUS Cabanis.   (κίστος, a shrub, rock-rose ; θοῦρος, leaping.)

a. Bill about half as long as head; no white superciliary line. (*Cistothorus.*)

1005. **C. stellaris** (Lichtenstein). SHORT-BILLED MARSH WREN. Dark brown, head and back darker ; entire upper parts with white streaks; lower parts buffy. L. 4½. W. 1¾. T. 1¾. E. U. S., in marshes ; rather rare. (Lat., starry.)

aa. Bill slender, about as long as head; a conspicuous white superciliary line. (*Telmatodytes* Cab.)

1006. **C. palustris** (Wilson). LONG-BILLED MARSH WREN. Clear brown ; back with a black patch containing white streaks; otherwise unstreaked above : crown blackish ; lower parts brownish white. L. 5. W. 2. T. 1¾. U. S., abundant in reedy swamps.

Family CLXXVI. CERTHIIDÆ. (The Creepers.)

Primaries 10, first less than half second. Bill slender, as long as head, without notch or bristles. Tarsus scutellate, shorter than middle toe. Claws all very long, curved and compressed. Wings about as long as tail; tail feathers pointed, with stiffened shafts, somewhat like the tail of a wood-pecker, and similarly used for support. Genera 5; species about 12, widely distributed. Habits similar to those of the Nuthatches, but the voice different, being small and fine. (The above diagnosis applies rather to the subfamily, *Certhiinæ*.)

a. Bill decurved, about as long as head. . . . . . . CERTHIA, 526.

526. CERTHIA Linnæus. (Lat., a creeper.)

1007. **C. familiaris** L. BROWN CREEPER. Plumage dark brown, above much barred and streaked with whitish; pale below; rump clear tawny. L. 5½. W. 2¾. T. 2¾. N. Am. A curious little bird. The E. American form (white below) is var. **americana** (Bonap.). (*Eu.*)

Family CLXXVII. PARIDÆ. (The Nuthatches and Titmice.)

Primaries 10, the first short. Bill not notched nor decurved: loral feathers bristly; nostrils concealed by dense tufts. Tarsus scutellate; plumage more or less lax, subject to few variations. Small birds, apparently allied to the jays on the one hand and to the wrens and thrushes on the other. Species 100 or more, in most parts of the world; insectivorous and usually not migratory.

a. Bill slender, as long as head; hind toe longer than middle toe: tail much shorter than wing. (*Sittinæ*.) . . . . SITTA, 527.
aa. Bill stoutish, much shorter than head; hind toe shorter than middle; tail not shorter than wing. (*Parinæ*.) . . . . PARUS, 528.

527. SITTA Linnæus. (σίττα. nuthatch.)

1008. **S. carolinensis** Latham. WHITE-BELLIED NUTHATCH. "SAP-SUCKER." Ashy blue above, white below; crissum with rusty brown; crown and nape black, unstriped; middle tail feathers like the back, others black, blotched with white; ♀ with less or no black on the head. L. 5½. W. 3½. T. 2. U. S., abundant everywhere. An active, nimble little bird, running up and down trees, and hanging in every conceivable attitude, the head down as often as up.

1009. **S. canadensis** L. RED-BELLIED NUTHATCH. Ashy blue, brighter than the preceding, rusty brown below; crown glossy black (♂), or bluish (♀), bordered by white and black stripes. L. 4½. W. 2¾. T. 1½. N. Am., chiefly N.

1010. **S. pusilla** Latham. BROWN-HEADED NUTHATCH. Ashy blue ; crown clear brown, a whitish spot on nape ; pale rusty below. L. 4. W. 2½. T. 1½. S. E. U. S., N. to Md. (Lat., weak.)

528. PARUS Linnæus. (Lat., a titmouse.)

a. Head conspicuously crested. (*Lophophanes* Kaup.)

1011. **P. bicolor** L. TUFTED TITMOUSE. Grayish ash, the forehead alone black ; whitish below ; sides washed with reddish. L. 6¼. W. 3¼ T. 3¼. E. U. S., southerly, N. to Mich. ; abundant in woodland and remarkable for its loud, cheerful whistle.

aa. Head not crested. (*Parus.*)

1012. **P. carolinensis** Audubon. SOUTHERN CHICKADEE. Similar to the next ; tertials and greater wing coverts without whitish edgings ; smaller : tail shorter. L. 4½. W. 2¼. T. 2⅛. Southern, N. to S. Pa. and Ind. ; often regarded as a winter resident variety of the next.

1013. **P. atricapillus** L. TITMOUSE. BLACK-CAPPED CHICKADEE. Grayish ash ; wings and tail plain, with whitish edgings ; crown, nape, chin, and throat black ; cheeks white ; no white superciliary line. L. 5. W. 2½. T. 2½. N. Am., S. to Ind. and Va., abundant : represented N. W. by var. **septentrionalis** Harris ; paler, with tail (2¾) longer than wings. (Lat., black-haired.)

1014. **P. hudsonicus** Forster. Olive brown ; crown browner ; some pale chestnut below ; throat black ; a white stripe through eye. L. 5. W. 2½. T. 2⅓. N. N. Am., S. to Mass.

FAMILY CLXXVIII. SYLVIIDÆ. (THE OLD WORLD WARBLERS.)

Diminutive Thrushes. Primaries 10, the first short. Bill slender, depressed at base, notched and decurved at tip. Rictus with bristles ; nostrils oval. Tarsus usually booted, scutellate in *Polioptilinæ.* Basal joint of middle toe attached its whole length externally, half way internally. A large family of nearly 600 species of small birds, chiefly of the Old World, where they fill the place taken in America by the *Mniotiltidæ.* The most famous of the group is the European nightingale (*Luscinia luscinia* L.).

a. Tarsus booted ; nostril with one or more minute feathers ; wings longer than tail. (*Regulinæ.*) . . . . REGULUS, 529.

aa. Tarsus scutellate ; wings not longer than tail. (*Polioptilinæ.*)

POLIOPTILA, 530.

529. REGULUS Cuvier. (Lat., dim. of *rex*, king — "of the wrens.")

a. Nostril hidden by a single tiny feather. (*Regulus.*)

1015. **R. satrapa** Lichtenstein. GOLDEN-CROWNED KINGLET. Olivaceous ; crown with a yellow patch, bordered with black, orange red in the centre in ♂; forehead and line over eye whitish; a vague dusky blotch at base of secondaries. L. 4. W. 2¼. T. 1¾. N. Am.; not rare. (σατρáπης, a ruler.)

a. Nostril with a tuft of small bristle-like feathers. (*Phyllobasileus* Cabanis.)

1016. **R. calendula** (L.). RUBY-CROWNED KINGLET. Olivaceous ; crown with a scarlet patch in both sexes, wanting the first year; no black about head. L. 4¼. W. 2⅓. T. 1¾. N. Am., common. (Lat., a little fire.)

530. **POLIOPTILA** Sclater. (πολιός, hoary ; πτῖλον, feather.)

1017. **P. cærulea** (L.). BLUE-GRAY GNAT-CATCHER. Clear ashy blue, brightest on head; whitish below ; ♂ with forehead and sides of crown black ; outer tail feathers chiefly white. L. 4½. W. 2. T. 2¼. U. S., chiefly southerly; N. to Mass. and L. Mich. A sprightly little bird with a squeaky voice, but really a fine singer.

FAMILY CLXXIX. **TURDIDÆ.**[1] (THE THRUSHES.)

Primaries 10, the first short or spurious ; bill generally rather long. not conical, usually with a slight notch near the tip ; nostrils oval, not concealed, but nearly or quite reached by the bristly frontal feathers ; rictus with bristles, which are well developed in most of our species ; tarsus always "booted," *i. e.*, enveloped in a continuous plate, formed by the fusion of all the scutella except 2 or 3 of the lowest. Toes deeply cleft, the inner one free, the outer united to the middle one, not more than half the length of the first basal joint.

A large family of about 300 species, found in most parts of the world, and embracing quite a wide variety of forms. Nearly all of them are remarkable for their vocal powers. Their food consists of insects and soft fruits.

a. Bill short, depressed, notched and slightly hooked at tip; gonys not more than ½ the commissure; tail about as long as wings. (*Myadestinæ*.)

MYADESTES, 531.

aa. Bill not depressed nor hooked; gonys more than ⅓ the commissure. (*Turdinæ*.)

  b. Wings moderate ; (no blue).

    c. Tarsus longer than middle toe with claw; nostrils exposed; nasal fossæ without feathers; bill notched near its tip; sexes similar.

      d. Bill much widened at base; (breast spotted). . . . TURDUS, 532.

      dd. Bill little widened at base; (breast in adult unspotted).

MERULA, 533.

---

[1] One of the most remarkable of the thrush-like birds is the Ouzel or Dipper (*Cinclus mexicanus* Swainson), an aquatic thrush which swims (or rather flies) freely under water, although not web-footed. It is a fine singer, living about mountain torrents in the Rocky Mountain regions ; a similar species (*C. merula*) occurs in Europe. They are now placed in a separate family, *Cinclidæ*.

cc. Tarsus not longer than middle toe with claw; nostrils partly concealed by feathers in the nasal fossæ; bill not notched; sexes unlike.

HESPEROCICHLA, 534.

bb. Wings long and pointed; (plumage partly blue). . . . SIALIA, 535.

531. MYADESTES Swainson. (μυῖα, fly; ἐδεστής, eater.)

1018. M. townsendi (Audubon). FLY-CATCHING THRUSH. TOWNSEND'S SOLITAIRE. Ashy gray, paler below; wing bands buffy; tail blackish; whitish ring about eye; young with reddish spots. L. 8. W. 4½. T. 4½. Rocky Mountains and westward, straying E. to Ill. (*Nelson.*) A most exquisite songster. (To J. K. Townsend.)

532. TURDUS Linnæus. (Lat., thrush.)

a. Wings never more than 3½ times tarsus; plain brownish above; spotted below. Wood-thrushes. (*Hylocichla* Baird.)

b. Reddish color of back most distinct on head.

1019. T. mustelinus Gmelin. WOOD THRUSH. Cinnamon brown, brightest on the head, shading into olive on the rump; breast with large, very distinct dusky spots. L. 8. W. 4¼. T. 3. E. U. S., in woodland; our largest and handsomest wood thrush. An exquisite songster. (Lat., weasel-colored.)

bb. Reddish color of back equally distinct from head to tail.

1020. T. fuscescens Stephens. VEERY. TAWNY THRUSH. WILSON'S THRUSH. Uniform reddish brown above; breast and throat washed with brownish or pinkish yellow, and marked with small indistinct brownish spots. L. 7½. W. 4¼. T. 3½. E. N. Am., in damp woods, frequent; a fine songster, superior to the wood-thrush in its range of notes. The Western variety, Ill. to Rocky Mts., var. **salicicolus** Ridgway, is russet olive, the cheeks paler, with broader markings. (Lat., dusky.)

bbb. Back entirely olive, with no reddish shade anywhere.

c. Sides of head without buffy shades.

1021. T. aliciæ Baird. GRAY-CHEEKED THRUSH. Very similar to the next, of which it may be a variety, but without buffy or whitish ring about eye, or any buffy tint about head. E. N. Am., ranging more northerly. A smaller form, with slenderer bill is var. **bicknelli** Ridgway, in Catskills and N. (To Alice Kennicott.)

cc. Sides of head more or less shaded with buffy.

1022. T. ustulatus Nuttall. OLIVE-BACKED THRUSH. Uniform olive above; breast and throat thickly marked with large, dusky olive spots; breast and sides of head strongly buffy-tinted; a conspicuous buffy orbital ring. L. 7¼. W. 4. T. 3. N. Am. The Western form (var. **ustulatus**) is russet brown above, rather than grayish olive as in the Eastern form, which is var. **swainsoni** Cabanis. (Lat., scorched.)

bbbb. Reddish color of back chiefly confined to the tail.

1023. **T. aonalaschkæ** Gmelin. HERMIT THRUSH. Olive brown above, becoming rufous on rump and tail; breast with numerous, rather distinct, dusky spots; a whitish orbital ring. L. 7. W. 3¼. T. 2½. N. Am., migrating early; a sweet singer. The Eastern bird, var. **pallasi** Cabanis, is more "smoky" in hue, the tail a little less red, the bill larger. (From Unalaska Island.)

533. MERULA Leach. (Lat., merle or blackbird.)

1024. **M. migratoria** (L.). ROBIN. AMERICAN RED BREAST. Olive gray above; head and tail blackish; throat white, with black streaks; under parts chestnut brown. L. 9¾. W. 5½. T. 4½. N. Am., everywhere abundant; a familiar, easy-going bird.

534. HESPEROCICHLA Baird. (ἑσπέρα, sunset; κίχλη, thrush.)

1025. **H. nævia** (Gmelin). OREGON ROBIN. Slate color, orange brown below; throat not streaked; ♂ with black collar. L. 9¾. W. 5. T. 4. Pacific slope, rarely straying E. (Lat., spotted.)

535. SIALIA Swainson. (σιαλίς, name of some bird; σίαλος, plump.)

1026. **S. sialis** (L.). COMMON BLUE BIRD. Bright blue above, throat and breast reddish brown ("the sky on its back and the earth on its breast"); belly white; ♀ usually duller, with a brownish tinge on back; young, as in others, spotted. L. 6¾. W. 4. T. 3. E. N. Am., abundant; breeds everywhere; one of our most attractive and familiar birds.

1027. **S. mexicana** Swainson. WESTERN BLUE BIRD. Head, neck all around and upper parts generally, deep bright blue; back with more or less chestnut; breast and sides reddish brown, throat bluish; size of last. Pacific slope, rarely E. to Iowa.

1028. **S. arctica** Swainson. ROCKY MOUNTAIN BLUE BIRD. Rich greenish blue; breast also blue; belly white; ♀ with pale drab instead of blue, on breast, etc.; size of others, or smaller. Rocky Mountains, E. to Missouri R.; the prettiest of thrushes and one of the most attractive of our birds.

With this beautiful bird we close the long series of feathered *Sauropsida.*

The next class, the *Mammalia,* is widely different from the birds, but its lowest forms, the Monotremes, approach the common reptilian stock from which both mammals and birds have probably sprung.

CLASS I. **MAMMALIA.** (THE MAMMALS.)

A Mammal is a warm-blooded, air-breathing vertebrate, having the skin more or less hairy (or rarely naked); viviparous, the embryo developed from a minute egg destitute of food-yolk (except in the *Monotremata*, in which group the eggs are large, as in Reptiles, and are developed outside the body); the young nourished for a time after birth by milk, secreted in the mammary glands of the mother; respiration never by means of gills, but after birth by lungs, suspended freely in the thoracic cavity, which is completely separated from the abdominal cavity by a muscular septum (the diaphragm); heart with four cavities; a complete double circulation; blood warm. Skeleton more firm than in other Vertebrates, the bones containing a larger proportion of salts of lime. Skull articulating with the atlas by means of two occipital condyles; bones of face immovably joined by sutures; each half of lower jaw of a single bone, articulating directly with the skull, the quadrate bone becoming one of the bones of the ear (the *malleus*). Brain case comparatively large, corresponding with the increased development of the brain. The numerous other peculiarities of the skeleton and the viscera need not be noticed in this connection.

The following analysis of the Orders of Mammals which occur within our limits is mostly taken from Professor Gill's "Arrangement of the Families of Mammals."

Orders of Mammalia.

a. Young developed within the uterus from a minute egg which is destitute of food-yolk; milk glands with nipples; no cloaca. (EUTHERIA.)
   b. Young born when of very small size and incomplete development, never connected by a placenta to the mother; brain small, its corpus callosum rudimentary. (Subclass DIDELPHIA.) . . MARSUPIALIA, XLVII.
   b'. Young not born until of considerable size and nearly perfect development, deriving its nourishment, before birth, from the mother through the intervention of a placenta; a well developed corpus callosum. (Subclass MONODELPHIA.)
     c. Brain with a relatively small cerebrum, which does not cover the other ganglia, much of the cerebellum being exposed behind, and in front much of the optic lobes. (*Ineducabilia.*)
      d. Canine teeth none; incisors $\frac{2}{2}$, rarely $\frac{1}{2}$, chisel-shaped; limbs adapted for walking. . . . . GLIRES, XLVIII.
      d'. Canine teeth present, in some form; incisors not $\frac{2}{2}$ nor $\frac{1}{2}$.

e. Anterior limbs not adapted for flight; ulna and radius not united; hand normal; mammæ usually abdominal.

INSECTIVORA, XLIX.

ee. Anterior limbs adapted for flight; ulna and radius united; bones of hand and fingers much elongated, supporting a thin, leathery skin, extending along sides of body to the posterior limbs; mammæ pectoral. . . . . CHIROPTERA, L.

cc. Brain with a relatively large cerebrum overlapping much, or all, of the cerebellum and optic lobes. (*Educabilia*.)

f. Posterior limbs absent, the pelvis rudimentary; anterior limbs reduced to broad flattened paddles, without distinct fingers or claws; no clavicles; tail with a broad, horizontally placed caudal fin; cervical vertebræ more or less grown together; carnivorous. . . . . CETE, LI.

ff. Posterior limbs and pelvis well developed; anterior limbs with hoofs, claws, or nails.

g. Femur and humerus not exserted beyond the common integuments of the body; clavicles more or less rudimentary; mammæ abdominal or inguinal.

h. Feet with hoofs; molars mostly with grinding surfaces; incisors various; no tusks; developed toes, 1 to 4; herbivorous. . . . . UNGULATA, LII.

hh. Feet with developed claws; canines specialized; molars, one or more, sectorial, adapted for cutting; incisors ⅔; carnivorous. . . . . FERÆ, LIII.

gg. Femur and humerus exserted; feet with distinct toes which are provided with nails · clavicles present; an inner digit of hand (thumb) opposable to the others; orbits encircled by bone and directed forwards; mammæ pectoral, two in number (rarely also an inguinal pair). PRIMATES, LIV.

ORDER XLVII. MARSUPIALIA. (THE MARSUPIALS.)

Young developed without a placenta, and born at a very early stage and incomplete condition of development. The young at birth are usually placed in an abdominal pouch formed by a fold of skin about the milk glands of the mother, where they remain for a considerable time. Reproductive organs in both sexes of peculiar structure, nearly all the parts being double in the female. Skeleton showing numerous peculiarities, the teeth usually more numerous than in the higher Mammals. Brain small, the corpus callosum rudimentary. Heart with two venæ cavæ. This large group is chiefly confined to Australia. It represents an early or primitive type of Mammalia, which has now become extinct in most parts of the world. The single non-Australian family approaches most nearly to ordinary Mammals. (Lat., *marsupium*, pouch.)

Families of Marsupialia.

a. Tail long, prehensile, nearly naked; feet plantigrade, 5-toed, the first toe thumb-like and without claw; teeth 50. . . . . DIDELPHIDIDÆ, 180.

FAMILY CLXXX. **DIDELPHIDIDÆ.** (THE OPOSSUMS.)

Marsupial mammals of small size, with the teeth i. $\frac{5}{4}:\frac{4}{4}$, c. $\frac{1}{1}:\frac{1}{1}$, pm. $\frac{3}{3}:\frac{3}{3}$, m. $\frac{4}{4}:\frac{4}{4}$. Feet five-toed, plantigrade, the claws 5–4. Tail usually very long, prehensile, nearly naked, covered by a scaly skin. with a few scattered hairs. Genera 2, species about 15 ; all American and chiefly belonging to the tropics. The common opossum is one of the largest of the group. All are sluggish animals, arboreal (*Didelphis*) or aquatic (*Chironectes*), and becoming very fat. They feed on insects and other small animals.

a. Arboreal; feet not webbed. . . . . DIDELPHIS, 536.

536. DIDELPHIS Linnæus. (δίς, two ; δελφύς, womb.)

1029. **D. virginiana** Shaw. COMMON OPOSSUM. Soiled yellowish, with some darker hairs ; ears black, leathery ; legs dark. L. 35. T. 15. N. Y. to Cal. and S. ; common.

ORDER XLVIII. **GLIRES.** (THE RODENTS OR GNAWERS.)

Mammals with the incisor teeth $\frac{4}{4}$ or $\frac{2}{2}$ in number, chisel-shaped, adapted for gnawing ; no canine teeth, a toothless space in the place of canines ; molar teeth adapted for grinding ; cerebrum small, little convoluted ; intestinal canal elongate ; ears and eyes usually well developed. Food chiefly vegetable.

The *Glires* or *Rodentia* is the largest order of Mammals, and in individuals by far more numerous than any other. Most of the species are of small size, the Beaver being one of the very largest of the forms now living.

"Though a feeble folk, comparatively insignificant in size and strength, they hold their own in legions against a host of natural enemies, rapacious beasts and birds, by their fecundity, their wariness and cunning, their timidity and agility, their secretiveness, each after the means by which it is provided for exercising its instinct of self-preservation, among which insignificance itself is no small factor." (*Coues.*) (Lat., *glis*, dormouse ; the Linnæan name *Glires* is much older than Cuvier's *Rodentia*.)

Families of Glires.

a. Incisors $\frac{2}{2}$, the median upper incisors large. vertically grooved, the outer small; teeth 28; tail very short; ears long; fibula united with the heel-bone. . . . . LEPORIDÆ, 181.
aa. Incisors $\frac{2}{2}$; tail well developed.
  b. Fur with stiff spine-like bristles; tibia and fibula separate.
                                            HYSTRICIDÆ, 182.
  bb. Fur more or less soft, without spines.
    c. Tibia and fibula united below.
      d. Tail and hind legs excessively elongated, the latter adapted for leaping; molars $\frac{4}{4}$ on each side. . . . . . . ZAPODIDÆ, 183.

dd. Tail and hind-legs not excessively elongated.

   e. Limbs very short, subequal, adapted for digging; fore-claws much enlarged; large, external cheek-pouches; body thick-set and heavy; molars ⁴⁄₄ on each side. . . . GEOMYIDÆ, 184.

   ee. Limbs moderate, not as above; cheek-pouches usually absent; molars ⅜ to ⅓ on each side. . . . . MURIDÆ, 185.

  cc. Tibia and fibula separate.

   f. Tail broad, flat, and scaly; feet webbed; molars ⁴⁄₄ on each side; body robust. . . . . CASTORIDÆ, 186.

   ff. Tail with fur; feet not webbed; molars ⁴⁄₄ or ⁵⁄₄ on each side.
                                       SCIURIDÆ, 187.

FAMILY CLXXXI. LEPORIDÆ (THE HARES.)

Incisors ⅖, the extra pair in upper jaw small, and placed behind the principal pair, which are grooved in front; molars ⁶⁻⁶⁄₅⁻₅; the teeth 28 in all; tail short, bushy, recurved; eyes large; ears long; soles furred. A single genus widely distributed, with about 30 species, among them the familiar Rabbit (*Lepus cuniculus* L.) of Europe, and several native species commonly called rabbits, but more properly hares.

537. LEPUS Linnæus. (Lat., a hare.)

a. Postorbital processes united with the skull; hind feet short; fur never white.

1030. **L. palustris** Bachman. MARSH HARE. Width of skull half its length. Yellowish brown; tail grayish, not cottony. L. 17. T. 1. Ear 2½. N. C. to S. Ill. and S., in swamps.

1031. **L. aquaticus** Bachman. WATER HARE. Width of skull not half its length. Yellowish brown, white below; tail white below, as in *L. sylvaticus.* L. 22. T. 2. Ear 3. S. Ill. to La. and S. W., in canebrakes and about lowland streams.

aa. Postorbital processes united with the skull.

 b. Fur never white; hind feet not longer than head.

1032. **L. sylvaticus** Bachman. GRAY RABBIT. COTTON-TAIL. Tail cottony-white; ears two-thirds length of head. Gray above, varied with black, and more or less tinged with yellowish brown; below white. L. 18. T. 2. Ear 2½. U. S., rather S., N. to Mass.; very abundant.

 bb. Fur becoming more or less white in winter; hind feet longer than head.

1033. **L. americanus** Erxleben. WHITE RABBIT. NORTHERN HARE. Ears about as long as head; fur, in summer, cinnamon brown, in winter, becoming white at the surface, plumbeous at base, with a median band of reddish brown. L. 20. T. 2½. Ear 3. Wooded districts. New England to Minn., and S. to Va., along the Alleghanies. The Eastern form var. **virginianus** Harlan,

has median brown band on fur *broad*. Further N. (Hudson's
Bay) occurs the European Hare, *L. timidus* L., a similar but
larger animal.

1034. **L. campestris** Bachman.   JACK-RABBIT.   PRAIRIE ⟵
HARE.   Ears much longer than head.   Fur pale yellowish gray
in summer. in winter white at surface and base. yellowish in
middle : tail long, all white.   L. 23.   T. 3¼.   Ear 5.   Kan. and
Dakota. to Oregon.   (Lat., of the fields.)

FAMILY CLXXXII. **HYSTRICIDÆ.**   (THE PORCUPINES.)

Molar teeth ¼ on each side ; fur more or less mixed with bristly
spines ; tip of muzzle with small hairs ; tibia and fibula distinct.
Genera 6 : species about 50, largely American.   The American
forms (*Sygnetherina*) differ in many respects from the Old World
allies of the European Porcupine (*Hystrix cristata* L.).   The
former are chiefly arboreal, and most of them have the tail prehensile.   Allied to this family is the South American group of
Caviidæ, represented by the Guinea (Guiana) Pig. (*Cavia aperea*).
(Lat. *hystrix*, porcupine .

a. Tail short, thick, not prehensile; claws 4–5, long, compressed, and
   curved; nostrils close together. . . . . ERETHIZON, 538.

   538. ERETHIZON Frédéric Cuvier.   (ἐρεθίζω, to irritate.)

   1035. **E. dorsatus** (L.)   CANADA PORCUPINE.   Dark brown, ⟵
spines tipped with yellowish white, and 4 to 6 inches long.   L. 40.
T. 6.   N. Am., from Me. to Mexico, formerly common.   (Lat.,
dorsum, back.)

FAMILY CLXXXIII. **ZAPODIDÆ.**   (THE JUMPING MICE.)

Hind legs greatly elongated, adapted for taking long leaps ; fore
legs short.   Tail very long.   Molars $\frac{4-4}{4}$; upper incisors compressed, grooved : molars rooted ; cheek pouches present ; toes
5–5 ; tibia and fibula united.   A single species, North American.

   539. ZAPUS Coues.   (ζά, an intensive particle ; ποῦς, foot.)

   1036. **Z. hudsonius** (Zimmermann).   JUMPING MOUSE.   Yel- ⟵
lowish brown ; fur coarse and rough ; soles naked.   L. 8.   T. 5.
Ear ¼.   U. S. chiefly N., scarce ; variable.

FAMILY CLXXXIV. **GEOMYIDÆ.**   (THE POUCHED
                                    GOPHERS.)

Cheek pouches large and distinct, opening outside of the mouth.
Molars $\frac{4-4}{4}$; incisors large and thick ; skull heavy ; temporal bones
enormously developed.   Limbs about equal, the fore claws five in
number, very large ; tibia and fibula united.   Body thick-set and
clumsy.   Genera 2, species 7 ; all North American, and chiefly

inhabiting the central plains; habits nocturnal and subterranean. Farther west occur numerous species of *Saccomyidæ* or Pocket-Mice, smaller than the Gophers, and with thin and papery skulls.

a. Upper incisors, each with a large groove near the middle; ears rudimentary; fore claws enormous. . . . . GEOMYS, 540.
aa. Upper incisors not grooved; ears distinct but very small; claws moderate.
THOMOMYS, 541.

540. GEOMYS Rafinesque. (γή. earth ; μῦς, mouse.)

1037. **G. bursarius** Shaw.) POCKET GOPHER. Reddish brown, with plumbeous tinge ; upper incisors with two grooves, the larger near the middle line ; tail and feet hairy. L. 11. T. 3. Prairies, Wis., Ill., and W. to Rocky Mts. (Lat., pouched.)

541. THOMOMYS Maximilian. (θωμός, heap; μῦς, mouse.)

1038. **T. talpoides** (Richardson). NORTHERN POCKET GOPHER. Dusky plumbeous ; tail, feet and breast mostly white ; ears in a dusky area. L. 9½. T. 2½. Minn. to Utah and N. W. (Lat., like a mole.)

FAMILY CLXXXV. MURIDÆ. (THE MICE.)

Incisors $\frac{2}{2}$; molars usually $\frac{3-3}{3-3}$; anteorbital foramen a vertical slit, widening above and bounded externally by a broad plate of the upper maxillary ; coronoid and condyloid processes of lower jaw well developed. Tibia and fibula united below. Genera 35 ; species 300. A large family, found in all parts of the globe, some of the species (*Mus*) being cosmopolitan, having accompanied man in all his migrations; all are of small size, the muskrat being one of the largest, and some are smaller than any other quadrupeds, except the Shrews. About one-third of the species belong to the the Old World genus, *Mus*.

a. Incisors broad, often broader than deep; molars rootless (except in *Evotomys*) with flat crown and serrate margin; (body heavy, eyes small, snout blunt, legs short, ears small). (*Arvicolinæ*.)
  b. Tail flattened, scant-haired; hind feet partly webbed. . . FIBER, 542.
  bb. Tail subterete.
    c. Upper incisors grooved; ears large. . . . . SYNAPTOMYS, 543.
    cc. Upper incisors not grooved.
      d. Molars rootless; ears concealed; coronoid process of lower jaw reaching level of condyle. . . . . ARVICOLA, 544.
      dd. Molars with roots; ears overtopping the fur; coronoid process of lower jaw not reaching level of condyle. . . EVOTOMYS, 545.
aa. Incisors narrow, compressed; molars rooted, tuberculate, with crenate margin ; (body slender; eyes and ears large; snout pointed, motions rapid). (*Murinæ*.)
  e. Molars of upper jaw with tubercles in two series ; palate ending opposite last molar. (American species.)

f. Mouse-like; length with tail 4 to 8 inches.
　g. Upper incisors grooved. . . . . OCHETODON, 546.
　gg. Upper incisors not grooved; ears very large. . CALOMYS, 547.
　ff. Rat-like; length, with tail, a foot or more. . . . NEOTOMA, 548.
　cc. [Molars of upper jaw with tubercles in three series; soles naked; tail
　　　long, scant-haired or scaly. (Introduced species).] . . . . MUS.

542. FIBER Cuvier. (Lat., beaver.)

1039. F. zibethicus (L.). MUSKRAT. Color dark brown. L.
22½. T. 11. N. Am.; everywhere. Largest of our *Muridæ;* building houses or burrows about streams and ponds. (Lat., *zibetha,* the
Civet, from the odor.)

543. SYNAPTOMYS Baird. (συνάπτω, to join ; μῦς, mouse.)

1040. S. cooperi Baird. LEMMING MOUSE. Mouse-color, grayish below: head very large with long whiskers; fur soft and long.
L. 4¾. T. ⅞. Indiana (Brookville, Nashville) to Oregon and
Alaska, a remarkable animal, between the field-mice and the lem-
mings (*Myodes*). (To Dr. J. G. Cooper.)

544. ARVICOLA Lacépède. FIELD MICE. (Lat., *arvum,* field ; *colo,* I inhabit.)

a. Posterior upper molar with one exterior triangle and a posterior trefoil;
　　middle upper molar with one internal triangle; front lower molar with
　　two internal and one external triangle.
　b. Fore claws larger than hinder; fur dense, silky, mole-like; size small.
　　(*Pitymys* McMurtrie.)

1041. A. pinetorum Le Conte. PINE MOUSE. Chestnut color,
ashy below. L. 4¾. T. ⅞. N. Y. to Ill. and S. (Lat., of the
pines.)

　bb. Fore claws not larger than hinder; fur coarse, not glossy; size medium.
　　(*Pedomys* Baird.)

1042. A. austerus Le Conte. Grizzly brownish, rusty plumbeous below. L. 5½. T. 1. Mich. to Dak. and La. (Lat., harsh.)
aa. Posterior upper molar with two external triangles and a posterior cres-
　　cent; middle upper molar with two internal triangles; front lower
　　molar with three internal and two or three lateral triangles; size large.
　　(*Mynomes* Rafinesque.)

1043. A. pennsylvanicus Ord. MEADOW MOUSE. Fore
claws not longer than hind claws. Grayish brown, blackish
mesially, hoary below. L. 5½. T. 1½. U. S., generally abundant; variable. (*A. riparius* Ord, a later name.)

545. EVOTOMYS Coues. (εὖ, well ; οὖς, ear ; μῦς, mouse.)

1044. E. rutilus (Pallas). LONG-EARED MOUSE. Color chestnut, median line of back rusty-red; sides yellowish, belly dull
white : fur full and soft; ears prominent. L. 4½. T. 1½. North-

ern regions. Var. **gapperi** Vigors, a little darker, with longer feet and tail, ranges S. to Mass.; the typical form circumpolar. (*Eu.*) (Lat., red-haired.)

546. OCHETODON Coues. (ὀχετός, channel ; ὀδών, tooth.)

1045. **O. humilis** (Audubon & Bachman.) HARVEST MOUSE. Mouse-color, the fur soft and silky ; whitish below. L. 4⅓. T. 2. Smallest of our mice. S. C. to Iowa, Utah and S. (Lat., humble.)

547. CALOMYS Waterhouse (1837). (*Hesperomys* Waterhouse 1839). WHITE-FOOTED MICE. (καλός, beautiful ; μῦς, mouse.)

a. Tail very long, scant-haired, about as long as head and body; ears rather small, closely hairy. (*Oryzomys* Baird.)

1046. **C. palustris** (Harlan). RICE-FIELD MOUSE. Blackish and ashy above, becoming paler below; fur harsh, but compact; soles perfectly naked ; a large, rat-like species. L. 8. T. 4. N. J. to Kan. and S.

aa. Tail rather long, closely hairy, about as long as head and body; ears large, rounded, scant-haired; feet and under parts white. (*Musculus* Rafinesque.)

1047. **C. michiganensis** (Audubon & Bachman.) Yellowish brown, a sooty dorsal band ; belly white ; feet not quite white ; tail bicolor; hind feet less than ¾ inch. L. 4⅓. T. 1⅓. Mich. to Ill. and Kan.

1048. **C. aureolus** (Audubon & Bachman.) RED MOUSE. Golden cinnamon, especially bright on ears; belly not pure white ; tail unicolor. L. 6. T. 2⅓. Pa. to Ill. and S. (Lat., golden.)

1049. **C. americanus** (Kerr). COMMON WHITE-FOOTED MOUSE. DORMOUSE. Yellowish brown, grayish or fawn color; belly and feet pure white ; tail distinctly bicolor; hind feet more than ¾ inch. L. 6⅓. T. 3¼. N. Am.; abundant everywhere. (*Hesperomys leucopus* Raf.)

aaa. Tail very short, closely hairy, not much longer than head. (*Onychomys* Baird.)

1050. **C. leucogaster** (Maximilian). Mouse color, snow-white below; ears high, furred. L. 5⅓. T. 1⅓. Minn. to Kas. and Montana. (λευκός, white ; γαστήρ, belly.)

548. NEOTOMA[1] Say & Ord. (νέος, new ; τομός, cutting, *i. e.*, rodent).

1051. **N. floridana** Say & Ord. WOOD RAT. Brownish gray ; the sides tawny ; belly and feet all white; tail scantily hairy. L. 13. T. 5. S. N. Y. to Col., Ariz., and S.

[1] Allied to *Neotoma* is the familiar Old World genus : MUS Linnæus. (μῦς, mouse.)
M. decumanus Pallas. Brown Rat. Wharf Rat. Tail nearly an inch shorter than head and body ; grayish brown above ; paler below ; feet dusky white ; fur mixed

FAMILY CLXXXVI. CASTORIDÆ. (THE BEAVERS.)

Aquatic rodents of large size, having the molars rootless, $\frac{1\cdot1}{1\cdot1}$; feet four-toed, the hind feet webbed; body stout and heavy; tail broad, flat and scaly; tibia and fibula distinct; no postorbital process. A single species now living, belonging to the northern hemisphere.

549. CASTOR Linnæus. (Lat., the Beaver.)

1052. **C. fiber** L. BEAVER. Reddish brown, grayish below. L. 40. T. 10. Weight 45 to 60 lbs. Northern regions, S. to Mexico; once abundant, now being rapidly exterminated. (*Eu.*) (Lat., the beaver or badger.)

FAMILY CLXXXVII. SCIURIDÆ. (THE SQUIRRELS.)

Molars rooted, $\frac{5\cdot5}{4\cdot4}$ (upper anterior often deciduous), the last 4 of nearly equal size; a distinct postorbital process of frontal bone; tibia and fibula distinct. Species of rather small size, in all parts of the world except Australia. Genera about 12, species 150. The variations in color, etc., are extremely great, and the number of well defined species is very much less than was once supposed.

a. Sides without membrane for "flying."
  b. Upper outline of skull nearly straight; frontal region depressed; cheek pouches rudimentary; thumb with a broad flat nail; tail short, bushy; ears small; fur coarse, heavy; body stout, clumsy. ARCTOMYS, 550.
  bb. Upper outline of skull more or less convex.
    c. Cheek pouches present; tail moderate.
      d. Skull strong and massive; ears rudimentary; thumb with well developed nail; body heavy, thickset. . . . . CYNOMYS, 551.
      dd. Skull comparatively thin.
        e. Thumb with rudimentary nail; (other characters drawn from the skull). . . . . SPERMOPHILUS, 552.
        ee. Thumb with well developed nail; skull narrowed anteriorly. TAMIAS, 553.
    cc. Cheek pouches wanting; tail very long and bushy, the hairs mostly on its sides; skull short, broad and rounded; thumb nail rudimentary; eyes well developed. . . . . SCIURUS, 554.
aa. Sides with a densely furred lateral membrane joining the anterior and posterior limbs: body and tail depressed; no cheek-pouches; ears large; molars subequal in size. . . . . SCIUROPTERUS, 555.

with stiff hairs: cosmopolitan; introduced into America about 1775, and now the commonest species, having nearly exterminated the next. (Lat., the tenth.)
M. rattus L. Black Rat. Tail not shorter than head and body: sooty black, plumbeous below; feet brown; introduced about 1544, but now supplanted by the preceding.
M. alexandrinus Geoffroy St. Hilaire. Roof Rat. White-bellied Rat. Introduced in the Southern States. (From Alexandria in Egypt.)
M. musculus L. Common House Mouse. Cosmopolitan; too well known. (Lat., a little mouse.)

SCIURIDÆ. — CLXXXVII. **323**

550. ARCTOMYS Schreber. MARMOTS. (ἄρκτος, bear; μῦς, mouse.)

1053. **A. monax** (L.). WOODCHUCK. GROUND HOG. Grizzly gray, varying to chestnut and blackish. L. 18. T. 5. Hudson's Bay to Va., W. to Neb.; common. (Lat., solitary.)

551. CYNOMYS Rafinesque. (κύων, dog; μῦς, mouse.)

1054. **C. ludovicianus** (Ord). MISSOURI PRAIRIE DOG. Reddish brown, varied with darker and gray; tawny below. L. 16½. T. 3½. Kas. to Montana, E. of Rocky Mts., living in large colonies. (Lat., of Louisiana.)

552. SPERMOPHILUS Cuvier. (σπέρμα, seed; φίλος, loving.)

a. Skull very long and narrow, the snout broad and very long; tail long; ear small. (*Ictidomys* Allen).

1055. **S. franklini** (Sabine). GRAY GOPHER. Yellowish brown, mottled with black. L. 15. T. 5½. Prairies; N. Ill. and N. W. (To Sir John Franklin.)

1056. **S tridecemlineatus** (Mitchill). STRIPED GOPHER. Dark brown, mixed with reddish, with 6 to 8 light stripes alternating with lines of dots, about 13 streaks in all; yellowish below, with a broad black stripe on each side. L. 10½. T. 3½. Prairies; Ark. to Ill. and N. W., common. (Lat., 13-lined.)

553. TAMIAS Illiger. (ταμίας, a steward.)

a. Premolars ⁴⁄₄.

1057. **T. striatus** (L.). CHIPMUNK. GROUND SQUIRREL. Reddish brown; back with 5 black stripes and 2 whitish ones; rump reddish. L. 11. T. 4½. Maine to Va. and W., abundant; the Southern form is var. **lysteri** (Pallas).

aa. Premolars ⁴⁄₄.

1058. **T. asiaticus** (Gmelin). MISSOURI CHIPMUNK. Back with 5 black stripes and 4 whitish ones; rump grayish. L. 8. T. 4. Wis. (*Hoy.*), N. W. to Asia. The E. form, var. **quadrivittatus** Say. has the sides bright rusty, the stripes on back well-defined. (*Eu.*)

554. SCIURUS Linnæus. (σκίουρος, squirrel; σκιά, shade; οὐρά, tail.)

1059. **S. hudsonicus** Erxleben. RED SQUIRREL. CHICKAREE. Yellowish gray, back with a median wash of bright rusty red; tail short and narrow, with a subterminal band of black. L. 14. T. 6½. N. Am., S. to Penn. and N. Ind.; abundant N.

1060. **S. carolinensis** Gmelin. GRAY SQUIRREL. BLACK SQUIRREL. Whitish gray, usually varied with tawny; middle of back brownish; ears not tufted; often entirely jet black. L. 20. T. 9. N. Am., E. of the plains, abundant. The common Northern form is var. leucotis Gapper, larger, the brownish band on back narrow. The typical **carolinensis** is Southern, N. to St. Louis. L. 17½. T. 8.

1061. **S. niger** L. Fox Squirrel. General color rusty gray, varying from almost white, through various shades of rusty red, to jet black, the latter color rare northward, reddish and orange shades predominating westward; tail very large and bushy. L. 26. T. 11. E. U. S., very abundant S. W., N. to Mass. Leading varieties are: var. **niger**, the Southern form, gray to black, with the ears and nose white, the belly reddish; var. **cinereus** L., the Eastern form, similar in color, the ears and nose not white, the ears short, scarcely longer than the fur; and var. **ludovicianus** Custis, the common Western Fox Squirrel, with high ears and a prevailing tinge of orange red; ears, feet and belly reddish.

555. **SCIUROPTERUS** Frédéric Cuvier. (σκίουρος, squirrel; πτερόν, wing.)

1062. **S. volans**[1] (L.). Common Flying Squirrel. Dull yellowish brown, creamy white below. L. 10. T. 4. N. Am., abundant. The Canadian form, var. **sabrinus** Shaw, is larger (L. about 12; T. 4½). with more dusky, especially on tail. (Lat., flying.)

" But we have reached the end of the chain of rodent beings of the earth, the water, and almost of the air, a cycle of mammalian life which circumscribes extraordinary diversity of form and function, revolving about a single central point of organization, namely, adze-like teeth, to gnaw wood with. The number of individuals which make a living in this way in a world of Malthusian strife is simply incalculable. . . . Yet they have one obvious part to play, that of turning grass into flesh, in order that carnivorous Goths and Vandals may subsist also, and in their turn proclaim, ' All flesh is grass.' " (*Coues.*)

ORDER XLIX. **INSECTIVORA.** (The Insect-eaters.)

Teeth of three kinds, molars, canines and incisors, all with enamel; brain small, the cerebrum without sylvian fissure; limbs well-developed and adapted for walking.

A large group of small animals, analogous to the *Carnivora* in many respects, but the individuals so small as to be unable to attack vertebrate animals, and therefore feeding chiefly on insects. But two of the numerous families are represented in our fauna.

a. Fur soft, without spines: sides of body without membrane for "flying;" canine teeth indistinct.

---

[1] This is *Mus volans* L. S. N. ed. x. p. 63. *Sciurus volans* L. (p. 64) is the European Flying Squirrel, *Pteromys volans.*

b. Fore-feet not enlarged; mnzzle elongate; external ear developed; appearance mouse-like. . . . . SORICIDÆ, 188.
bb. Fore feet very broad, with stout claws adapted for digging; no external ear. . . . . TALPIDÆ, 189.

FAMILY CLXXXVIII. **SORICIDÆ.** (THE SHREWS.)

Small Insectivora, mouse-like in appearance, with the eyes and external ears developed. Muzzle elongate. Feet normal, not fossorial ; the fore-feet mostly smaller than the hind ones. Teeth $\frac{16 \text{ to } 20}{12}$; canines obsolete. The most abundant and widely distributed family of the Insectivora, comprising more than half the known species, arranged in 8 genera. The number of species of *Blarina* and *Sorex* is still uncertain, and the geographical distribution of the species has been little studied.

a. Feet very long, with a fringe of hairs; ears valvular (to exclude water); concha (external ear) directed backwards; tail long; teeth $\frac{32}{32}=32$.
NEOSOREX, 556.
aa. Feet not fringed.
b. Ears large, the concha turned backward. . . . . . . SOREX, 557.
bb. Ears small, not visible externally, the concha directed forwards, so as to hide the opening; tail short, not longer than head. . BLARINA, 558.

556. NEOSOREX Baird. (*νέω*, to swim ; *Sorex.*)

1063. **N. palustris** (Richardson). WATER SHREW. Hoary black; belly ashy gray ; largest of our shrews. L. 6. T. 2¼. Mass. to Rocky Mts. and N. *Aquatic.*

557 SOREX Linnæus. (Lat., a field-mouse.)

a. Teeth colored, 32 = $\frac{32}{32}$; feet large. (*Sorex.*)
b. Third upper premolar larger than fourth.

1064. **S. forsteri** (Richardson). Ears small ; tail ⅔ length of head and body; snout slender. L. 4½. T. 1¾. N. U. S., S. to Penn. (To John Reinhold Forster.)

1065. **S. richardsoni** Bachman. Ears rather small; tail scant-haired. L. 4. T. 1¼. Wis. and N. (To Sir John Richardson.)

1066. **S. platyrhinus** (DeKay). COMMON SHREW. SHREW MOUSE. Ears very large for a Shrew ; tail short, scant-haired ; color chestnut-brown. L. 3¾. T. 1½. N. U. S., rather common. (*πλατύς*, broad ; *ῥίν*, snout.)

bb. Third upper premolar equal to fourth and smaller than first and second.

1067. **S. cooperi** Bachman. Ears large; chestnut brown. L. 3¾. T. 1½. Mass. to Neb. and N. (To Dr. J. G. Cooper.)

1068. **S. personatus** Geoffroy St. Hilaire. Ears large : chestnut brown. L. 2¾. T. 1. Smallest of our shrews ; Penn. and S. (Lat., masked.)

aa. Teeth 30 = $\frac{30}{30}$; feet small. (*Microsorex* Baird.)

1069. **S. hoyi** Baird. Very small and slender; ears large; olive brown. L. 3. T. 1¼. Wis. to Nova Scotia and N. (To Dr. P. R. Hoy.)

558. BLARINA Gray. (A coined name.)

a Teeth 32 = $\frac{?}{?}$. (*Blarina*.)

1070. **B. brevicauda** (Say). MOLE-SHREW. Size large for a Shrew; fur short and coarse; color dark ashy gray. L. 4½. T. 1. Mass. to Va. and Dak., generally common. (Lat., short-tail.)

1071. **B. carolinensis** (Bachman). Smaller; leaden gray. L. 3¼. T. $\frac{?}{?}$. Mo. to N. C. and S.

1072. **B. angusticeps** Baird. Skull unusually narrow; uniform plumbeous; tail as long as head. L. 3½. T. 1. Vermont. (Lat., narrow head.)

aa. Teeth 30 = $\frac{1.5}{.}$; tail bicolor. (*Soriciscus* Coues.)

1073. **B. parva** (Say). Body stout; iron gray, with brown gloss. L. 3¼. T. $\frac{?}{?}$. Penn. to Ga. and S., not rare.

1074. **B. exilipes** Baird. Fur full; feet very small; hoary olive. L. 2½. T. $\frac{2}{3}$. Va. to Ill. and S. (Lat., *exilis*, slender; *pes*, foot.)

FAMILY CLXXXIX. TALPIDÆ. (THE MOLES.)

Body stout, thick, and clumsy, without distinct neck. Eyes rudimentary, sometimes concealed. No external ears. Limbs very short; feet greatly expanded and provided with strong claws, adapted for digging; anterior limbs much larger than posterior. Scapula as long as humerus and radius together. Canines usually distinct. Fur compact, soft, and velvety. Genera 7; found throughout the Northern hemisphere; most of them digging elaborate burrows. (Lat., *talpa*, mole.)

a. Snout elongated, not star-shaped at tip; tail shorter than head.

b. Teeth $\frac{?}{?}$ = 36; nostrils partly superior; tail nearly naked.
                                                               SCALOPS, 559.

bb. Teeth $\frac{?}{?}$ = 44; nostrils lateral; tail densely hairy. . SCAPANUS, 560.

aa. Snout elongated, fringed at tip with a circle of long fleshy projections; nostrils terminal; tail much longer than head; teeth $\frac{?}{?}$ = 44.
                                                               CONDYLURA, 561.

559. SCALOPS Cuvier. (σκάλοψ, mole, from σκάλλω, to dig.)

1075. **S. aquaticus** (L.). COMMON MOLE. Dark plumbeous, paler below; feet full webbed; palms broader than long; eye not wholly covered by skin. L. 5½. T. 1. Mass. to Ind., and S., very abundant.

1076. **S. argentatus** Audubon & Bachman. PRAIRIE MOLE. Silvery plumbeous; palms scarcely broader than long; eyes covered by skin; larger and more silvery than the preceding. L. 6¼. T. 1¼. Mich. to La. and W., chiefly in the prairie region.

560. SCAPANUS Pomel. (σκαπάνη, hoe.)

1077. **S. americanus** (Bartram). HAIRY-TAILED MOLE.
Dark plumbeous, with brown gloss; palms narrow; tail densely
hairy. L. 5. T. 1. Mass. to Ohio.

561. CONDYLURA Illiger. (κόνδυλος, node; οὐρά, tail.)

1078. **C. cristata** (L.). STAR-NOSED MOLE. Blackish; skull ⌣ ⌢
long and slender. L. 6¾. T. 2¾. Nova Scotia to Ind., and N.
(Lat., crested.)

ORDER L. **CHIROPTERA.** (THE BATS.)

Flying Insectivora. Mammals with the anterior limbs modified
for flight by the elongation of the fore arm, and especially of four
of the fingers, all of which are connected by a thin leathery membrane, which also includes the hind feet and the tail; humerus and
femur not included in the common integument of the body; teeth
with enamel, the three sorts differentiated; mammæ pectoral. The
Bats are chiefly nocturnal in their habits, going into retirement in
day-time, and hanging, head downward, by their hind claws. Most
of them are insectivorous, a few in tropical regions feeding on fruits.
About 400 species are known, chiefly of small size. The order is
very sharply defined, but it has probably sprung from the same
stock as the Insectivora. (χείρ, hand; πτερόν, fin.)

a. Insectivorous; ears large; no leaf-like appendage to snout; hairs with imbricated scales arranged in spirals. . . . . . VESPERTILIONIDÆ, 190.

FAMILY CXC. **VESPERTILIONIDÆ.** (THE COMMON
BATS.)

Insectivorous Bats with the snout not appendaged, or merely with
two lateral excrescences; wing membranes ample; tail completely
enclosed in the interfemoral membrane or only the last joint exserted; fur of peculiar structure, each hair with a series of minute
imbricated scales arranged in spiral. The largest family of bats,
with about 16 genera; especially abundant in temperate regions.

a. Nostrils simple, at tip of snout; ears moderate; forehead not grooved.
  b. Incisors ⅔.⅔.
    c. Teeth 38; muzzle narrow, hairy in front of eyes; ears as long as head;
      slender species with thin wings and ears. . . VESPERTILIO, 562.
    cc. Teeth 32 to 36; muzzle nearly naked before eyes; ears shorter than
      head; stout species with thick wings and ears. VESPERUGO, 563.
  bb. Incisors ½.½.
    d. Teeth 30; upper incisors small; wings and interfemoral membranes
      nearly naked. . . . . NYCTICEJUS, 564.
    dd. Teeth 32, upper incisors stout; interfemoral membranes hairy above,
      the wings with furry patches. . . . . . ATALAPHA, 565.
aa. Nostril margined behind by grooves and glandular prominences; cheeks
  with large excrescences; ears very large (an inch high); teeth 36.
  PLECOTUS, 566.

562. VESPERTILIO Linnæus. (Lat., bat, from *vesper*, evening.)

1079. **V. subulatus** Say. LITTLE BROWN BAT. Face small, fox-like, with high forehead and pointed snout; ears large, oval, twice the height of the erect tragus; wings naked; interfemoral membrane naked except at base; face whiskered; color dull olive-brown. L. 3. E. 9. T. 1½. E. N. Am., abundant everywhere; very variable. (Lat., awl-shaped.)

563. VESPERUGO Keyserling & Blasius. (Lat, *vesper*, evening.)

a. Teeth 36; molars ⅔⅔. (*Vesperides* Coues.)

1080. **V. noctivagans** (Le Conte). SILVER BLACK BAT. Tragus almost as broad as high, scarcely one-third height of ear; femoral membrane entirely though scantily furred; fur long and silky, black, usually with silvery tips to the hairs. L. 3⅓. E. 12. T. 1½. U. S. generally. (Lat., *nox*, night; *vagans*, wandering.)

aa. Teeth 34; molars ⅔⅔. (*Vesperugo.*)

1081. **V. georgianus** (F. Cuvier). Tragus slender, erect, half the height of the ear; upper incisors about equal in size; femoral membrane one-third furred; dark reddish brown, brighter forwards. L. 3. E. 9. T. 1½. Maine to Texas; chiefly southward.

aaa. Teeth 32; molars ⅔⅔. (*Vesperus* Coues.)

1082. **V. serotinus** (Schreber). LONG-EARED BAT. Tragus never pointed, nearly half as high as ear; wings naked; interfemoral membrane furred at base; ear more or less turned outward; upper lateral incisors small, scarcely visible. L. 3 to 4. E. 12. T. 1½. Northern hemisphere, widely diffused; the American form, var. **fuscus** Beauvais, is said to be rather smaller than the European. (*Eu.*). (Lat. of evening.)

564. NYCTICEJUS Rafinesque. (*νύξ*, night.)

1083. **N. crepuscularis** (Le Conte.) TWILIGHT BAT. Ears small, wide apart; a small wart above eye; fur rather scanty. Dark fawn color above, passing into brownish below. L. 3⅓. E. 9. T. 1⅓. Penn. to Mo. and S. W., common. (Lat., of twilight.)

565. ATALAPHA Rafinesque.

1084. **A. noveboracensis** (Erxleben). RED BAT. Fur long and silky, reddish brown, mostly white at tip; lips and ears not edged with black; a whitish tuft at base of thumb. L. 3¾. E. 12. T. 1¾. U. S. everywhere, very abundant; known by its reddish color. (Lat., of New York.)

1085. **A. cinerea** (Beauvais). HOARY BAT. Rich chocolate-brown, overlaid with white; lips and ears marked with black. L. 5. E. 14. T. 2¼. U. S., rather northward, rare. (Lat., ashy.)

566. PLECOTUS Geoffroy St. Hilaire. (πλέκω, to fold ; οὖς, ear.)

a. Nostril without " rose-leaf " or disk-like appendage. (*Corynorhinus* H. Allen.)

1086. **P. macrotis** (Le Conte). BIG-EARED BAT. Blackish; fur soft and long. L. 3½. E. 11. T. 1¾. Va. to Dak. and S. (μακρός, large ; οὖς, ear.)

ORDER LI. **CETE.** (THE CETACEANS.)

Mammals of the sea, more or less fish-like in form, and adapted for life in the open ocean. Bones of the neck short, more or less fused; posterior limbs wanting; pelvis rudimentary; anterior limbs developed as broad, flattened paddles, without distinct fingers and without nails. Nostrils developed as spiracles, and opening usually on top of head, thus enabling the animals to breathe without raising the head from the water; eyes small ; no external ear; skin nearly or quite destitute of hair; tail ending in a broad horizontal fin or paddle; back sometimes with a dorsal fin. Skin thick and tough; beneath it a thick layer of fat (blubber), which protects the animal from the cold. Species numerous; found in all seas, some of them being the largest of all animals. The nearest relationships of the whales are perhaps with the seals, among living forms, but the differentiation is now very wide. Of the numerous species occasionally straying to our coasts, the following seem properly to belong to our fauna. The nomenclature and analysis of genera is chiefly taken from True's paper on " Collecting specimens of Cetaceans," in Rept. U. S. F. C. for 1883–1885. I have also made considerable use of MSS. lists of species, kindly given me by Mr. F. W. True, and by Professor Cope. (κῆτος, whale.)

Families of Cete.

a. Upper jaw without whalebone; spiracles coalescent into one; lower jaw much less thick than upper; skull unsymmetrical. (*Denticete.*)
  b. Upper jaw with teeth (except in the adult of one genus); eye inserted behind angle of mouth and not much above it; snout more or less sharp at tip; lower jaw with numerous (6 to 120) teeth.
      DELPHINIDÆ, 191.
  bb. Upper jaw toothless; eye decidedly above angle of mouth.
    c. Lower jaw with 2 to 4 teeth, or apparently toothless; snout more or less sharp at tip. . . . . ZIPHIIDÆ, 192.
    cc. Lower jaw with 18 to 50 teeth; snout not sharp, sometimes truncate at tip. . . . . PHYSETERIDÆ, 193.
aa. Upper jaw with long strips of baleen or whalebone; no teeth; spiracles separate; eye very small, close to angle of mouth, between mouth and pectorals; lower jaw very thick and deep, nearly as deep as upper, the cleft of mouth curved. (*Mysticete.*) . . . . . BALÆNIDÆ, 194.

Family CXCI. DELPHINIDÆ. (The Dolphins.)

Cetaceans with well developed teeth in both jaws (deciduous in the upper jaw in one genus) and a single, somewhat complicated nasal tube. Genera 17, species about 40, including the smaller of the Cetacea, but many of the most active and voracious of the species.

 .. Head with an elongate beak; a distinct dorsal fin.
    b. Teeth in each jaw about 44; truncate at tip; palate without lateral
        grooves. . . . .   Tursiops, 567.
    bb. Teeth in each jaw 80 to 120.
        c. Palate without lateral grooves. . . . . . . Prodelphinus, 567 (b).
        xx. Palate with deep lateral grooves. . . . . . Delphinus, 568.
  aa. Head with a very short beak or none.
    c. Teeth in both jaws persistent.
        d. Teeth flattened; dorsal present. . . . . . . Phocæna, 569.
        dd. Teeth terete.
            e. Dorsal fin well developed.
               f. Teeth in each jaw 44 to 46; dorsal fin falcate.
                                      Lagenorhynchus, 570.
               ff. Teeth in each jaw 16 to 24.
                  g. Dorsal moderate; head almost globular; P. long and narrow;
                    teeth rather weak, none in corner of mouth.
                                  Globicephalus, 571.
                  gg. Dorsal very high, sword-shaped, its height greater than length
                    of pectorals; teeth very strong; skull massive. Orca, 572.
           ee. Dorsal fin obsolete; pectoral short; teeth few.
                                Delphinapterus, 573.
  cc. Teeth in upper jaw feeble, disappearing with age; 6 to 14 bluntish
      teeth in lower jaw; dorsal fin low, rather posterior. Grampus, 574.

567. TURSIOPS Gervais. (Lat., *tursio*, porpoise.)

1087. **T. tursio** (L.). Bottle-nose Dolphin. Gray above, pure white below; beak short and stout; teeth $\frac{44}{44}$. Vertebræ $7 + 18 + 37$. L. 11 feet. N. Atl., common; caught in numbers at Cape May. (*Eu.*)

1088. **T. erebennus** (Cope.) Black Dolphin. Wholly black; premaxillaries forming an elevated, rounded ridge; teeth $\frac{44}{44}$. Vertebræ $7 + 11 + 16 + ?$. L. 8 feet. Atlantic Coast of U. S. (ἐρεβεννός, black as Erebus.)

567 (b). PRODELPHINUS Gervais.

1089. **P. plagiodon** (Cope) Spotted Dolphin. Form of *D. delphis;* dorsal high, recurved; P. broad; beak stout. Dark purplish slate-color, above; white below; back and fins spotted with pale; lower parts spotted with dark slate. L. 10 feet. N. Atl., S. to N. J. (πλάγιος, oblique; ὀδών, tooth.)

1090. **P. doris** (Gray). Outline from foramen to crest curved; cranium rounded; occiput broad, rounded; muzzle short; its length 2⅓ times its width at premaxillary notch. N. Atl. (not American?) (*Eu.*) (δορίς, a sacrificial knife.)

568. DELPHINUS Linnæus. (δελφίς, dolphin.)

1091. **D. delphis** L. COMMON DOLPHIN. Snout narrow, sharp; occiput short, rounded. L. 10 feet. N. Atl., scarce on our coast. (*Eu.*)

569. PHOCÆNA Cuvier. (φώκαινα, porpoise, from φώκη, seal.)

1092. **P. phocæna** (L.). COMMON HARBOR PORPOISE. PUF-FING PIG. SNUFFER. Color nearly plain dusky above, paler below. L. 5 feet. N. Atl. and N. Pac.; very common in surf and near shore, ascending rivers. (*Eu.*) (*P. brachycium* Cope, the E. American form, and *P. lineata* Cope, a form with a reddish brown lateral band on sides separating the blackish back from the white belly, are probably varieties of *P. phocæna*.)

570. LAGENORHYNCHUS Gray. (λάγηνος, flagon; ρύγχος, snout.)

1093. **L. acutus** Gray. SKUNK PORPOISE. BAY PORPOISE. Sides with broad stripes of white and yellow. L. 10 to 15 feet. Coast of N. E. U. S., and in the open seas in large schools; like other porpoises, often swimming alongside of ships as if racing with them. Common, used for bait. Two or three other species occur in the N. Atl. (*Eu.*)

571. GLOBICEPHALUS Lesson. (Lat., *globus*, globe; κεφαλή, head.)

1094. **G. melas** (Traill). BLACK FISH. PILOT WHALE. GRIND WHALE. Black; P. about 4 in length. L. 20 feet. N. Atl., common, in large schools, S. to N. J. (μέλας, black.) (*Eu.*)

1095. **G. brachypterus** Cope. Black; P. 6 in length. Coast of N. J. and S. (βραχύς, short; πτερόν, fin.)

572. ORCA Gray. (Latin name.)

1096. **O. orca** (L.). KILLER. SWORD GRAMPUS. Black, white below. A most persistently voracious and destructive cetacean, attacking all large sea animals, tunnies, sword-fishes, seals, and all whales, even the largest, to the great annoyance of fishermen. L. 20 feet or more. D. 6 feet high. Atl. (*Eu.*) (*O. gladiator* Bonnaterre.)

573. DELPHINAPTERUS Lacépède. (δελφίς, dolphin; a. privative; πτερόν, fin.)

1097. **D. leucas** (Pallas). WHITE WHALE. BELUGA. Creamy

white. young dusky.   N. Atl., S. to Cape Cod.   L. 15 feet.
(*Eu.*)

574. GRAMPUS Gray.   (A corruption of the French
"grand poisson.")

1098. **G. griseus** (Cuvier).   GRAMPUS.   COW FISH.   Slate
color, with white scratches.   L. 15 to 20 feet.   N. Atl., not rare.
(*Eu.*)   (Lat., gray.)   •

FAMILY CXCII. **ZIPHIIDÆ.**   (THE BOTTLED-NOSED
WHALES.)

This group is intermediate between the Sperm Whales and the
Dolphins.   It is distinguished from the former chiefly by the very
small number of teeth, usually not more than four developed in the
lower jaw, these fitting into pits in the upper; these teeth are mostly
developed only in the male.   Dorsal small, posterior; pectoral short,
ovate, placed low, with five fingers enclosed in thick skin; snout
more or less produced, the forehead rising abruptly in the adult.
Genera 4; species about 10, mostly of the Southern seas.

a. Teeth in ♂ evident.
   b. Visible teeth two, in tip of lower jaw. . . . . ZIPHIUS, 575.
   bb. Visible teeth two, in side of lower jaw. . . . . MESOPLODON, 576.
aa. Teeth in both jaws wanting or concealed; beak long.
                                    HYPEROODON, 577.

575. ZIPHIUS Cuvier.   (An old name, from ξίφος, sword?)

1099. **Z. cavirostris** Cuvier.   L. 20 feet.   Atl.   (*Eu.*).   (Lat.,
concave-snout.)

576. MESOPLODON Gervais.   (μέσος, middle ; ὅπλον,
armature ; ὀδών, tooth.)

1100. **M. sowerbiensis** (Blainville).   COW-FISH.   N. Atl.,
scarce.   L. 20 feet.   (*Eu.*)

577. HYPEROODON Lacépède   (ὑπερῴα, palate ; ὀδών, tooth.)

1101. **H. rostratus** (Müller).   BOTTLE-NOSED WHALE.
SPERM-WHALE PORPOISE.   Beak distinct in young, obscured in
adult by the development of bony crests which give the head the
shape of a trunk or chest.   N. Atl.   L. 25 feet.   (*Eu.*)   (Lat.,
long-nosed.)   (*H. bidens* Owen ; *Z. semijunctus* Cope.)

FAMILY CXCIII. **PHYSETERIDÆ.**   (THE SPERM
WHALES.)

Teeth numerous, in lower jaw only ; lower jaw very thin and
flat ; upper jaw heavy ; eye placed high, much above angle of
mouth.   Two genera, with 3 or 4 species, in warm seas.

a. Dorsal fin present, behind middle of back; teeth 18 to 30, very sharp; head bluntish but not truncate; spiracles on top of head; length about 20 feet. . . . . Kogia, 578.

aa. Dorsal fin wanting: teeth 40 to 50, large and blunt; head very long and deep, truncate in front, the cavity of the snout filled with oil and spermaceti; spiracles in front of head; length 60 to 80 feet.

Physeter, 579.

578. KOGIA Gray.

1102. K. breviceps (Blainville). Pigmy Sperm Whale. Warm seas, occasional on our coast. (Lat., short-headed.) (*Eu.*) (*K. grayi, goodei,* etc.)

579. PHYSETER Linnæus. (φυσητήρ, a whale, from φυσάω, to blow.)

1103. P. macrocephalus L. Sperm Whale. Cachalot. Blackish, paler below. Open sea, commonest far S.; one of the most valuable of the whales. L. ♂ 80 feet; ♀ much smaller. (*Eu.*) (μακρός, long; κεφαλή, head.)

Family CXCIV. BALÆNIDÆ. (The True Whales.)

Teeth disappearing before birth, their place taken in the upper jaw by an array of parallel plates with fringed edges, known as baleen or whalebone. Eye very small, placed close to angle of mouth. Spiracles separate, comparatively simple in structure; lower jaw very large and thick, its edge convex upward. Genera 8, species about 20; huge creatures, mostly of the colder seas, feeding chiefly on small animals and sought by man for the sake of the oil (blubber), and the whalebone.

a. Belly with conspicuous longitudinal furrows; pectorals shorter than head.
   b. Dorsal fin well developed.
      c. Dorsal fin much nearer tail than head.
         d. Whalebone slaty or pale; beak rather acuminate. Physalus, 580.
         dd. Whalebone black; beak rather obtuse. . . Sibbaldius, 581.
      cc. Dorsal fin nearly median. . . . . Balænoptera, 582.
   bb. Dorsal fin obsolete; back with a fleshy hump; belly with furrows; pectoral as long as head. . . . . Megaptera, 583.
aa. Belly without furrows; dorsal fin obsolete.
   e. Whalebone short and white (1 to 2 feet long). . . Agaphelus, 584.
   ee. Whalebone very long and blackish (7 to 12 feet long). Balæna, 585.

580. PHYSALUS Gray. (φύσαλος, whale, from φυσάω, to blow.)

1104. P. physalus (L.). Common Rorqual. Finner. Fin-back. Razor-back. Black above, paler below. L. 70 feet. N. Atl. (*Eu.*) (*P. antiquorum* Gray; *Balænoptera boops,* and *musculus* L.)

581. SIBBALDIUS[1] Gray. (To Robert Sibbald, who described the whales of Scotland in 1773.)

1105. **S. borealis** (Cuvier). SILVER-BOTTOM WHALE. Steel gray. L. 90 feet. N. Atl. (*Eu.*) (Allied to this species is the Pacific "Sulphur Bottom," *Sibbaldius sulfureus* Cope, the largest of all animals, reaching a length of more than 100 feet.)

1106. **S. tuberosus** Cope. Back with a series of humps or tuberosities along the median line from dorsal to tail; uniform black above; P. without band. L. 40 to 50 feet. N. Atl.

1107. **S. tectirostris** Cope. FIN-BACK WHALE. L. 60 feet. Coast of E. U. S., the most common large whale in Mass. Bay. (Lat., *tectus*, covered ; *rostrum*, snout.)

582. BALÆNOPTERA Lacépède. (*Balæna ;* πτερόν, fin.)

1108. **B. rostrata** (Müller). PIKED WHALE. L. 30 feet. N. Atl. (*Eu.*) (Lat., long-nosed.)

583. MEGAPTERA Gray. (μέγας, large ; πτερόν, fin.)

1109. **M. longimana** (Rudolphi). HUMP-BACK WHALE. Body short, thick, with humps and protuberances; skin often covered with barnacles. L. 50 to 75 feet; color usually black. N. Atl., formerly common. An American form has been described as *M. osphyia* Cope, on account of the shorter head and fins, and higher neural spines. (*Eu.*) (Lat., *longus*, long ; *manus*, hand.)

584. AGAPHELUS Cope.

1110. **A. gibbosus** (Erxleben). SCRAG WHALE. N. Atl., a rare or very doubtful species, sometimes thought to be the young of the Right Whale. (*Eu.*)

585. BALÆNA Linnæus. (Lat., whale.)

1111. **B. cisarctica** Cope. RIGHT WHALE. BLACK WHALE. The common large whale of our Eastern coasts and the North Atlantic generally, occasionally S. to S. C. Color black. L. 40 feet. (*Eu.*) (Lat., this side of Arctic.) (*B. biscayensis* Gervais, the American name older.)

In the Arctic seas occurs the great Bowhead, *B. mysticetus* L., the most valuable of the whales, reaching a length of 50 or 60 feet, yielding 200 to 300 barrels of oil and from 1 to 2 tons of whalebone.

---

[1] According to Professor Cope, this genus is a doubtful one, *S. tectirostris* being perhaps not really different from *Balænoptera (Physalus) physalus.*
Probably *Physalus* and *Sibbaldius* should be united to *Balænoptera.* (*True.*)

ORDER LII. **UNGULATA.** (THE HOOFED MAMMALS.)

Herbivorous mammals provided with 1 to 4 enlarged and thickened claws or hoofs on each foot ; molar teeth adapted for grinding. The anatomical characters of this well-known and varied group are too numerous to be here summarized. The order is usually subdivided into the *Perissodactyli*, or odd-toed ungulates, and the *Artiodactyli* or even-toes. The former group is exemplified by the Horse (*Equus caballus* L.), the Ass (*Asinus asinus* L.), the Rhinoceros and the Tapir. The *Artiodactyli* are again subdivided into the non-ruminating, omnivorous, hornless, naked or bristly allies of the Common Hog and Wild Boar (*Sus scrofa*), and the group of *Pecora* (Ruminants). To the latter belong all the living ungulates occurring within our limits. (Lat., *ungulatus*, hoofed.)

Families of Ungulata.

a. Feet bifid; first toe wanting; second and fourth rudimentary. (*Artiodactyli.*)

   b. Stomach compound, of 3 or 4 compartments; horns usually present. (*Pecora.*)

      c. Upper jaw without incisors, in the adult.

        d. Horns solid, usually branching, deciduous. . . . CERVIDÆ, 195.

        dd. Horns permanent, hollow, each enclosing a process of the frontal bone. . . . . BOVIDÆ, 196.

FAMILY CXCV. **CERVIDÆ.** (THE DEER.)

Horns deciduous, solid, developed from the frontal bone, more or less branched, covered at first by a soft, hairy integument, known as " velvet "; when the horns attain their full size, which they do in a very short time, there arises at the base of each a ring of tubercles known as the " burr;" this compresses and finally obliterates the blood-vessels supplying the velvet. which dries up and is stripped off, leaving the bone hard and insensible ; the horns or " antlers " are shed annually, the separation of the " beam " from its " pedicel " taking place just below the burr : antlers are wanting in the female (excepting in the Reindeer) but they are present in the male of nearly all species. Stomach in four divisions, of the ordinary ruminant pattern. Dental formula. i. $\frac{0}{3}\frac{0}{3}$; c. (usually) $\frac{0:0}{1:1}$; pm. $\frac{3:3}{3:3}$; m. $\frac{3:3}{3:3}$. A widely distributed family of about 13 genera.

a. Horns present in males only.

   b. Horns rounded more or less; rarely sub-palmated . nose naked and moist.

      c. Horns small, curving forward, the first snag short, at some distance above the base, and like the others curving upward; tail rather long; hoofs rather elongate. . . . . CARIACUS, 586.

      cc. Horns large, curving backward, with the snags all directed forward, one of them immediately above the burr, tail very short; hoofs broad and rounded. . . . . CERVUS, 587.

bb. Horns very broadly palmated to the tip; nose very broad, entirely
hairy except a small naked spot between nostrils. . . ALCE, 588.
aa. Horns (present in both sexes) broadly palmated at tip; nose entirely
hairy. . . . . RANGIFER, 599.

586. CARIACUS Gray. (Old name.)

1112. **C. virginianus** (Boddaert). VIRGINIA DEER. RED
DEER. General color chestnut red, grayish in winter; tail white
below. Maine to Rocky mountains and S., formerly very common,
and still abundant in wild districts.

1113. **C. leucurus** (Douglas). WHITE TAILED DEER. Yellowish gray, waved with dusky; lower side of tail, etc., white;
chin mostly white; size of preceding. Neb. to Texas and W.
(λευκός, white: ουρά, tail.)

1114. **C. macrotis** (Say). MULE DEER. Larger; ears very
long, nearly as long as tail. Ashy brown, a darker dorsal stripe.
Neb. to Ore. (μακρός, long; ους, ear.)

587. CERVUS Linnæus. (Lat., deer.)

1115. **C. canadensis** Erxleben. WAPITI. Chestnut red, gray-
ish in winter; size nearly equal to that of the Moose. Va. to Wis.,
Dak. and W., now becoming rare; commonly and wrongly called
"Elk" in America.

588. ALCE Hamilton Smith. (From Elk.)

1116. **A. alces** (L.). MOOSE. TRUE ELK. Tawny above,
yellowish below; ears large; profile of snout very convex. Largest of our *Cervidæ*, reaching the size of a horse. Maine and N.
N. Y. to Oregon and N. (*Eu.*) The American form is var.
americanus Jardine.

589. RANGIFER Hamilton Smith. (Old name.)

1117. **R. tarandus** (L.). REINDEER. Brownish, grayer in
winter. American varieties of the Reindeer, or possibly distinct
species, are the Woodland Caribou, var. **caribou** (Kerr), found from
Maine to L. Superior and N., and the Barren Ground Caribou,
var. **grœnlandicus** (Kerr), smaller and confined to the treeless
Arctic regions. (*Eu.*) (Lat., reindeer.)

FAMILY CXCVI. BOVIDÆ. (THE CATTLE.)

Ruminants with the horns, if present, simple or nearly so, hollow, permanent, each enclosing a process of the frontal bone.
Teeth i. $\frac{0}{0}$; c. $\frac{0}{0}$; m. $\frac{6.6}{6.6}$ = 32. Genera about 35; species 80 or
more, inhabiting warm regions, and most abundant in the Old
World. The ox (*Bos taurus* L.), the sheep (*Ovis aries*), and the
goat (*Capra hircus*), are familiar members of the family.

a. Horns erect, compressed at base, with a short branch or flattened process in front, the end conical, recurved; nose hairy at tip, except along the central line; tail very short; false hoofs obsolete.

ANTILOCAPRA, 590.

aa. Horns simple.

  b. Nose almost entirely hairy, not broad at tip; horns curving spirally; nose convex in profile; no beard on chin; a gland and duct between hoofs; false hoofs present; tail long. . . . . OVIS, 591.

  bb. Nose naked at tip and very broad; horns curved, the base directed outwards; hoofs broad; tail long; forehead broader than long; body highest at shoulders; anterior parts with a long, shaggy mane.

BISON, 592.

590. ANTILOCAPRA Ord. (Antilope + Capra.)

1118. **A. americana** (Ord). PRONG-HORN. CABREE. ROCKY MOUNTAIN "ANTELOPE." Yellowish brown, marked with brown and white. L. about 5 feet. T. 7 inches. Height 3 feet. Dak. to Tex. and W.

591. OVIS Linnæus. (Lat., sheep.)

1119. **O. montana** Cuvier. BIG-HORN. ROCKY MOUNTAIN SHEEP. Larger than a domestic sheep; ♂ with immense horns; ♀ with smaller ones. Grayish brown. Dak. to Col. and W.

592. BISON Audubon & Bachman. (Lat., a wild ox or buffalo.)

1120. **B. bison** (L.). BUFFALO. BISON. Brown; the snout, hoofs, horns, etc., black. U. S. generally; formerly abundant, but now extinct, except a few herds in the Yellowstone region.

ORDER LIII. **FERÆ.** (THE FLESH-EATERS OR CARNIVORA.)

Canine teeth distinct, conical; molars more or less adapted for cutting; clavicles imperfect or wanting; toes provided with claws; skin covered with hair or fur; alimentary canal short. General structure in accordance with the predatory life led by all these animals. (Lat., *ferus*, a wild beast; the name *Feræ* of Linnæus, is much older than *Carnivora*.)

Families of Feræ.

a. Limbs short, unfitted for walking, the toes united in a flat paddle, from which only the claws project; tail very short; eyes very large; incisors often less than ⅔. (*Pinnipedia.*)

  b. Hind limbs directed backwards, used only in swimming; claws strong; neck short. . . . . PHOCIDÆ, 197.

aa. Limbs fitted for walking, the toes distinct; incisors ⅗. (*Fissipedia.*)

  b. Hind feet with 5 toes.

    c. Feet fully plantigrade; sectorial teeth and the molars behind them all tuberculate.

338 MAMMALIA : FERÆ. — LIII.

d. Tail well developed; body rather slender, the snout sharp.
Procyonidæ, 198.
dd. Tail rudimentary; body very robust; snout not acuminate.
Ursidæ, 199.
   cc. Feet sub-plantigrade or digitigrade; only one tuberculate molar, the
sectorial premolar of typical form. . . . . . . Mustelidæ, 200.
bb. Hind feet with 4 toes.
  c. Teeth 42; claws not retractile; snout more or less produced.
Canidæ, 201.
   cc. Teeth 28 to 30; claws retractile into a sheath; snout short, the
head broad. . . . . Felidæ, 202.

Family CXCVII. PHOCIDÆ.[1] (The Earless Seals.)

Seals with the fore-limbs well forward; neck short; hind limbs directed backward, useless on land; hand and foot hairy; nails usually well developed; no external ear. Other characters further distinguishing these seals from the Fur Seals and Sea-Lions (*Otariidæ*), and the Walruses (*Rosmarida*), are drawn from the skeleton. Genera 11, species 17, found on most coasts, swimming freely in the water and feeding chiefly on fishes, resting and sunning in the rocks on the shore.

a. Incisors usually ³⁄₃; interorbital region very narrow; nails of all digits well-developed; (other characters drawn from the skull). (*Phocinæ.*)
b. Snout narrow; incisors simple, conical. . . . . . . Phoca, 593.

593. PHOCA Linnæus. (φώκη, seal.)

1121. **P. vitulina** L. Harbor Seal. Yellowish gray, usually blotched with darker; variable. L. 3 to 5 feet; weight 50 to 65 pounds. Northern shores, S. to N. J., common N. (Lat., calf-like.) (*Eu.*) Several other seals occur N. of Newfoundland.

Family CXCVIII. PROCYONIDÆ. (The Raccoons.)

Plantigrade Carnivora of moderate size, with the body comparatively slender and the tail well developed. Teeth i. ³⁄₃; c. ¹⁄₁; pm. ⁴⁄₄; m. ²⁄₂ = 40. Sectorial tooth broad, tubercular. Snout more or less elongated; no cæcum. Genera 2, — *Nasua* of Mexico, and the following, all American.

a. Tail not prehensile; snout moderate, not flexible. . . . Procyon, 594.

594. PROCYON Storr. (προκύων, before the dog.)

1122. **P lotor** (L.). Common Raccoon. Grayish white; hairs black-tipped; tail with black rings; a black cheek-patch; body rarely entirely black. L. 33. T. 10½. U. S., abundant. (Lat., washer.)

[1] For a full account of the seals, see Allen's admirable "Monograph of the Pinnipedes." 1880.

Family CXCIX. URSIDÆ. (The Bears.)

Plantigrade Carnivora having the body thick and clumsy. Tail rudimentary. Teeth 42 ; molars broad and tuberculated, according with the omnivorous diet. Genera 5 ; species few and widely distributed, — in North America, there are probably but three, although many have been described, the Polar Bear, *Thalarctos maritimus,* and the following.

a. Snout depressed, so that the profile does not form a straight line ; soles not fully furred ; claws moderate. . . . . Ursus, 595.

595. URSUS Linnæus. (Lat., bear.)

1123. **U. americanus** Pallas. Brown, Black or Cinnamon Bear. Color black or brownish, exceedingly variable, a fact which has given rise to numerous nominal species, but the several forms or varieties intergrade perfectly. N. Am., abundant, where not exterminated.

1124. **U. horribilis** [1] Ord. Grizzly Bear. Grizzly gray or brownish. Largest of the bears, reaching a length of 9 feet and a weight of 800 lbs. Neb. to Cal., in the mountains. Very near the European bear, *U. arctos* L. but larger.

Family CC. MUSTELIDÆ. (The Weasels.)

Carnivora either plantigrade or digitigrade, with the toes 5-5. Molars $\frac{4\cdot4}{3\cdot3}$ (rarely $\frac{4\cdot4}{4\cdot4}$) ; the upper and the last lower one tubercular ; sectorial premolar without tubercles ; no cæcum. Most species provided with glands near the anus which secrete a fetid liquid. Some are strictly carnivorous while others are rather omnivorous. Size usually median or small. *Mustelidæ* are found in all parts of the earth excepting the Australian region. Some of the species are aquatic, and one (*Euhydris,* the Sea Otter) presents numerous analogies with the Seals.

a. Skull with the cerebral portion swollen backwards and outwards, the snout short, high and truncate forwards ; toes webbed, the feet adapted for swimming ; teeth 36. Aquatic. (*Lutrinæ.*) . . . . . Lutra, 596.
aa. Skull with the cerebral portion posteriorly somewhat compressed, the snout produced, attenuate and transversely convex above ; feet scarcely webbed ; mostly not aquatic.
  b. Auditory bulla little inflated, constricted ; last upper molar above quad-

[1] "Coward, — of heroic size
In whose lazy muscles lies
Strength we fear and yet despise ;
Savage, whose relentless tusks
Are content with acorn husks ;
Robber, whose exploits ne'er soared
O'er the bee's or squirrel's hoard ;
Whiskered chin and feeble nose,
Claws of steel on baby's toes." (*Bret Harte.*)

rangular, very large, with an outer cutting ridge; claws non-retractile; fore-claws lengthened, fossorial. (*Mephitinæ.*)

  c. Teeth 34; tail long and bushy; anal secretion very strong-scented.

MEPHITIS, 597.

  bb. Auditory bulla much inflated, not constricted.

    d. Last molar of upper jaw enlarged, sub-triangular; toes straight, with long, non-retractile claws. (*Melinæ.*)

      e. Body robust; tail very short; teeth 32. . . . TAXIDEA, 598.

    dd. Last molar of upper jaw short, transverse; toes short, arched, the claws retractile. (*Mustelinæ.*)

      f. Feet sub-plantigrade; body stout; tail very full and bushy; teeth 38. . . . . GULO, 599.

      ff. Feet digitigrade; body slender.

        g. Teeth 38; sectorial tooth with an internal tubercle.

MUSTELA, 600.

        gg. Teeth 34; sectorial tooth without internal tubercle.

PUTORIUS, 601.

596. LUTRA Linnæus. (Lat., otter.)

1125. **L. hudsonica** (Lacépède). AMERICAN OTTER. Liver-brown. L. 45. T. 15. N. Am., aquatic. (*Mustela canadensis* Turton, not of Schreber.)

597. MEPHITIS Cuvier. (Lat., a bad odor.)

a. Skull not depressed, its upper outline irregularly convex. (*Mephitis.*)

1126. **M. mephitica** (Shaw). COMMON SKUNK. Usually black, with tip of tail, dorsal stripes and nuchal patch white; sometimes all black or even nearly all white. L. 28. T. 9. Hudson's Bay to Mexico, abundant, its odor " mephitic " beyond comparison.

aa. Skull depressed, its upper outline nearly straight. (*Spilogale* Gray.)

1127. **M. putorius** (L.). LITTLE STRIPED SKUNK. Black, with white patch on forehead; four parallel dorsal stripes, broken behind; tail black, with white pencil at tip. Size very small. L. 15. T. 4. Wis. (*Hoy*) to Ga., and S. W. (Lat., ill-scented.)

598. TAXIDEA Waterhouse. (*Taxus*, a related genus; εἶδος, form.)

1128. **T. americana** (Boddaert). AMERICAN BADGER. Chiefly grayish. L. 27. T. 5. Wis. to Tex. and W., formerly E. to Ohio.

599. GULO Storr. (Lat., glutton.)

1129. **G. gulo** (L.). WOLVERENE. Blackish; a pale lateral band meeting its fellow above root of tail; forehead pale; fur shaggy. L. 30. T. 8. N. Y. to Col. and N. (*Eu.*)

600. MUSTELA Linnæus. (Lat., weasel.)

1130. **M. americana** Turton. SABLE. PINE MARTEN. Brown, not darker below than above, usually a tawny throat-patch. Ears high, sub-triangular. L. 24. T. 8. Me. to Ore., and N.

1131. **M. pennanti** (Erxleben). PEKAN. BLACK CAT. Blackish, paler anteriorly, darkest below; no throat-patch; ears low, semicircular. L. 35. T. 14. Penn. to Hudson's Bay, and W. (To Thomas Pennant, author of Arctic Zoölogy.)

601. PUTORIUS Cuvier. (Lat., *putor*, a bad odor.)

a. Species of large size (length to base of tail over 12).
b. Toes somewhat webbed; pads of palm coalescent; tail bushy; ears low. (*Lutreola* Wagner).

1132. **P. vison** (Schreber). MINK. Dark chestnut-brown, uniform or varied with whitish below. L. 28. T. 8. N. Am.; common, aquatic. (Lat., a scout.)

bb. Toes not webbed; pads of palm separate; tail short, slender; ears high, round. (*Cynomyonax* Coues.)

1133. **P. nigripes** Audubon & Bachman. BLACK-FOOTED FERRET. Pale brown; feet, tip of tail and bar across face black. L. 23. T. 4. Neb. and W., in "Prairie-dog towns," feeding on the rodents. An allied species is the European Ferret, *P. putorius* L., trained to hunt rats. (Lat., black-foot.)

aa. Species of small size (length to base of tail less than 12); body attenuate: neck long; ears conspicuous, orbicular; tail slender; toes cleft: pads on feet separate; coloration bicolor, reddish brown, yellowish or white below, the fur usually becoming snow-white in winter. (*Gale* Wagner.)
c. Tail black at tip.

1134. **P. longicauda** Bonaparte. LONG-TAILED WEASEL. Belly tawny or salmon-yellow; black tip of tail reduced to a terminal pencil. L. 16½. T. 6. Minn. to Ariz. and N.

1135. **P. erminea** (L.). WEASEL. ERMINE. STOAT. Belly sulphur-yellow; black of tail not confined to tip; fur snow-white in winter. L. 11. T. 3. Northern regions, S. to Kan., common N. (*Eu.*) (From ermine.)

cc. Tail pointed, scarcely black at tip.

1136. **P nivalis** (L.). LEAST WEASEL. Mahogany-brown, white, rarely yellowish below; white in winter. L. 10. T. 2. Northern regions, S. to Penn. (Lat., snow-white.) (*Eu.*)

FAMILY CCI. **CANIDÆ.** (THE DOGS.)

Digitigrade Carnivora with blunt, non-retractile claws; toes 5–4. Muzzle more or less elongated. Dentition typically i. $\frac{3\text{-}3}{3\text{-}3}$; c. $\frac{1\text{-}1}{1\text{-}1}$; pm. $\frac{4\text{-}4}{4\text{-}4}$; m. $\frac{2\text{-}2}{3\text{-}3}$ = 42; canines large, rather blunt. Genera about 5. Species widely distributed, all of them more or less dog-like or fox-like in habit.

aa. Pupil elliptical; tail long and bushy; upper incisors scarcely lobed; body rather slender.
b. Tail with soft fur and long hair; muzzle long. . . . VULPES, 602.

bb. Tail with a concealed mane of stiff hairs, and without soft fur; muzzle shorter. . . . . UROCYON, 603.
aa. Pupil circular; tail moderate; upper incisors distinctly lobed. CANIS, 604.

602. VULPES Brisson. (Lat., fox.)

1137. **V. vulpes** (L.). RED FOX. Chiefly reddish gray, with black feet and ears; tip of tail white. The American form is var. *fulvus* (Desmarest). The Cross Fox is var. *decussatus* Desm., with a dark cross on back; the Black or Silver Fox is var. *argentatus* Shaw. These forms fully intergrade with the Common Fox. L. 45. T. 15. Northern regions, S. to Texas. (*Eu.*)

1138. **V. velox** Say. KIT FOX. Smaller, with closer fur; yellowish gray, ears not black. L. 33. T. 9. Iowa to Ore. (Lat., swift.)

603. UROCYON Baird. (οὐρά, tail; κύων, dog.)

1139. **U. cinereo-argentatus** (Schreber). GRAY FOX. Chiefly gray; fur dusky or tawny, hairs hoary at tip; tip of tail usually dark. L. 40. T. 14. Penn. to Texas and S. W. (Lat., ashy-silvery.)

604. CANIS Linnæus. (Lat., dog.)

1140. **C. latrans** Say. COYOTE. PRAIRIE WOLF. Yellowish gray, clouded with black; fur coarse; snout sharp. L. 55. T. 11. Wis. to Texas and W., common on the plains, burrowing in the ground. A vagabond dog-like animal, "half bold and half timid, yet lazy all through." (Lat., barking.)

1141. **C. lupus** L. WOLF. Color exceedingly variable; chiefly gray, becoming whitish northward (var. **occidentalis**), southward more and more blackish and reddish, till in Florida black wolves (var. **ater** Richardson) predominate, and in Texas red ones (var. **rufus** Aud. & Bach.), while on the plains is the dusky wolf (var. **nubilus** Say). L. 65. T. 15. Northern regions, common where not exterminated. Allied to the wolf, or perhaps descended from it, and more or less mixed with other *Canidæ* is the Dog, *Canis familiaris* L. (*Eu.*)

FAMILY CCII. FELIDÆ. (THE CATS.)

Digitigrade Carnivora with the toes 5–4; claws compressed, very sharp, retractile; palms and soles hairy, with naked pads under each toe and the ball of the foot. Body compact; head short, broad and rounded. Dentition i. $\frac{3-3}{3-3}$; c. $\frac{1-1}{1-1}$; pm. $\frac{3-3}{2-2}$ or $\frac{2-2}{2-2}$; m. $\frac{1-1}{1-1} = 30$ or 28; canine teeth long and sharp; teeth all strongly trenchant; tongue with short, retrorse papillæ. General aspect cat-like. Species about 50, found in all parts of the world excepting Australia and its islands, "the fiercest, strongest and most terrible of beasts," "brave when hungry and in the dark, cowardly or lazy in the day-time, and magnanimous when not in need of food." The Common House Cat, *Felis domesticus* Schreber, one of the smallest of the *Felidæ*, is a familiar representative of the group.

a. Premolars ⅔, (anterior upper one wanting); tail less than half length of body proper; ears triangular, tufted. . . . . **LYNX, 605.**

aa. Premolars 2.3, anterior upper one very small; tail at least half as long as the body (exclusive of head and neck); fur compact and glossy; ears not tufted. . . . . **FELIS, 606.**

605. LYNX Rafinesque. (λύγξ, a wild cat.)

1142. **L. canadensis** (Desmarest). CANADA LYNX. Feet very large, densely furred beneath in winter, concealing the small, naked patches. Grayish hoary, waved with black ; tail black at tip ; no distinct bars on inner side of legs ; larger than the next, with larger feet and longer fur. L. 39. T. 4½. N. Am.

1143. **L. rufus** (Guldenstädt). AMERICAN WILD CAT. Reddish, overlaid by grayish ; inner sides of legs with dark cross-bands ; tail with a black patch at end above, preceded by half rings. L. 35. T. 7. N. Am.

606. FELIS Linnæus. CATS. (Lat., cat.)

1144. **F. concolor** L. AMERICAN PANTHER. COUGAR. PUMA. Above tawny brownish yellow ; a wash of darker along dorsal line ; dirty white below ; kittens spotted, their tails ringed ; larger than a sheep. L. 90. T. 32. America, N. to Canada. (Lat., one-color.)

ORDER LIV. PRIMATES. (THE ANTHROPOID MAMMALS.)

Both limbs nearly or quite outside of the common integument of the body : fingers and toes usually 5, the thumb sometimes wanting, when present opposable to the others ; great toe with a depressed nail ; teeth various, usually with distinct incisors, canines and molars : clavicles completely developed ; shoulders distinct, well-separated ; brain large, the cerebrum and cerebellum highly developed ; parts of the brain well differentiated. Mammæ pectoral, except in some lemurs. A large and varied order, the highest among animals, comprising men, apes, baboons, monkeys, and lemurs. The lemurs diverge in many respects from the other primates, and should perhaps stand as a separate order (*Prosimii*). The structural peculiarities of man are not numerous, and are mostly correlated with the great development of the brain, the chief peculiarity characteristic of the *Hominidæ*. (Lat., *primatus*, the chief place.)

Families of Primates.

a. Hair on body little developed, except in certain specialized areas ; body erect in locomotion ; great toe not opposable ; dentition i. ²⁄₂ ; c. ¹⁄₁ ; pm. ²⁄₂ ; m. ³⁄₃ on each side ; no gaps between the teeth. . . . . HOMINIDÆ, 203.

Family CCIII. **HOMINIDÆ.** (The Men.)

The most prominent characters of the *Hominidæ* are "derived from the distribution of hair on the body, which is subject to wide modification in the different races, from the fact that locomotion is easiest in the erect posture, owing to the relative shortness of the arms; from the greater length and mobility of the thumb and the comparative immobility of the great toe. Well-marked skeletal peculiarities are the possession of 12 rib-bearing vertebræ, the rounded skull in which the muscular ridges are little prominent, and the great capacity of the cranium. This is, of course, in adaptation to the relatively enormous development of the cerebral hemispheres, which much exceed in bulk those of other primates and to which man owes his specific name." It is apparent that different races have arrived at different stages of evolution in the development of the brain, "as well as in the employment of articulate speech, to which man owes the power of transmitting to others the results of his experience and his position as the 'highest animal.'" (*R. Ramsay Wright.*) As usually understood, this family contains but a single species, cosmopolitan, and highly variable.

607. **HOMO** Linnæus. (Lat., man.)

1145. **H. sapiens** L. Man. This species is now split up into many sub-species or races, the native man of this continent, or "American Indian," being var. **americanus** L. Other races now naturalized in America are the Caucasian race, var. **europæus** L., the Mongolian race, var. **asiaticus** L., and the Negro race, var. **afer** L. The first of these is an immigrant from Europe, the second from Asia, and the third was brought hither from Africa by representatives of var. *europæus* to be used as slaves. The wild man, or typical var. *sapiens*, as described by Linnæus ("Homo diurnus: varians cultura, loco. tetrapus, mutus, hirsutus"), seems to be now extinct. (Lat., knowing.) (*Eu.*)

"*Sic vivimus ut immortales et morimur ut mortales.*" (Seneca.)

GLOSSARY OF TECHNICAL TERMS.

Abdomen. Belly.

Abdominal. Pertaining to the belly, — said of the ventral fins of fishes when inserted considerably behind the pectorals, the pelvic bones to which the ventral fins are attached having no connection with the shoulder girdle.

Abortive. Remaining or becoming imperfect.

Acuminate. Tapering gradually to a point.

Acute. Sharp-pointed.

Adipose fin. A peculiar, fleshy, fin-like projection behind the dorsal fin, on the backs of Salmons, Cat Fishes, etc.

Adult. A mature animal.

Ægithognathous. Having the peculiar palate of Passerine birds.

Air-bladder. A sac filled with air, lying beneath the back-bone of fishes, corresponding to the lungs of the higher vertebrates.

Allantois. An organ of the embryo.

Altrices. Birds hatched in an immature condition, reared in the nest and fed by the parents.

Altricial. Having the nature of *Altrices.*

Alula. The feathers attached to the "thumb" (rather the "index finger") of a bird.

Alveolar surface. A portion of the jaw of a turtle, where the teeth-sockets (alveola) are developed in other reptiles.

Amnion. An organ of the embryo.

Amphicœlian. Double concave, — said of vertebræ.

Anadromous. Running up, — said of marine fishes which run up rivers to spawn.

Anal. Pertaining to the anus or vent.

Anal fin. The fin on the median line, behind the vent, in fishes.

Anal plate. The plate, immediately in front of the vent, in serpents, often divided in two by a median suture.

Anchylosed. Grown firmly together.

Anteorbital plate. The plate (one or two) in front of the eye in serpents, with its longest diameter vertical; also called preocular.

Antrorse. Turned forwards.

Anus. The external opening of the intestines; the vent.

Arboreal. Living in trees.

Arterial bulb. The muscular swelling, at the base of the great artery, in fishes.

Articulate. Jointed.

Artiodactylous. Even-toed (toes 2 or 4).

Atrophy. Non-development.

Attenuate. Long and slender, as if drawn out.

Auricle. The large lobe of the external ear; also, one of the chambers of the heart.

Barbel. An elongated fleshy projection, usually about the head, in fishes.

Basal. Pertaining to the base; at or near the base.

Basipterygoid. Bones developed in the palatine arch in some birds.

Beak. The bill of birds, or (in other animals) any beak-like structure.

Bend of Wing. Angle at the carpus when the wing is folded.

Bicolor. Two-colored.

Bicuspid. Having two points.

Booted. Said of the tarsus in birds, when its scales coalesce and form a continuous envelope, as in the robin.

Branchiæ. Gills; respiratory organs of fishes.

Branchial. Pertaining to the gills.

Branchihyals. Small bones at base of gill arches.

Branchiostegals. The bony rays supporting the branchiostegal membranes, under the head of a fish, below the opercular bones, and behind the lower jaw.

Bristle. A stiff hair, or hair-like feather.

Buccal. Pertaining to the mouth.

Caducous. Falling off early.

Cæcal. Of the form of a blind sac.

Cæcum. An appendage of the form of a blind sac, connected with the alimentary canal.

Calcareous. Containing or composed of carbonate of lime

Canines. The teeth behind the incisors, — the "eye-teeth"; in fishes, any conical teeth in the front part of the jaws, longer than the others.

Carapace. The upper shell of a turtle, usually composed of bony plates covered by horny scales.

Cardiform (teeth). Teeth coarse and sharp, like wool-cards.

Carinate. Keeled; having a ridge along the middle line.

Carotid. The great artery running to the head.

Carpus. The wrist.

Caudal. Pertaining to the tail.

Caudal fin. The fin on the tail of fishes and whales.

Caudal peduncle. The region between the anal and caudal fins in fishes.

Cavernous. Containing cavities, either empty or filled with a mucous secretion.

Cephalic fins. Fins on the head of certain rays a detached portion of the pectoral.

Cere. Fleshy, cutaneous, or membranous covering of the base of the bill in certain birds, particularly the Owls, Hawks, and Parrots.

Cervical. Pertaining to the neck.

Chiasma. Crossing of the fibres of the optic nerve.

Chin. The space between the rami of the lower jaw.

Ciliated. Fringed with eye-lash like projections.

Cirri. Fringes.

Claspers. Organs attached to the ventral fins in the male of sharks, skates, etc.

Clavicle. The collar bone, or lower anterior part of shoulder girdle, not entering into socket of arm.

Cloaca. A common opening of genital, urinary, and alimentary canals.

Commissure. The line on which the mandibles of a bird are closed.

Compressed. Flattened laterally.

Condyle. Articulating surface of a bone.

Conirostral. Said of a bill like that of a Sparrow; conical in form and with the commissure angulated.

Coracoid. The principal bone of the shoulder girdle in fishes; otherwise a bone or cartilage on the ventral side, helping to form the arm-socket.

Costal folds. Folds of the skin (of a Salamander) showing the position of the ribs (costæ).

Coverts. Small feathers hiding the bases of quills.

Crest. In birds, any lengthened feathers about the head; elsewhere, any elevated or crest-like projection.

Crissum. The under tail coverts, in birds.

Ctenoid. Rough-edged, said of scales when the posterior margin is minutely spinous or pectinated.

Culmen. The middle line or ridge of the upper mandible in birds.

Cuneate. Wedge-shaped; said of a bird's tail when the middle feathers are longest and the rest regularly shorter.

Cycloid. Smooth-edged; said of scales not ctenoid, but concentrically striate.

Deciduous. Temporary; falling off.

Decurved. Curved downward.

Dentary. The principal or anterior bone of the lower jaw, usually bearing the teeth.

Dentate. With tooth-like notches.

Denticle. A little tooth.

Dentirostral. Having the bill notched near its tip.

Depressed. Flattened vertically.

Depth. Vertical diameter (usually of the body of fishes).

Dermal. Pertaining to the skin.

Desmognathous. United palate, as in the lower water-birds (Loons, Gulls, etc.).

Diaphanous. Translucent.

Diaphragm. Muscular septum between thorax and abdomen.

Diapophysis. Transverse process of a vertebra.

Digitigrade. Walking on the toes, like a dog.

Distal. Remote from point of attachment.

Dorsal. Pertaining to the back.

Dorsal fin. The fin on the back of fishes.

Emarginate. Slightly forked or notched at the tip; abruptly narrowed or notched toward the tip (said of quills).

Endoskeleton. The skeleton proper, — the inner bony framework of the body.

Epignathous. Upper mandible hooked over tip of lower.

Erectile. Susceptible of being raised or erected.

Eustachian tubes. Tubes connecting the inner ear with the pharynx.

Even (tail). Having all the feathers of equal length.

Exoskeleton. Hard parts (scales, scutes, feathers, hairs) on the surface of the body.

Exserted. Projecting beyond the general level.

Extra-limital. Beyond the limits (of this book).

Facial. Pertaining to the face.

Falcate. Scythe-shaped; long, narrow, and curved.

Falciform. Curved, like a scythe.

Fauna. The animals inhabiting any region, taken collectively.

Femoral. Pertaining to the femur, or proximal bone of the hinder leg.

Fibula. The small outer leg bone.

Filament. Any slender or thread-like structure.

Filiform. Thread-form.

Fissirostral. Having the bill very deeply cleft, beyond the base of the horny part, as in the Swallows.

Fontanelle. An unossified space on top of head covered with membrane.

Foramen. A hole or opening.

Forehead. Frontal curve of head.

Forficate. Deeply forked; scissors-like.

Fossæ (nasal). Grooves in which the nostrils open.

Fossorial. Adapted for digging.

Frontal bone. Anterior bone of top of head.

Fulcra. Rudimentary spine-like projections extending on the anterior rays of the fins of ganoid fishes.

Furcate. Forked.

Fusiform. Spindle-shaped; tapering toward both ends but rather more abruptly forward.

Ganglion. A nerve centre.

Ganoid. Scales or plates of bone covered by enamel.

Gape. Opening of the mouth.

Gastrosteges. Band-like plates along the belly of a serpent; ventral plates.

Gills. Organs for breathing the air contained in water.

Gill arches. The bony arches to which the gills are attached.

Gill openings. Openings leading to or from the branchiæ.

Gill rakers. A series of bony appendages variously formed along the inner edge of the anterior gill arch.

Glabrous. Smooth.

Gonys. The middle line of the lower mandible.

Gorget. Throat patch of peculiar feathers.

Graduated (spines). Progressively longer backward; the third being as much longer than second as second is longer than first.

Graduated (tail). One in which the outer feathers are regularly shorter from the middle.

Granulate. Rough with small prominences.

Gular. Pertaining to the *gula*, or upper fore-neck.

Hæmal spine. The lowermost spine of a caudal vertebra, in fishes.

Hæmopophyses. Appendages on the lower side of abdominal vertebræ, in fishes.

Hallux. The great toe, — in birds, the hind toe.

Height. Vertical diameter.

Heterocercal. Said of the tail of a fish, when unequal, — the back-bone evidently running into the upper lobe.

Hirsute. With shaggy hairs.

Homocercal. Said of the tail of a fish when not evidently unequal; the back-bone apparently stopping at the middle of the base of the caudal fin.

Humerus. Bone of the upper arm.

Hyoid. Pertaining to the tongue.

Hypognathous. Having the lower mandible longer than the upper, as in the Black Skimmer.

Imbricate. Overlapping, like shingles on a roof.

Imperforate. Not pierced through.

Inarticulate. Not jointed.

Incisors. The front or cutting teeth.

Infraoral. Below the mouth.

Interfemoral membrane. The membrane connecting the posterior limbs of a bat.

Interhæmals. Bones to which anal rays are attached in fishes.

Intermaxillaries. The premaxillaries; the bones forming the middle of the front part of the upper jaw, in fishes.

Internasals. Plates on the forehead of the snake on the line connecting the two nostrils.

Interneurals. Bones to which dorsal rays are attached in fishes.

Interopercle. Membrane bone between the preopercle and the branchiostegals.

Interorbital. Space between the eyes.

Interspinals. Bones to which fin-rays are attached (in fishes); inserted between neural spines above and hæmal spines below.

Isocercal (tail). Last vertebræ progressively smaller and ending in median line of caudal fin, as in the Cod-fish.

Jugular. Pertaining to the lower throat, — said of the ventral fins, when placed in advance of the attachment of the pectorals.

Keeled. Having a ridge along the middle line.

Labials. Plates forming the lip of a serpent.

Lamellæ. Plate-like processes inside of the bill of a duck.

Lamellate. Said of a bill provided with lamellæ, as in a duck.

Larva. An immature form, which must undergo change of appearance before becoming adult.

Lateral. To or towards the side.

Lateral line. A series of muciferous tubes forming a *raised line* along the sides of a fish.

Laterally. Sidewise.

Lobate. Furnished with membranous flaps, — said of the toes of birds.

Longitudinal. Running lengthwise.

Loral plate. Plate between eye and nostril of a serpent, before and below preocular when this is present; its longest diameter horizontal.

Lores. Space between eye and bill.

Lunate. Form of the new moon; having a broad and rather shallow fork.

Mammary glands. Glands secreting milk.

Mandible. Under jaw (or in birds, either jaw).

Maxilla. Upper jaw.

Maxillaries. Outermost or hindmost bones of the upper jaw, in fishes; they are joined to the premaxillaries in front, and usually extend farther back than the latter.

Metacarpus. The hand proper, exclusive of the fingers.

Metamorphosis. A decided change in form.

Metatarsus. The foot proper.

Molars. The grinding teeth; posterior teeth in the jaw.

Monogamous. Pairing; said of birds.

Muciferous. Producing or containing mucus.

Myocomma. A muscular band.

Nape. Upper part of neck, next to the occiput.

Nares. Nostrils, anterior and posterior.

Nasal. Pertaining to the nostrils.

Nasal plate. Plate in which the nostrils are inserted.

Neural spine. The uppermost spine of a vertebra.

Nictitating membrane. The third or inner eye-lid, of birds, sharks, etc.

Notochord. A cellular cord, which in the embryo precedes the vertebral column.

Nuchal. Pertaining to the nape or *nucha.*

Obscure. Scarcely visible.

Obsolete. Faintly marked; scarcely evident.

Obtuse. Blunt.

Occipital. Pertaining to the occiput.

Occipital plates. Plates on the head of a serpent, behind the vertical plate.

Occiput. Back of the head.

Ocellate. With eye-like spots, generally roundish and with a lighter border.

Oid (suffix). Like, — as *Percoid*, perch-like.

Opercle, or *operculum.* Gill cover; the posterior membrane bone of the side of the head, in fishes.

Opercular bones. Membrane bones of the side of the head, in fishes.

Opercular flap. Prolongation of the upper posterior angle of the opercle, in Sun-fishes.

Opisthocœlian. Concave behind only; said of vertebræ which connect by ball and socket joints.

Orbicular. Nearly circular.

Orbit. Eye socket.

Oscine. Musical.

Osseous. Bony.

Ossicula auditus. Bones of the ear in fishes.

Osteology. Study of bones.

Oviparous. Producing eggs which are developed after exclusion from the body, as in all birds.

Ovoviviparous. Producing eggs which are hatched before exclusion, as in the Dog-fish and Garter Snake.

Ovum. Egg.

Palate. The roof of the mouth.

Palatines. Membrane bones of the roof of mouth; one on each side extending outward and backward from the vomer.

Palmate. Web-footed; having the anterior toes full-webbed.

Palustrine. Living in swamps.

Papilla. A small, fleshy projection.

Papillose. Covered with papillæ.

Paragnathous. Having the two mandibles about equal in length.

Parasphenoid. Bone of roof of mouth behind the vomer.

Parotoid. A glandular body behind the ear, in Batrachians.

Parietal. Bone of the side of head above.

Pectinate. Having teeth like a comb.

Pectoral. Pertaining to the breast.

Pectoral fins. The anterior or uppermost of the paired fins, in fishes, corresponding to the anterior limbs of the higher Vertebrates.

Pelage. The hair of a Mammal, taken collectively.

Pelagic. Living on or in the high seas.

Pelvis. The bones to which the hinder limbs (ventral fins in fishes) are attached.

Perforate. Pierced through; said of nostrils when without a septum.

Perissodactylous. Odd-toed (toes 1, 3, or 5).

Peritoneum. The membrane lining the abdominal cavity.

Phalanges. Bones of the fingers and toes.

Pharyngeal bones. Bones behind the gills and at the beginning of the œsophagus of fishes, of various forms, almost always provided with teeth; usually one pair below and four pairs above. They represent a fifth gill-arch.

Pharyngognathous. Having the lower pharyngeal bones united.

Physoclistous. Having the air-bladder closed.

Physostomous. Having the air-bladder connected by a tube with the alimentary canal.

Pigment. Coloring matter.

Pineal body. A small ganglion in the brain; a rudiment of an optic lobe, which in certain lizards (and in extinct forms) is connected with a third or median eye.

Pituitary body. A small ganglion in the brain.

Planta. Sole of foot.

Plantigrade. Walking on the sole of the foot, as do men and bears.

Plastron. Lower shell of a turtle.

Plicate. Folded; showing transverse folds or wrinkles.

Plumage. The feathers of a bird, taken collectively.

Plumbeous. Lead-colored, — dull bluish gray.

Poller. Thumb; in birds, the digit which bears the alula, — corresponding to the index finger.

Polygamous. Mating with more than one female.

Post-frontal (plates). The ones before the vertical plate.

Post-orbital. Behind the eye.

Post-temporal. The bone, in fishes, by which the shoulder girdle is suspended to the cranium.

Præcoces. Birds able to run about and feed themselves at once when hatched.

Præcocial. Having the nature of *Præcoces.*

Præcoracoid. A portion of coracoid more or less separated from the rest.

Præcoracoid arch. An arch in front of the coracoid in most soft-rayed fishes.

Prefrontal (plates). Those in front of post-frontal.

Premaxillaries. The bones, one on either side, forming the front of the upper jaw in fishes. They are usually larger than the maxillaries and commonly bear most of the upper teeth.

Premolars. The small grinders; the teeth between the canines and the true molars.

Preocular. Before the eye.

Preopercle. The membrane bone lying in front of the opercle and more or less nearly parallel with it.

Preorbital. The large membrane bone before the eye in fishes.

Primary. Any one of the ten (or nine) of the large, stiff quills growing upon the pinion or hand-bones of a bird; as distinguished from the secondaries, which grow upon the fore-arm.

Primary wing coverts. The coverts overlying the bases of the primaries.

Procœlian. Concave in front only.

Procurrent (fin). With the lower rays inserted progressively farther forward.

Projectile. Capable of being thrust forward.

Protractile. Capable of being drawn forward.

Proximal. Nearest.

Pseudobranchiæ. Small gills developed on the inner side of the opercle, near its junction with the preopercle.

Pterygoids. Bones of roof of mouth in fishes, behind the palatines.

Pubis. Anterior lower part of pelvis.

Pulmonary. Pertaining to the lungs.

Punctate. Dotted with points.

Pyloric cœca. Glandular appendages in the form of blind sacs opening into the alimentary canal of most fishes at the *pylorus* or passage from the stomach to the intestine.

Quadrate. Nearly square ; a bone of the lower jaw in lower vertebrates.

Quill. One of the stiff feathers of the wing or tail of a bird.

Quincunx. Set of five arranged alternately, thus * * *

Radius. Outer bone of fore-arm.

Ray. One of the cartilaginous rods which support the membrane of the fin of a fish.

Rectrices. Quills of the tail of a bird.

Recurved. Curved upward.

Remiges. Quills of the wing of a bird.

Reticulate. Marked with a network of lines.

Retractile. Susceptible of being drawn inward, as a cat's claw.

Retrorse. Turned backward.

Rhachis. Shaft of a quill.

Rictal. Pertaining to the rictus, as rictal bristles.

Rictus. Gape of the mouth.

Rostral. Pertaining to the snout, as rostral plate.

Rudimentary. Undeveloped.

Ruff. A series of modified feathers.

Rugose. Rough with wrinkles.

Sacral. Pertaining to the *sacrum*, or vertebræ of the pelvic region.

Saurognathous. Having the peculiar ("lizard-like") structure of the palate found in Woodpeckers.

Scansorial. Capable of climbing.

Scansorial tail. Tail feathers sharp and stiff, as in the scansorial birds (Woodpeckers).

Scapula. Shoulder blade ; in fishes, the bone of the shoulder girdle below the post-temporal.

Scapular arch. Shoulder girdle.

Schizognathous. Split palate, as in the Heron and similar birds.

Scute. Any external bony or horny plate.

Scutellate. Provided with scutella ; said of the tarsus when covered with broad plates in a regular vertical series, and separated by regular lines of impression.

Scutellum. One of the tarsal plates or scutella.

Secondaries. The quills growing on the fore arm.

Second dorsal. The posterior or soft part of the dorsal fin, when the two parts are separated.

Sectorial tooth. One of the premolars of carnivora, adapted for cutting.

Semipalmate. Half-webbed ; having the anterior toes more or less connected at base by a webbing which does not extend to the claws.

Septum. A thin partition.

Serrate. Notched, like a saw.

Sessile. Without a stem or peduncle.

Setaceous. Bristly.

Setiform. Bristle-like.

Shaft. Stiff axis of a quill.

Shoulder girdle. The bony girdle posterior to the head, to which the anterior limbs are attached (post-temporal; scapula, and coracoid or clavicle).

Soft dorsal. The posterior part of the dorsal fin in fishes, when composed of soft rays.

Soft rays. Fin-rays which are articulate and usually branched.

Spatulate. Shaped like a spatula.

Sphenoid. Basal bone of skull.

Spine. Any sharp projecting point; in fishes those fin-rays which are unbranched, inarticulate, and usually, but not always, more or less stiffened.

Spinous. Stiff or composed of spines.

Spinous dorsal. The anterior part of the dorsal fin when composed of spinous rays.

Spiracles. Openings in the head or neck of some fishes and Batrachians.

Spurious. Said of the first primary when less than about one-third the length of the second. (The student will notice that in *Oscines* the *presence* of a short or spurious quill indicates *ten* primaries; its *absence, nine.*)

Stellate. Star-like; with radiating ridges.

Sternal fontanelle. A pit at the top of the sternum.

Sternum. The breast bone.

Striate. Striped or streaked.

Sub (in composition). Less than; somewhat; not quite; under, etc.

Sub-caudal. Under the tail.

Sub-opercle. The bone immediately below the opercle (the suture connecting the two often hidden by scales).

Sub-orbital. Below the eye.

Sub-orbital stay. A bone extending from one of the sub-orbital bones in certain fishes, across the cheeks, to or towards the preopercle.

Subulate. Awl-shaped.

Suffrago. Heel joint; joint of tibia and tarsus.

Superciliary. Pertaining to the region of the eyebrow.

Supplemental maxillary. A small bone lying along upper edge of the maxillary.

Supraoccipital. The bone at posterior part of skull in fishes, usually with a raised crest above.

Supra-oral. Above the mouth.

Supra-orbital. Above the eye.

Supra-scapula. The post-temporal or bone by which the shoulder girdle in fishes is joined to the skull.

Suspensory bones. Bones by which the lower jaw, in fishes, is fastened to the skull.

Symphysis. Point of junction of the two parts of lower jaw; tip of chin.

Symplectic. The bone in fishes that keys together the hyomandibular and quadrate posteriorly.

Syndactyle. Having two toes immovably united for some distance, — as in the Kingfisher.

Synonym. A different word having the same or a similar meaning.

Synonomy. A collection of different names for the same group, species, or thing; " a burden and a disgrace to science." (*Cones.*)

Tail. In mammals, the vertebræ, etc., posterior to the sacrum ; in birds, the tail-feathers or rectrices, taken collectively ; in serpents, the part of the body posterior to the vent ; in fishes (usually), the part of the body posterior to the anal fin. (Often used more or less vaguely.)

Tail coverts. The small feathers overlapping the bases of the rectrices.

Tarso-metatarsus. The correct name for the so-called tarsus of birds ; the bone reaching from the tibia to the toes, composed chiefly of the metatarsus, but having at its top one of the small tarsal bones confluent with it.

Tarsus. The ankle-bones collectively ; in birds, commonly used for the shank-bone, lying between the tibia and the toes, the *tarso-metatarsus.*

Tectrices. The wing and tail coverts.

Temporal. Pertaining to the region of the temples.

Tenuirostral. Slender-billed.

Terete. Cylindrical and tapering.

Terminal. At the end.

Tertials. The quills attached to the humerus.

Tessellated. Marked with little checks or squares, like mosaic work.

Thoracic. Pertaining to the chest ; ventral fins are thoracic when attached immediately below the pectorals, as in the perch, the pelvic bones being fastened to the shoulder girdle.

Tibia. Shin-bone ; inner bone of leg between knee and heel.

Tomium. Cutting edge of the bill.

Totipalmate. Having all *four* toes connected by webbing.

Tragus. The inner lobe of the ear; the lobe opposite the auricle.

Transverse. Crosswise.

Trenchant. Compressed to a sharp edge.

Truncate. Abrupt, as if cut squarely off.

Tubercle. A small excrescence, like a pimple.

Tympanum. Drum of the ear; external in some Batrachia, etc.

Typical. Of a structure the most usual in a given group.

Ulna. The inner or posterior bone of the fore-arm.

Unguiculate. Provided with claws.

Ungulate. Provided with hoofs.

Unicolor. Of a single color.

Ultimate. Last or farthest.

Urosteges. The plates underneath the tail of a serpent.

Vent. The external opening of the alimentary canal.

Ventral. Pertaining to the abdomen.

Ventral fins. The paired fins behind or below the pectoral fins in fishes, corresponding to the posterior limbs in the higher vertebrates.

Ventral plates. In serpents, the row of plates along the belly between throat and vent.

Ventricle. One of the thick-walled chambers of the heart.

Versatile. Capable of being turned either way.

Vertebra. One of the bones of the spinal column.

Vertical. Up and down.

Vertical fins. The fins on the median line of the body; the dorsal, anal, and caudal fins.

Vertical plate. Central plate on the head of a serpent.

Villiform. Said of the teeth of fishes when slender and crowded into velvety bands.

Viscous. Slimy.

Viviparous. Bringing forth living young.

Vomer. In fishes, the front part of the roof of the mouth; a bone lying immediately behind the premaxillaries.

Web. The vane of a feather, on either side of the rhachis or " stem;" also, the membrane connecting the toes.

Xiphisternum. Tip of the sternum.

Zygodactyle. Yoke-toed; having the toes in pairs, — two in front, two behind.

Zygoma. The malar or cheek bone.

EXPLANATION OF SIGNS AND ABBREVIATIONS.

I. FISHES.

L. = Total length in inches of a well-grown example.

D. = Dorsal fin.

2d D. = Second dorsal fin.

P. = Pectoral fins.

V. = Ventral fins.

A. = Anal fin.

C. = Caudal fin.

B. = Branchiostegals.

Vert. = Vertebræ. The number is usually divided into abdominal and caudal vertebræ; the latter having the hæmapophyses united, forming hæmal spines. Thus Vert. 10 + 14, the usual number in typical fishes, means 10 abdominal and 14 caudal vertebræ.

♂ = Male.

♀ = Female.

Roman numerals used with abbreviations for the fins indicate the number of *spines* or *inarticulate* rays in a fin. *Arabic* numerals indicate the number of *soft rays.* In a fin containing both spines and soft rays, a *comma* (,) separating the numerals indicates that the two kinds of rays are *continuous,* or more or less connected. A *dash* (—) indicates

their *separation*. Thus, "D. X, 12," describes a single dorsal fin with
10 spines and 12 soft rays; "D. X — 12," indicates two dorsal fins —
the first of 10 spines, the second of 12 soft rays; "D. X — 1, 12,"
would indicate the presence of a single spine in the second dorsal.

The posterior soft ray of the dorsal and anal fins is usually split to
the base. It should be counted as *one* ray and not as *two*.

" *Gill rakers* 5 + 15," indicates 5 above and 15 below angle of gill-
arch; rudiments not counted. When the number above the angle is
uncertain or non-essential, it is indicated as "x."

Lat. l. = Lateral line, *i. e.*, the number of scales contained in its course.
When the lateral line is obsolete, "lat. l." signifies the number
of scales in a row from the head to the base of the caudal fin.
Thus, "lat. l. 36" means that there are 36 scales in a row along
the sides from the head to the caudal.

" *Scales* 5–36–10" indicates the presence of 36 scales in the lateral
line itself; 5 scales in a vertical series between front of dorsal and
lateral line, and 10 scales between lateral line and vent.

In all cases the number of rays or scales, as given in the descriptions, is intended to represent a fair average, and a variation of one-
sixth, or even more, in either direction need not surprise the student.
Generally the spines and scales are more constant in their numbers
than the soft rays, and the fewer of either, the less variable.

Length, as used in proportionate measurements, is distance along the
side from tip of snout to end of last vertebra. It *does not include* the
caudal fin.

Depth in length. = The greatest depth of the body as contained in the
distance along the side from the snout to the base of the caudal.

Head in length. = The distance from the snout along the cheeks to the
extremity of the opercle, as contained in the distance from the
snout to the base of the caudal.

Eye in head. = Its longitudinal diameter as contained in the length of
the side of the head.

As above stated, these measurements, as given in the descriptions,
are *intended* to be the *average* of *living adults*, and must be applied to
young specimens or preserved ones with caution.

Young fishes are usually but not always more elongate than adults,
and the eye is proportionally much larger.

A fin is said to be "*long*" when it has a long base, or is many-rayed.
A "*high*" fin is one in which the individual rays are elongated.

II. Reptiles.

L. = Length in inches of an adult example, from tip of snout to tip of
tail.

Sc. or *Scales.* = Number of longitudinal rows of scales exclusive of the
ventral series.

V. P. — Number of ventral plates, or gastrosteges, counted along the belly, from the throat to the vent. The figures given in the descriptions are intended to be *average*, the actual number being somewhat variable.

S. C. P. = Number of pairs of sub-caudal plates, or urosteges, counted from the vent to the tip of the tail.

III. BIRDS.

L. = Length in inches (along back from tip of bill to end of longest tail feather); thus, " L. 7¼ " means, length 7¼ inches.

E. = Extent (spread of wing) measured in inches.

W. = Length of wing (from bend of closed wing — carpal joint — to tip of longest feather) in inches.

T. = Length of tail in inches (*i. e., actual* length of the longest tail feather).

B. = Length of bill in inches (measured along middle line of culmen to tip of bill).

Hd. = Length of head in inches (measured with dividers from base of bill to nape.)

Ts. = Length of Tarsus in inches (measured in front).

Tel. = Length of middle toe with its claw.

The measurements given in the descriptions are understood to represent a fair *average adult male;* a variation of one-sixth, or more, in *absolute* length is nothing unusual; *relative* lengths, as of wings and tail, are much more constant. To save space I have usually preferred to say " L. 6," to saying " L. 5½ to 6½."

♂ = Male.
♀ = Female.
Yg. = Young.
> = More than, longer than, or more than equivalent to.
< = Less than, in its various senses.
= = Sign of equivalence.

The toes are numbered 1, 2, 3, 4; 1 being the hind toe, or hallux; 2 the inner anterior toe; 3 the middle toe; and 4 the outer toe.

IV. MAMMALS.

L. = Length in inches from tip of snout to tip of last vertebra of tail.

T. = Length of tail in inches (exclusive of hairs).

i. = Incisor teeth.

c. — Canines.

pm. — Premolars.

m. = Molars.

Thus, " i. $\frac{2}{7} : \frac{2}{1}$ " indicates two incisor teeth on each side in the upper jaw, and one on each side in the lower.

" Toes 5–4 " implies fore feet five-toed, hind feet four-toed.

NOTE. — As authority for names of species in this work, the original describer of the species is alone given. The name is written in full except in case of Linnæus, abbreviated as " L."

In case the original combination of general and specific name is still retained, the name of the author is printed without parentheses. In case, however, the original describer placed the species in question in a genus different from the one here adopted, the author's name is enclosed in parentheses.

Thus (page 277), " *Corvus corax* L." means that Linnæus placed his species *corax* in *Corvus*, where it still remains.

" *Melanerpes erythrocephalus* (L.)," indicates that the species (*Picus erythrocephalus* of Linnæus) is now placed in a genus different from the one in which it originally stood. *Melanerpes* is a modern subdivision of *Picus*, which formerly included all Woodpeckers.

" *Eu.*" indicates that the species in question is also found in Europe.

ADDITIONAL NOTE ON UPSILONPHORUS (page 156). — Two species of this genus are found on our coast : 443. **U. y-græcum**, described in the text, and 443 (b). **U. guttatus** (Abbott), with the white spots less distinct, and the Y on top of head short and broad, its basal part about as broad as long. In *U. y-græcum* the Y is slender, its basal part long. *U. guttatus* is the more common northward.

INDEX.

INCLUDING COMMON NAMES AND NAMES OF GENERA, SUBGENERA, FAMILIES, AND HIGHER GROUPS.

Abastor, 191.
Acantharchus, 115.
Acanthis, 284.
Acanthopteri, 26, 101.
Accipiter, 257.
Achirus, 168.
Acipenser, 34.
Acipenseridæ, 33.
Acris, 183.
Actitis, 248.
Actochelidon, 221.
Actodromas, 246.
Æchmophorus, 215.
Ægialitis, 249.
Æsalon, 260.
Æstrelata, 224.
Aëtobatidæ, 22.
Aëtobatis, 23.
Agaphelus, 334.
Agelaius, 279.
Agkistrodon, 199.
Agonidæ, 151.
Ailurichthys, 38.
Aix, 231.
Ajaja, 236.
Alaudidæ, 275.
Albacora, 106.
Albacore, 106.
Albatross, 223.
Albula, 70.
Albulidæ, 70.
Alburnops, 57.
Alca, 217.
Alce, 336.
Alcedinidæ, 265.
Alcidæ, 216.

Alectis, 109.
Alewife, 72.
Alle, 218.
Alligator, 187.
Alligator-fish, 151.
Alligator Gar, 35.
All-mouth, 172.
Allosomus, 79.
Alopias, 17.
Alopiidæ, 17.
Alosa, 73.
Alvordius, 126.
Amber-fish, 110.
Ambloplites, 115.
Amblyopsidæ, 82.
Amblyopsis, 83.
Amblystoma, 177.
Amblystomatidæ, 177.
Ameiurus, 39.
Amia, 37.
Amiatus, 37.
Amiidæ, 36.
Ammocœtes, 10.
Ammocrypta, 122.
Ammodramus, 286.
Ammodytes, 101.
Ammodytidæ, 101.
Ampelidæ, 293.
Ampelis, 293.
Amphioxus, 8.
Amphiuma, 176.
Amphiumidæ, 176.
Amyda, 206.
Anacanthini, 161.
Anarrhichadidæ, 159, 160.
Anarrhichas, 160.

Anas, 230.
Anatidæ, 227.
Anchovy, 74.
Ancylocheilus, 247.
Ancylopsetta, 166.
Angel-fish, 19, 146.
Angler, 172.
Anguidæ, 201.
Anguilla, 90.
Anguillidæ, 89.
Anhinga, 225.
Anhingidæ, 225.
Anolis, 202.
Anorthura, 308.
Anser, 234.
Anseres, 227.
Antelope, 337.
Antennariidæ, 172.
Anthropoids, 342.
Anthus, 306.
Antilocapra, 337.
Antrostomus, 269.
Apeltes, 98.
Aphoristia, 168.
Aphredoderidæ, 112.
Aphredoderus, 113.
Aphrizidæ, 250.
Aplodinotus, 144.
Apodes, 26, 89.
Apomotis, 117.
Aprionodon, 14.
Aquila, 259.
Archibuteo, 259.
Archosargus, 140.
Arcifera, 182.
Arctomys, 323.

Ardea, 238.
Ardeidæ, 237.
Ardetta, 238
Arenaria, 250.
Argentinidæ, 76.
Argyrosomus, 73.
Ariopsis, 39.
Aristonetta, 232.
Arius, 39.
Aromochelys, 207.
Arquatella, 246.
Artediellus, 150.
Artiodactyli, 335.
Arvicola, 320.
Ascidians, 5.
Asinus, 335.
Asio, 262.
Aspidonectes, 206.
Aspidophoroides, 151.
Ass, 335.
Astragalinus, 285.
Astroscopus, 156.
Astur, 258.
Asturina, 259.
Atalapha, 328.
Atherinidæ, 99.
Athlennes, 92.
Auk, 216, 217, 218.
Auxis, 106.
Aves, 212.
Avocet, 243, 244.
Aythya, 231.

Bachelor, 115.
Badger, 340.
Bairdiella, 143.
Balæna, 334.
Balænidæ, 333.
Balænoptera, 334.
Balanoglossus, 7.
Bald Eagle, 259.
Baldpate, 231.
Balistes, 169.
Balistidæ, 168.
Baltimore Oriole, 279.
Bank Swallow, 293.
Barb, 144.
Bar-fish, 115.
Barnacle Goose, 235.
Barn-door Skate, 21.
Barn Owl, 261.
Barn Swallow, 292.
Barracuda, 100.

Barred Owl, 262.
Barren Ground Caribou, 336.
Bartramia, 248.
Bascanion, 195.
Bashaw, 41.
Basking Shark, 19.
Bass, 120, 137.
Bat, 327, 323.
Bat-fish, 172.
Batrachia, 174.
Batrachidæ, 154.
Batrachus, 154.
Bay-breasted Warbler, 302.
Bay Porpoise, 331.
Bead Snake, 198.
Bear, 339.
Beaver, 322.
Bee-Martin, 274.
Bellows-fish, 172.
Beluga, 331.
Bergall, 146.
Big-horn, 337.
Bill-fish, 92, 104.
Birds, 212.
Birds of Prey, 255.
Bishop Ray, 23.
Bison, 337.
Bittern, 237, 238.
Black and White Creeper, 298.
Black and Yellow Warbler, 302.
Black-backed Gull, 220.
Black Bass, 120.
Blackbird, 279, 280.
Black-capped Warbler, 305.
Black Cat, 341.
Black Dolphin, 330.
Black Duck, 230.
Black-fish, 137, 146, 331.
Black Fox, 342.
Black Hawk, 258, 259.
Black-headed Gull, 220.
Black-Horse, 46.
Black Moccasin, 199.
Black-nosed Dace, 63.
Black-poll Warbler, 302.
Black Rat, 322.
Black Sea Bass, 137.
Black Snake, 195.
Black Squirrel, 323.

Black - throated Blue Warbler, 301.
Black-throated Bunting, 290.
Black - throated Green Warbler, 303.
Black Whale, 334.
Blarina, 326.
Blenniidæ, 157.
Blenny, 157.
Blind-fish, 83.
Blob, 149.
Blowing Viper, 197.
Blue-back Trout, 81.
Blue-bill, 232.
Blue-Bird, 313.
Blue-fin, 78.
Blue-fish, 111.
Blue-gills (*Lepomis pallidus*), 118.
Blue Goose, 234.
Blue Grosbeak, 290.
Blue-headed Vireo, 296.
Blue Heron, 238.
Blue Jay, 277.
Blue Perch, 146.
Blue-Racer, 195.
Blue Shark, 17.
Blue-Stocking, 244.
Blue Sun-fish, 118.
Blue-tailed Lizard, 201.
Blue-winged Teal, 230.
Blue - winged Warbler, 299.
Blue Yellow-backed Warbler, 299.
Bobolink, 278.
Bob-White, 252.
Bohemian Waxwing, 293.
Boleichthys, 134.
Boleosoma, 123.
Bonasa, 252.
Bone-fish, 70.
Bonito, 106.
Bonnet-head Shark, 16.
Borer, 9.
Bos, 336.
Botaurus, 237.
Bottle-nosed Dolphin, 330.
Bottle-nosed Whale, 332.
Bovidæ, 336.
Bow-fin, 36.
Bowhead, 334.
Box Turtle, 207, 210.

Brachyotus, 262.
Branch Herring, 72.
Branchiostoma, 8.
Branchiostomatidæ, 8.
Branta, 235.
Brant Goose, 235.
Bream, 68, 119.
Brevoortia, 73.
Brook Trout, 80.
Brosmius, 162.
Brown Bat, 328.
Brown Bear, 339.
Brown Creeper, 309.
Brown Lark, 306.
Brown Thrush, 307.
Brown Trout, 79.
Bubo, 263.
Bubonidæ, 261.
Buffalo, 337.
Buffalo-fish, 44.
Buffle-head, 233.
Bufo, 182.
Bufonidæ, 182.
Bug-fish, 73.
Bull Bat, 270.
Bull-Frog, 186.
Bullhead, 40.
Bullhead Minnow, 54.
Bull Snake, 196.
Bull-Trout, 80.
Bumper, 110.
Bunting, 286.
Burbot, 162.
Burgomaster, 220.
Burr-fish, 170.
Burrowing Owl, 263.
Butcher Bird, 294.
Buteo, 258.
Butorides, 239.
Butter-Ball, 233.
Butter-fish, 112, 158.
Butterfly Ray, 22.
Buzzard, 258.

CACHALOT, 333.
Cærulean Warbler, 302.
Calamospiza, 291.
Calcarius, 285.
Calico Bass, 115.
Calidris, 247.
Calomys, 321.
Campephilus, 267.
Campostoma, 52.

Camptolaimus, 233.
Canada Warbler, 305.
Canary, 285.
Canidæ, 341.
Canis, 342.
Canvas-back Duck, 232.
Capelin, 76.
Cape May Warbler, 301.
Capra, 336.
Caprimulgidæ, 269.
Carangidæ, 107.
Caranx, 108.
Carassius, 50.
Carcharhinus, 17.
Carcharias, 18.
Carchariidæ, 18.
Carcharodon, 19.
Cardinal Grosbeak, 289.
Cardinalis, 289.
Carduelis, 285.
Cariacus, 336.
Caribou, 336.
Carinatæ, 212, 213.
Carnivora, 337.
Carolina Wren, 308.
Carp, 50.
Carphophiops, 191.
Carpiodes, 45.
Carpodacus, 284.
Carp Sucker, 45.
Carrion Crow, 256.
Castor, 322.
Castoridæ, 322.
Cat, 342, 343.
Cataphracti, 146.
Cat-Bird, 307.
Cat-fish, 38, 39.
Catharista, 256.
Cathartes, 256.
Cathartidæ, 255.
Catostomidæ, 43.
Catostomus, 46.
Caucasian, 344.
Cavalla, 109.
Cave Fish, 82.
Cave Salamander, 180.
Cavia, 318.
Cecomorphæ, 214.
Cedar Bird, 293.
Cemophora, 197.
Centrarchidæ, 113.
Centrarchus, 114.
Centrocyllium, 14.
Centropristis, 137.

Centroscymnus, 14.
Centurus, 268.
Ceophlœus, 268.
Cephalacanthidæ, 151.
Cephalacanthus, 152.
Cephalochordata, 6.
Cepphus, 217.
Ceratichthys, 63.
Cerna, 138.
Certhia, 309.
Certhiidæ, 309.
Cervidæ, 335.
Cervus, 336.
Ceryle, 266.
Cestreus, 142.
Cetaceans, 329.
Cete, 321.
Cetorhinidæ, 19.
Cetorhinus, 19.
Chænobryttus, 115.
Chætodipterus, 146.
Chætura, 270.
Chain Snake, 196.
Chameleon, 202.
Channel Bass, 143.
Channel Cat, 39, 40.
Characinidæ, 43.
Charadriidæ, 249.
Charadrius, 249.
Charitonetta, 233.
Charr, 80.
Chasmodes, 158.
Chat, 305.
Chatterer, 293.
Chaulelasmus, 230.
Chelidon, 292.
Chelonia, 205.
Cheloniidæ, 205.
Chelopus, 210.
Chelydra, 207.
Chelydridæ, 206.
Chen, 234.
Chenomorphæ, 227.
Cherry Bird, 293.
Chestnut-sided Warbler, 302.
Chewink, 289.
Chickadee, 310.
Chickaree, 323.
Chicken Hawk, 258, 259.
Chicken-Snake, 195.
Chilomycterus, 170.
Chimæra, 24.
Chimæridæ, 24.

Chimney Swallow, 270.
Chipmunk, 323.
Chipping Sparrow, 288.
Chippy, 288.
Chiroptera, 327.
Chloroscombrus, 110.
Chogset, 146.
Chologaster, 83.
Chondestes, 287.
Chondrostei, 25.
Chondrotus, 179.
Chordata, 6.
Chordeiles, 270.
Chorophilus, 183.
Chrœcocephalus, 220.
Chrosomus, 53.
Chrysemys, 210.
Chub, 66, 140.
Chub-Mackerel. 107.
Chub-Sucker, 46.
Chuckle-headed Cat, 39.
Chuck-will's Widow, 269.
Ciconiidæ, 237.
Cigar-fish, 108.
Cinclus, 311.
Circus, 257.
Cirrostomi, 8.
Cisco, 78.
Cistothorus, 308.
Cistudo, 210.
Citharichthys, 167.
Clamatores, 271.
Clam-Cracker, 22.
Clangula, 233.
Clapper Rail, 241.
Cliff Swallow, 292.
Cling-fish, 155.
Clinostomus, 67.
Cliola, 54.
Clivicola, 293.
Clupea, 72.
Clupeidæ, 71.
Cnemidophorus, 201.
Coal-fish, 163.
Cobia, 103.
Coccothraustes, 283.
Coccyges, 265.
Coccyzus, 265.
Cod-fish, 161, 163.
Codling, 162.
Colaptes, 268.
Colinus, 252.
Coluber, 194.
Colubridæ, 188.

Columba, 254.
Columbæ, 253.
Columbidæ, 254.
Columbigallina, 254.
Colymbus, 215.
Compsothlypis, 299.
Condylura, 327.
Conger, 91.
Conger Eel, 91.
Congo Snake, 176.
Connecticut Warbler, 304.
Contopus, 274.
Conurus, 264.
Coot, 242.
Copperhead, 199.
Coregonus, 77.
Cormorant, 225, 226.
Corn-Snake, 194, 196.
Corvidæ, 276.
Corvina, 143.
Corvus, 277.
Corynorhinus, 329.
Coryphæna, 112.
Coryphænidæ, 112.
Cottidæ, 147.
Cottogaster, 125.
Cotton-mouth, 199.
Cotton-tail, 317.
Cottus, 148.
Coturnicops, 241.
Coturniculus, 281.
Coturnix, 252.
Couesius, 65.
Cougar, 343.
Cow-Bird, 278.
Cow-fish, 332.
Cow-nosed Ray, 23.
Coyote, 342.
Crab-eater, 103.
Craig-fluke, 168.
Crake, 241.
Cramp-fish, 22.
Crane, 240.
Craniota, 6.
Crappie, 115.
Crawl-a-Bottom, 46, 126.
Creciscus, 241.
Creek Chub, 66.
Creek-fish, 46.
Creeper, 300.
Crevallé, 109.
Cricket Frog, 183.
Cristivomer, 80.
Croaker, 143.

Crocodilia, 187, 211.
Crocodilus, 187.
Crossbill, 284.
Cross Fox, 342.
Crotalidæ, 198.
Crotalus, 199.
Crow, 277.
Crow Blackbird, 280.
Crymophilus, 243.
Cryptacanthodes, 159.
Cryptacanthodidæ, 159.
Cryptobranchidæ, 176.
Cryptobranchus, 177.
Crystallaria, 123.
Ctenolabrus, 146.
Cuckoo, 265.
Cuculidæ, 265.
Cunner, 146.
Curlew, 248, 249.
Cusk, 162.
Cutlass-fish, 104.
Cut-lips, 49, 54.
Cutwater, 222.
Cyanocitta, 277.
Cycleptus, 46.
Cyclophis, 195.
Cyclopteridæ, 154.
Cyclopterus, 154.
Cyclostomi, 9.
Cynomyonax, 341.
Cynomys, 323.
Cynoperca, 135.
Cynoscion, 142.
Cyprinella, 58.
Cyprinidæ, 49.
Cyprinodon, 84.
Cyprinodontidæ, 83.
Cyprinus, 50.
Cypselurus, 95.

DAB, 165, 167.
Dab-chick, 215.
Dace, 58.
Daddy Sculpin, 150.
Dafila, 231.
Darter (bird), 225.
Darter (fish), 121.
Dasyatidæ, 22.
Dasyatis, 22.
Day Owl, 263.
Dscactylus, 46.
Decapterus, 108.
Deer, 335, 336.

Delphinapterus, 331.
Delphinidæ, 330.
Delphinus, 331.
Dendragapus, 252.
Dendroica, 299.
Dermochelydidæ, 204.
Dermochelys, 204.
Desmognathidæ, 180.
Desmognathus, 180.
Diadophis, 196.
Diamond-back, 209.
Diapterus, 145.
Dichromanassa, 238.
Dick-sissel, 290.
Didelphia, 314.
Didelphididæ, 316.
Didelphis, 316.
Diedapper, 215.
Diemyctylus, 181.
Diodontidæ, 170.
Diomedeidæ, 223.
Dionda, 53.
Diplesion, 125.
Dipnoi, 12, 173.
Dipper, 233, 311.
Discocephali, 102.
Diver, 216.
Diving-birds, 214.
Dog, 341, 342.
Dog-fish, 15, 37, 87.
Dog-Shark, 16.
Dolichonyx, 278.
Dollar-fish, 112.
Dolly Varden Trout, 80.
Dolphin (fish), 112.
Dolphin (mammal), 330, 331.
Donzella, 160.
Dorosoma, 74.
Dough Bird, 249.
Dove, 253, 254.
Dovekie, 218.
Dowitcher, 245.
Drum, 141, 144.
Dryobates, 267.
Dublin Trout, 81.
Duck, 227, 229.
Duck-bill Cat, 33.
Duck Hawk, 260.
Dunlin, 247.
Dytes, 215.

EAGLE, 259.
Eagle Ray, 22, 23.

Echelidæ, 90.
Echelus, 90.
Echeneididæ, 102.
Echeneis, 102.
Echinorhinus, 14.
Ectopistes, 254.
Educabilia, 315.
Eel, 89, 90.
Eel-back Flounder, 167.
Eel-Pout, 160.
Eft, 181.
Egret, 238.
Eider Duck, 233.
Elacate, 103.
Elacatidæ, 103.
Elagatis, 111.
Elanoides, 257.
Elanus, 257.
Elapidæ, 198.
Elaps, 198.
Elassoma, 113.
Elassomatidæ, 113.
Electric Ray, 21.
Elk, 336.
Elopidæ, 70.
Elops, 70.
Empidonax, 275.
Emydidæ, 208.
Emys, 210.
Engraulis, 74.
Engystoma, 184.
Engystomatidæ, 184.
Enneacanthus, 116.
Enteropneusta, 6.
Epelasmia, 146.
Ephippidæ, 146.
Epinephelus, 138.
Equus, 335.
Eremophila, 276.
Erethizon, 318
Eretmochelys, 205.
Ereunetes, 247.
Ericosma, 128.
Ericymba, 62.
Erimyzon, 46.
Erionetta, 233.
Erismatura, 234.
Ermine, 341.
Esocidæ, 88.
Esox, 88.
Esquimaux Curlew, 249.
Etheostoma, 121, 131.
Etropus, 167.
Etrumeus, 71.

Eucalia, 97.
Eulamia, 17.
Euleptorhamphus, 93.
Eumeces, 201.
Eumesogrammus, 159.
Eumicrotremus, 154.
Eupomotis, 119.
Eurhipiduræ, 212.
Eutainia, 192.
Eutheria, 314.
Euthynnus, 106.
Evening Grosbeak, 283.
Eventognathi, 26, 42.
Evet, 181.
Evotomys, 320.
Exocœtidæ, 91.
Exocœtus, 93.
Exoglossum, 54.

FALCO, 260.
Falcon, 256, 260.
Falconidæ, 256.
Fall-fish, 59, 66.
Farancia, 191.
Felidæ, 342.
Felis, 343.
Feræ, 337.
Ferret, 341.
Fiber, 320.
Field-Mouse, 320.
Field Sparrow, 288.
File-fish, 169.
Fin-back Whale, 333,334.
Finch, 280.
Fine-scaled Sucker, 46.
Finner, 334.
Fire Bird, 279.
Firmisternia, 182.
Fish Crow, 277.
Fish-Duck, 228, 229.
Fishes, 12.
Fish-Hawk, 261.
Fishing-frog, 172.
Fissipedia, 337.
Fistulariidæ, 97.
Flannel - mouthed Cat, 39.
Flasher, 138.
Flat-fish, 163, 167.
Flat-head Cat, 41.
Flat-headed Chub, 65.
Flicker, 268.
Flocking Fowl, 232.

Florida, 238.
Flounder, 164, 166.
Flycatcher, 273, 274, 275.
Fly-catching Thrush,312.
Flying-fish, 93, 94.
Flying Gurnard, 152.
Flying Squirrel, 324.
Fool-fish, 169.
Fork-tailed Gull, 221.
Fox, 342.
Fox-Shark, 17.
Fox Snake, 195.
Fox Sparrow, 289.
Fox Squirrel, 324.
Fratercula, 217.
Fregata, 227.
Fregatidæ, 226.
Friar, 100.
Frigate Mackerel, 106.
Fringillidæ, 280.
Frog, 184, 185, 186.
Frog-fish, 172.
Frost Bird, 249.
Frost-fish, 76, 163.
Fulica, 242.
Fuligula, 232.
Fulmar, 223.
Fulmarus, 223.
Fundulus, 84.

GADIDÆ, 161.
Gadus, 163.
Gadwall, 230.
Gaff-Topsail, 38.
Gale, 341.
Galeichthys, 39.
Galeocerdo, 16.
Galeorhinidæ, 16.
Galeoscoptes, 307.
Galeus, 16.
Gallinæ, 251.
Gallinago, 245.
Gallinula, 241.
Gallinule, 241.
Gallus, 251.
Gambusia, 87.
Gannet, 225.
Ganoidei, 25, 32.
Gar-fish, 92.
Gar-pike, 35.
Garrot, 232.
Garter Snake, 192, 193.
Garzetta, 238.

Gaspereau, 72.
Gaspergou, 144.
Gasterosteidæ, 97.
Gasterosteus, 98.
Gavia, 219.
Gelochelidon, 221.
Gennaia, 260.
Geomyidæ, 318.
Geomys, 319.
Geothlypis, 304.
Gerres, 145.
Gerridæ, 144.
Giant Salamander, 176.
Ginglymodi, 25, 35.
Gizzard Shad, 74.
Glaniostomi, 25, 33.
Glass-Snake, 202.
Glaucionetta, 232.
Glires, 316.
Globicephalus, 331.
Glut Herring, 72.
Glutton, 340.
Glyptocephalus, 168.
Gnatcatcher, 311.
Gnawers, 316.
Goat, 336.
Goatsucker, 269.
Gobiesocidæ, 155.
Gobiesox, 155.
Gobiidæ, 156.
Gobiosoma, 157.
Gobius, 157.
Goby, 156.
Godwit, 247.
Goggle Eye, 115.
Goggler, 108.
Golden-crowned Thrush, 304.
Golden Eagle, 259.
Golden-Eye, 232.
Golden Robin, 279.
Golden Shiner, 68.
Golden-winged Warbler, 298.
Golden - winged Wood-pecker, 268.
Goldfinch, 285.
Gold-fish, 50.
Goody, 143.
Goosander, 229.
Goose, 234, 235.
Goose-fish. 172.
Gopher, 323.
Gopher Turtle, 211.

Goshawk, 258, 259.
Gourd-seed Sucker, 46.
Grackle, 280.
Grallæ, 242.
Grampus, 332.
Grand-Écaille, 71.
Graptemys, 208.
Grass-Bass, 115.
Grasshopper Sparrow, 286.
Grass-Snake, 195.
Grass-Sparrow, 286.
Gray-cheeked Thrush, 312.
Gray Fox, 342
Gray Gopher, 323.
Gray Hawk, 259.
Gray Jay, 277.
Grayling, 79.
Gray Owl, 262.
Gray Rabbit, 317.
Gray Snapper, 139.
Gray Squirrel, 323.
Gray Wolf, 342.
Great Auk, 218.
Grebe, 215.
Green Heron, 239.
Greenlet, 295.
Green Snake, 195.
Green Turtle, 205.
Green-winged Teal, 230.
Grind-Whale, 331.
Grizzly Bear, 339.
Gronias, 41.
Grosbeak, 283, 289, 290.
Ground-Bird, 286.
Ground Dove, 254.
Ground Hog, 323.
Ground Snake, 191.
Ground Squirrel, 323.
Grouper, 138.
Grouse, 251-253.
Grubby, 150.
Gruidæ, 240.
Grus, 240.
Guara, 236.
Guillemot, 217.
Guinea Hen, 251.
Guinea Pig, 318.
Guiraca, 290.
Gull, 219, 220.
Gull-billed Tern. 221.
Gulo, 340.
Gurnard, 152.

Gymnacanthus, 151.
Gymnosarda, 106.
Gyrfalcon, 260.
Gyrinophilus, 179.

HABIA, 290.
Haddock, 163.
Hadropterus, 128.
Hæmatopodidæ, 250.
Hæmatopus, 250.
Hag-fish, 9.
Hake, 162, 163.
Haldea, 192.
Halecomorphi, 26, 36.
Half-beak, 93.
Haliæëtus, 259.
Halibut, 165.
Halocypselus, 93.
Hammer-head, 46.
Hammer-headed Shark, 16.
Haplodoci, 154.
Haplomi, 26, 82.
Harbor Porpoise, 331.
Harbor Seal, 338.
Hard-tail, 109.
Hare, 317.
Hare-lip Sucker, 49.
Harlequin Duck, 233.
Harlequin Snake, 198.
Harporhynchus, 307.
Harvest-fish, 112.
Harvest Mouse, 321.
Hawk, 257-259.
Hawk Owl, 263.
Hawksbill Turtle, 205.
Head-fish, 171.
Heath Hen, 253.
Helinaia, 298.
Hellbender, 177.
Helminthophila, 298.
Helmitherus, 298.
Helodromas, 248.
Hemibranchii, 26, 96.
Hemichordata, 6.
Hemidactylium, 179.
Hemiramphus, 93.
Hemitremia, 55.
Hemitripterus, 148.
Hen, 251.
Hen Hawk, 258.
Hermit Thrush, 313.
Herodias, 238.

Herodiones, 235.
Heron, 237-239.
Herring, 71, 72.
Herring Gull, 220.
Hesperiphona, 283.
Hesperocichla, 313.
Hesperomys, 321.
Heterodon, 197.
Heterosomata, 26, 163.
Hiatula, 146.
Hickory Shad, 74.
Hierofalco, 260.
High-Holer, 268.
Himantopus, 244.
Hiodon, 69.
Hiodontidæ, 69.
Hippocampus, 96.
Hippoglossoides, 165.
Hippoglossus, 165.
Hirundinidæ, 291.
Histrionicus, 233.
Hog, 335.
Hog-choker, 168.
Hog-fish, 126.
Hog Molly, 46.
Hog-nosed Snake, 198.
Hog Sucker, 46.
Holbrookia, 203.
Holocephali, 24.
Holostei, 25.
Hominidæ, 344.
Homo, 344.
Hooded Warbler, 305.
Hoop Snake, 191.
Horned Dace, 66.
Horned Lark, 276.
Horned Owl, 263.
Horned Pout, 40.
Horned Toad, 203.
Horn-fish, 135.
Horn Snake, 191.
Horny-Head, 65.
Horse, 335.
Horse-fish, 109.
Horse-head, 110.
Horse-mackerel, 108.
Hound-fish, 92.
Hound Shark, 16.
House Mouse, 322.
House Snake, 197.
House Sparrow, 284.
House Wren, 308.
Hudsonius, 57.
Humming Bird, 270.

Hump-back Whale, 334.
Hybognathus, 53.
Hybopsis, 63.
Hyborhynchus, 54.
Hydranassa, 238.
Hydrochelidon, 222.
Hydrophlox, 59.
Hyla, 184.
Hylatomus, 268.
Hylidæ, 183.
Hylocichla, 312.
Hypentelium, 46.
Hyperoartia, 10.
Hyperoodon, 332.
Hyperotreta, 9.
Hypleurochilus, 158.
Hystricidæ, 318.

IBIDIDÆ, 236.
Ibis, 236.
Ice-fish, 76.
Ice Gull, 220.
Ichthyomyzon, 10.
Ichthyopsida, 6.
Ictalurus, 39.
Icteria, 305.
Icteridæ, 277.
Icterus, 279.
Ictidomys, 323.
Ictinia, 257.
Ictiobus, 44.
Iguanidæ, 202.
Innostoma, 125.
Indian Hen, 237.
Indigo-Bird, 290.
Ineducabilia, 314.
Insectivora, 324.
Ioa, 123.
Ionornis, 241.
Iridoprocne, 292.
Isesthes, 158.
Isogomphodon, 14.
Isospondyli, 26, 69.
Istiophoridæ, 103.
Istiophorus, 104.
Isuropsis, 18.
Isurus, 18.
Ivory-billed Woodpecker, 267.
Ivory Gull, 219.

JACK-CURLEW, 249.
Jack-Rabbit, 318.

Jack-Snipe, 246.
Jæger. 218, 219.
Jay. 277.
Jerker, 65.
John A. Grindle, 37.
Johnny, 124.
Johnny Darter, 121.
Joint-Snake. 202.
Jumping Mouse, 318.
Jump-rocks, 48.
Junco. 288.
Jurel, 109.

Kentucky Warbler, 304.
Kildeer, 249.
Killer, 331.
Killifish, 83-85.
King-bird. 274.
King Eider. 233.
King-fish. 105, 144.
Kingfisher, 265, 266.
Kinglet, 310, 311.
King Salmon, 79.
Kinosternidæ, 207.
Kinosternon, 207.
Kite, 257.
Kit Fox, 342.
Kittiwake Gull, 220.
Knot, 246.
Kogia, 333.
Kyphosus, 140.

Labidesthes, 100.
Labridæ, 145.
Lacertilia, 200.
Lady-fish, 70.
Lagenorhynchus, 331.
Lagocephalus, 170.
Lagochila, 49.
Lagodon, 140.
Lagopus, 252.
Lake Herring. 78.
Lake Sturgeon, 34.
Lake Trout, 80.
Lamellirostres, 227.
Lamna, 18.
Lamnidæ, 18.
Lamprey, 10, 11.
Lancelet, 8.
Land-locked Salmon, 80.
Land Tortoise, 211.
Lanier, 260.

Laniidæ, 294.
Lanius, 294.
Lanivireo, 296.
Lant, 101.
Lapland Longspur, 285.
Laridæ, 219.
Lark, 275.
Lark Bunting. 291.
Lark Sparrow, 287.
Larus. 220.
Laughing Gull, 220.
Lawyer, 162, 244.
Least Bittern, 238.
Leather Carp, 50.
Leather-Jacket, 108, 169.
Leather-Turtle, 204, 206.
Leiostomus, 143.
Leirus, 111.
Leopard Frog, 185.
Lepibema, 137.
Lepidosteus, 35.
Lepisosteidæ, 35.
Lepisosteus, 35.
Lepomis, 116.
Leporidæ, 317.
Leptoblennius, 159.
Leptocardii, 8.
Leptocephalus, 90.
Leptops, 41.
Lepus, 317.
Leucosticte, 284.
Limanda, 167.
Limicolæ, 242.
Limosa, 247.
Ling, 162.
Linnet. 284.
Liopeltis, 195.
Liopsetta, 167.
Liparididæ, 153.
Liparis, 153.
Lizard. 200, 203.
Lizard-fish. 75.
Lobotes, 138.
Lobotidæ, 138.
Log-cock, 268.
Loggerhead-Shrike. 294.
Loggerhead-Turtle, 205.
Log-Perch, 126.
Longe, 80.
Long-eared Bat, 323.
Longipennes, 218.
Long-nosed Dace, 63.
Long-shanks, 244.
Longspur, 285, 286.

Look-down, 110.
Loon, 216.
Lophiidæ, 172.
Lophius, 172.
Lophobranchii, 26, 95.
Lophodytes, 229.
Lophophanes, 310.
Lords-and-Ladies, 233.
Lota, 162.
Loxia, 284.
Lucania, 86.
Lump-fish. 154.
Lump Sucker, 154.
Luscinia, 310.
Lutjanus, 139.
Lutra, 340.
Lutreola, 341.
Luxilus, 58.
Lycodidæ, 160.
Lynx, 343.
Lythrurus, 60.

Mackerel, 104, 106.
Mackerel Shark, 18.
Mackinaw Trout, 80.
Macrochelys, 207.
Macrochires, 269.
Macrorhamphus, 245.
Mademoiselle, 143.
Mad Tom (*Noturus insignis*), 42.
Magpie, 276.
Malaclemmys, 208.
Mallard Duck, 230.
Mallotus, 76.
Malthe, 172.
Malthidæ, 171.
Mammalia, 314.
Mammals, 314.
Man, 344.
Man-eater Shark, 19.
Mangrove Snapper, 139.
Man-o'-War Bird, 226, 227.
Manta, 23.
Mantidæ, 23.
Map-Turtle, 208.
Mareca, 230.
Marlin, 247.
Marmot, 323.
Marsh Hare, 317.
Marsh Harrier, 257.
Marsh-hen, 241.
Marsh-Robin, 289.

Marsh-Wren, 308.
Marsupialia, 315.
Marten, 340.
Martin, 292.
Maryland Yellow-throat, 305.
Maskinongy, 89.
Massasauga, 199.
Mattowacca, 72.
May-fish, 84.
Meadowlark, 279.
Meadow Mouse, 320.
Megalestris, 218.
Megalops, 71.
Megaptera, 334.
Megascops, 263.
Melanerpes, 268.
Melanitta, 234.
Melanogrammus, 163.
Meleagris, 253.
Melospiza, 289.
Menhaden, 73.
Menidia, 100.
Menobranchus, 175.
Menomonee White-fish, 77.
Menticirrhus, 143.
Merlin, 260.
Mephitis, 340.
Merganser, 229.
Merluccius, 163.
Merula, 313.
Mesogonistius, 116.
Mesoplodon, 332.
Methriopterus, 307.
Michigan Grayling, 79.
Michigan Herring, 78.
Microgadus, 163.
Microgobius, 157.
Micropalama, 245.
Microperca, 134.
Micropodidæ, 270.
Micropogon, 143.
Micropterus, 120.
Microsorex, 325.
Milk Snake, 197.
Miller's Thumb, 148, 149.
Milvulus, 274.
Mimus, 307.
Miniellus, 56.
Mink, 341.
Minnilus, 55.
Minnow, 49, 55.
Minytrema, 47.

Mirror Carp, 50.
Mississippi Cat, 39.
Missouri Skylark, 306.
Missouri Sucker, 46.
Mniotilta, 298.
Mniotiltidæ, 296.
Moccasin, 194, 199.
Mocking-bird, 307.
Mocking Wren, 308.
Moharra, 144.
Mola, 171.
Mole, 326.
Molidæ, 171.
Molothrus, 278.
Monacanthus, 169.
Mongolian, 344.
Mongrel Buffalo, 44.
Mongrel White-fish, 78.
Moniana, 57.
Monkey, 343.
Monk-fish, 19.
Monodelphia, 314.
Monotremata, 314.
Moon-eye, 69, 78.
Moon-fish, 109, 110.
Moose, 336.
Morone, 137.
Mossbunker, 73.
Motacillidæ, 306.
Mother-of-Eels, 160.
Mourning Dove, 254.
Mourning Warbler, 304.
Mouse, 319.
Mouse-fish, 172.
Moxostoma, 47, 173.
Mud Cat, 41.
Mud Eel, 175.
Mud-fish, 37, 85.
Mud Hen, 242.
Mud Minnow, 87.
Mud Puppy, 175.
Mud Shad, 74.
Mud Sun-fish, 115.
Mud Turtle, 207, 210.
Muffle-jaw, 149.
Mugilidæ, 99.
Mule Deer, 336.
Mullet, 99.
Mullet Sucker, 47.
Mullidæ, 141.
Mullus, 141.
Mummichog, 85.
Murænoides, 158.
Muridæ, 319.

Murre, 217.
Mus, 319, 320.
Musculus, 320.
Muskallunge, 89.
Muskrat, 320.
Musk Turtle, 207.
Mustela, 340.
Mustelidæ, 339.
Mustelus, 16.
Mutton-fish, 160.
Myadestes, 312.
Myiarchus, 274.
Myliobatis, 23.
Mynomes, 321.
Myxine, 9.
Myxinidæ, 9.
Myzonts, 9.

Nanemys, 210.
Nanostoma, 129.
Nashville Warbler, 299.
Necturus, 175.
Needle-fish, 91, 92.
Negro, 344.
Nematognathi, 26, 37.
Neocorys, 306.
Neosorex, 325.
Neotoma, 321.
Nettion, 230.
Newt, 181.
Night Hawk, 270.
Night Heron, 239.
Nightingale, 310.
Night Jar, 269.
Nocomis, 65.
Nomonyx, 234.
Nonpareil, 290.
Notemigonus, 68.
Nothonotus, 130.
Notropis, 55.
Noturus, 41.
Numb-fish, 22.
Numenius, 248.
Numida, 251.
Nuthatch, 309, 310.
Nuttallornis, 274.
Nyctala, 262.
Nyctanassa, 239.
Nyctea, 263.
Nycticejus, 328.
Nycticorax, 239.

Oceanites, 224.
Oceanodroma, 224.

24

Ochetodon, 321.
Ochthodromus, 249.
Odontoglossæ, 227.
Oidemia, 234.
Old Squaw, 233.
Oligocephalus, 132.
Oligoplites, 108.
Oligosoma, 201.
Olive-backed Thrush, 312.
Olor, 235.
Oncorhynchus, 79.
Onychomys, 321.
Ophibolus, 196.
Ophidia, 187.
Ophidiidæ, 160.
Ophidion, 160.
Ophisaurus, 202.
Opisthonema, 73.
Oporornis, 304.
Opossum, 316.
Opsopœodus, 68.
Orange-crowned Warbler, 299.
Orange-throated Warbler, 303.
Orbidus, 170.
Orca, 331.
Orchard Oriole, 279.
Oregon Robin, 313.
Oriole, 279.
Orthopristis, 140.
Oryzomys, 321.
Osceola, 197.
Oscines, 271.
Osmerus, 76.
Osprey, 261.
Otocoris, 276.
Otter, 340.
Ouzel, 311.
Oven-bird, 304.
Ovis, 337.
Owl, 261, 262.
Ox, 336.
Ox-bird, 247.
Ox-eye, 249.
Oxyechus, 249.
Oxygeneum, 52.
Oyster-Catcher, 250.
Oyster-fish, 146, 155.

PADDLE-FISH, 33.
Painted Bunting, 290.

Painted Turtle, 210.
Paludicolæ, 239.
Pandion, 261.
Panther, 343.
Paralichthys, 166.
Parexocœtus, 93.
Paridæ, 309.
Paroquet, 264.
Parrot, 263.
Partridge, 252.
Parula, 299.
Parus, 310.
Passenger Pigeon, 254.
Passer, 284.
Passerculus, 286.
Passerella, 289.
Passeres, 271.
Passerina, 290.
Pavo, 251.
Pavoncella, 248.
Peabody Bird, 288.
Peacock, 251.
Pea-lip Sucker, 49.
Pecora, 335.
Pediculati, 27, 171.
Pediocætes, 253.
Pedomys, 320.
Peep, 246.
Pekan, 341.
Pelecanidæ, 226.
Pelecanus, 226.
Pelican, 226.
Pelidna, 246.
Pelionetta, 234.
Pelobatidæ, 182.
Perca, 134.
Percesoces, 26.
Perch, 134.
Percidæ, 121.
Perciformes, 113.
Percina, 126.
Percopsidæ, 82.
Percopsis, 82.
Perisoreus, 277.
Perissodactyli, 335.
Perissoglossa, 301.
Petrel, 223, 224.
Petrochelidon, 292.
Petromyzon, 10.
Petromyzontidæ, 10.
Peucæa, 289.
Pewee, 274, 275.
Phalacrocoracidæ, 225.
Phalacrocorax, 226.

Phalarope, 243.
Phalaropodidæ, 243.
Phalaropus, 243.
Pharyngognathi, 145.
Phasianidæ, 253.
Pheasant, 252, 253.
Phenacobius, 62.
Philohela, 245.
Phoca, 338.
Phocæna, 331.
Phocidæ, 338.
Phœbe, 274.
Phoxinus, 66.
Phrynosoma, 203.
Phycis, 162.
Phyllobasileus, 311.
Physalus, 333.
Physeter, 333.
Physeteridæ, 333.
Physoclysti, 26, 37, 91.
Physostomi, 26, 37.
Pica, 276.
Picariæ, 264.
Pici, 266.
Picidæ, 266.
Pickerel, 88, 89.
Picoides, 267.
Pigeon-Hawk, 257, 260.
Pig-fish, 140.
Pigmy Sperm Whale, 333.
Pike, 88, 89.
Piked Whale, 344.
Pike Perch, 135.
Pilot-fish, 77, 110.
Pilot Snake, 194.
Pilot Whale, 331.
Pimelepterus, 140.
Pimephales, 54.
Pine-creeping Warbler, 303.
Pine Grosbeak, 283.
Pine Marten, 340.
Pine Mouse, 320.
Pine Siskin, 285.
Pine Snake, 196.
Pin-fish, 140.
Pinicola, 283.
Pinnated Grouse, 252.
Pinnipedia, 337.
Pin-tail, 231.
Pipe-fish, 96.
Pipilo, 289.
Piping Plover, 250.

Pipit, 306.
Piranga. 291.
Pirate-Perch, 113.
Pisces, 12.
Pituophis, 196.
Pitymys. 320.
Placopharynx, 48.
Plagusia, 168.
Platalcidæ, 236.
Platophrys, 166.
Platygobio. 65.
Platypodon, 17.
Platysomatichthys, 165.
Plautus, 218.
Plecotus, 329.
Plectognathi, 27, 168.
Plectospondyli, 42.
Plectrophenax, 285.
Plegadis, 236.
Plethodon, 179.
Plethodontidæ, 179.
Pleurodelidæ, 181.
Pleurolepis, 122.
Pleuronectes, 166.
Pleuronectidæ, 164.
Plover, 249.
Pneumatophorus, 107.
Pochard, 231.
Podicipidæ, 215.
Podilymbus, 215.
Pœcilichthys, 129.
Pogonias, 144.
Polar Bear, 339.
Polioptila, 311.
Pollachius, 163.
Pollack, 163.
Polyodon, 33.
Polyodontidæ, 33.
Pomatomidæ, 111.
Pomatomus, 111.
Pomolobus, 72.
Pomoxis, 115.
Pompano, 110.
Pond-fish, 119.
Pond Turtle, 208.
Poocætes, 286.
Porbeagle, 18.
Porcupine, 318.
Porcupine-fish, 170.
Porgie, 138, 140.
Poronotus, 112.
Porpoise, 331.
Porzana, 241.
Potamocottus, 149.

Pouched Gopher, 318, 319.
Prairie Chicken, 252.
Prairie Dog, 323.
Prairie Hare, 318.
Prairie Mole, 326.
Prairie Rattlesnake, 199.
Prairie Warbler, 303.
Prairie Wolf, 342.
Primates, 343.
Prionotus, 152.
Pristididæ, 20.
Pristis, 20.
Procellaria, 224.
Procellariidæ, 223.
Procyon, 338.
Procyonidæ, 338.
Prodelphinus, 330.
Progne, 292.
Prong Horn, 336.
Prosopium, 77.
Proteida, 175.
Proteidæ, 175.
Protonotaria, 298.
Pseudemys, 209.
Pseudopleuronectes, 167.
Pseudotriacis, 14.
Psittaci, 263.
Psittacidæ, 264.
Ptarmigan, 252.
Pterophryne, 172.
Pteroplatea, 22.
Puffer, 170.
Puffin, 217.
Puffing Pig, 331.
Puffinus, 223.
Puma, 343.
Pumpkin-seed, 119.
Purple Finch, 284.
Putorius, 341.
Pygopodes, 214.
Pygosteus, 97.

Qua-Bird, 239.
Quail, 252.
Quassilabia. 49.
Querquedula, 230.
Quill-back, 45.
Quinnat, 79.
Quiscalus, 280.

Rabbit, 317.
Rabbit-fish, 170.

Rabbit-mouth Sucker, 49.
Raccoon, 338.
Raft Duck, 232.
Raiæ, 19.
Rail, 240, 241.
Rainbow Trout, 79.
Rain Crow, 265.
Rain-water Fish, 86.
Raja, 21.
Rajidæ, 20.
Rallidæ, 240.
Rallus, 241.
Rana, 184.
Rangeley Trout, 81.
Rangifer, 336.
Ranidæ, 184.
Raptores, 255.
Rat, 321.
Rattlesnake, 198, 199.
Raven, 277.
Ray, 19.
Razor-backed Buffalo, 44.
Razor-billed Auk, 217.
Recurvirostra, 244.
Recurvirostridæ, 243.
Red Bat, 328.
Red-bellied Minnow, 53.
Red-bellied Snake, 192.
Red-bellied Terrapin, 209.
Red-bellied Woodpecker, 268.
Red Bird, 289, 291.
Red-breast, 313.
Red Deer, 336.
Red Eft, 181.
Red Eye, 115.
Red-eyed Vireo, 295.
Red-fin, 58, 60.
Red-fish, 143.
Red Fox, 342.
Red Grouper, 138.
Red Head, 231.
Red-headed Woodpecker, 268.
Red Horse, 47, 173.
Red Mouse, 321.
Red-mouthed Buffalo, 44.
Redpoll, 284.
Red-poll Warbler, 303.
Red-shafted Flicker, 268.
Red Snake, 196.
Red Snapper, 139.
Red Squirrel, 323.

Redstart, 305.
Red-winged Blackbird, 279.
Reed Bird, 278.
Regina, 193.
Regulus, 310.
Reindeer, 346.
Remora, 102.
Remoropsis, 102.
Reniceps, 16.
Reptiles, 187.
Reptilia, 187.
Rhinichthys, 63.
Rhinonemus, 162.
Rhinoptera, 23.
Rhombochirus, 102.
Rhombus, 112
Rhynchodon, 200.
Rhynchophanes, 283.
Rhynchopidæ, 222.
Rhynchops, 222.
Riband Snake, 192.
Rice Bird, 278.
Rice Mouse, 321.
Right Whale, 334.
Ring-billed Gull, 220.
Ring-necked Duck, 232.
Ring-neck Plover, 250.
Ring-neck Snake, 196.
Rissa, 220.
River Chub, 65.
River Duck, 228.
Roach, 66.
Robin, 313.
Robin Snipe, 246.
Roccus, 136.
Rock, 137.
Rock Bass, 115.
Rock-fish, 137, 147.
Rockling, 162.
Rock Sturgeon, 34.
Rocky Mountain Sheep, 337.
Rodentia, 316.
Rorqual, 333.
Rose-breasted Grosbeak, 290.
Rose-fish, 147.
Rosy Gull, 221.
Rough-legged Hawk, 259.
Rough-winged Swallow, 293.
Round-fish, 77.
Round Herring, 71.

Round Pompano, 110.
Round Robin, 108.
Ruby-throat, 270.
Rudder-fish, 110, 111, 140.
Ruff, 248.
Ruffed Grouse, 252.
Ruminants, 335.
Runner, 108.
Rusty Blackbird, 279.

SABLE, 340.
Sail-fish, 104.
Sailor's Choice, 140.
Salamander, 175–177.
Salientia, 181.
Salmo, 79.
Salmon, 76, 79.
Salmonidæ, 76.
Salmon Trout, 80.
Salt-marsh Turtle, 203.
Salvelinus, 80.
Sand Darter, 122.
Sanderling, 247.
Sand-hill Crane, 240.
Sand Lance, 191.
Sand Martin, 293.
Sandpeep, 247.
Sand Pike, 135.
Sandpiper, 246–248.
Sand Shark, 18.
Sap Sucker, 267.
Sarda, 106.
Sauger, 135.
Sault White-fish, 78.
Saurel, 108.
Sauropsida, 7, 187.
Saury, 93.
Savannah Sparrow, 286.
Saw-fish, 20.
Saw-whet Owl, 263.
Sayornis, 274.
Scabbard-fish, 104.
Scad, 108.
Scalops, 326.
Scapanus, 327.
Scaphiopus, 183.
Scaphirhynchops, 34.
Scaphirhynchus, 34.
Scarlet Tanager, 291.
Scaup Duck, 232.
Sceloporus, 203.
Schilbeodes, 41.
Sciæna, 143.

Sciænidæ, 141.
Scincidæ, 200.
Scissor-tail, 274.
Sciuridæ, 322.
Sciuropterus, 324.
Sciurus, 323.
Sclerognathus, 44.
Scolecophagus, 279.
Scoliodon, 17.
Scolopacidæ, 244.
Scolopax, 245.
Scomber, 106.
Scomberesox, 93.
Scomberomorus, 105.
Scombridæ, 104.
Scombriformes, 103.
Scorpænidæ, 147.
Scorpion, 201.
Scoter, 234.
Scotiaptex, 262.
Scrag Whale, 334.
Screech Owl, 263.
Sculpin, 147, 150.
Scup, 140.
Scuppaug, 140.
Scyphobranchii, 155.
Sea Bass, 135.
Sea Cat-fish, 38, 39.
Sea Coot, 234.
Sea Devil, 23.
Sea Duck, 228.
Sea Horse, 96.
Seal, 338.
Sea Raven, 148.
Sea Robin, 152.
Sea-side Finch, 287.
Sea Snail, 153, 154.
Sea Squirt, 7.
Sebastes, 147.
Seiurus, 304.
Selachii, 14.
Selachostomi, 25, 33.
Selene, 109.
Semotilus, 66.
Sergeant-fish, 103.
Serinus, 285.
Seriola, 110.
Serranidæ, 135.
Serraria, 127.
Serpent, 187.
Setophaga, 305.
Shad, 73.
Shad-waiter, 77.
Shag, 226.

Sharks, 14.
Sharp-shinned Hawk, 257.
Shearwater, 223, 224.
Sheep, 336.
Sheepshead, 140, 144.
Sheldrake, 229.
Shiner, 58, 68.
Shore-birds, 242.
Shore Lark, 276.
Shore Sparrow, 236.
Shoveller, 231.
Shovel-nosed Sturgeon, 34.
Shrew, 325.
Shrew Mouse, 325.
Shrike, 294.
Sialia, 313.
Sibbaldius, 334.
Sicklebill, 248.
Siluridæ, 38.
Silver-bottom Whale, 334.
Silver Eel, 104.
Silver-Fin, 58.
Silver-fish, 71.
Silver Fox, 342.
Silver Gar, 92.
Silver Hake, 163.
Silver Perch, 143.
Silversides, 99, 100.
Silver Whiting, 144.
Siphostoma, 96.
Siredon, 177.
Siren, 175.
Sirenidæ, 175.
Siskin, 285.
Sistrurus, 199.
Sitta, 309.
Skate, 21.
Skim-back, 45.
Skimmer, 222.
Skink, 200.
Skip-jack. 111.
Skipper, 93.
Skua Gull, 218.
Skunk, 340.
Skunk Porpoise, 331.
Skylark, 276.
Sleeper, 15.
Slow-worm, 201.
Smelt, 76.
Snake, 188.
Snake-bird, 225.

Snake-fish, 75.
Snapper, 139.
Snapping Turtle, 206, 207.
Snipe, 244–246.
Snow Bird, 283.
Snow Bunting, 285.
Snow Goose, 234.
Snowy Owl, 263.
Snuffer, 331.
Soft-shelled Turtle, 205, 206.
Sole, 168.
Solitaire, 312.
Somateria, 233.
Somniosidæ, 15.
Somniosus, 15.
Song Sparrow, 289.
Sora, 241.
Sorex, 325.
Soricidæ, 325.
Soriciscus, 326.
South-Southerly, 233.
Spade-fish, 146.
Spade-foot, 183.
Spanish Mackerel, 105.
Sparidæ, 138.
Sparrow. 284.
Sparrowhawk, 260.
Sparrow Owl, 262.
Spatula, 231.
Spawn-eater, 57.
Spear-fish, 104.
Speckle-bill, 234.
Speckled Tortoise, 210.
Speckled Trout, 80.
Spelerpes, 180.
Speotyto, 263.
Spermophilus, 323.
Sperm Whale, 333.
Sperm-whale Porpoise, 332.
Sphyræna, 100.
Sphyrænidæ, 100.
Sphyrapicus, 268.
Sphyrna, 16.
Sphyrnidæ, 15.
Spilogale, 340.
Spinus, 285.
Spiny-rayed Fishes, 101.
Spirit Duck, 233.
Spiza, 290.
Spizella, 288.
Split-mouth, 49.

Spoon-bill (fish), 33.
Spoon-bill (bird), 236.
Spoon-bill Duck, 231.
Spot, 143.
Spotted Adder, 197.
Spotted Dolphin, 330.
Spreading Adder, 197.
Sprig-tail, 231.
Spruce Partridge, 252.
Squali, 14.
Squalidæ, 15.
Squalius, 66.
Squalus, 15.
Squatarola, 249.
Squatina, 19.
Squatinidæ, 19.
Squawk, 239.
Squeteague, 142.
Squirrel, 322, 323.
Stake Driver, 237.
Star-gazer, 156.
Star-nosed Mole, 327.
Steganopodes, 224.
Steganopus, 243.
Stelgidopteryx, 293.
Stenotomus, 140.
Stercorariidæ, 218.
Stercorarius, 219.
Sterna, 221.
Stichæus, 159.
Stickleback, 97, 98.
Stilt, 244.
Stilt Sandpiper, 245.
Stingaree, 22.
Sting-ray, 22.
Stizostedion, 135.
Stoasodon, 23.
Stock-fish, 163.
Stolephoridæ, 74.
Stolephorus, 74.
Stone Cat, 41.
Stone Lugger, 46, 52.
Stone Roller, 46, 52.
Stone Snipe, 247.
Stone Toter, 46, 54.
Storeria, 192.
Stork, 237.
Storm Petrel, 224.
Strawberry Bass, 115.
Strigidæ, 261.
Striped Bass, 137.
Striped Gopher, 323.
Striped Snake, 193.
Striped Sucker, 47.

Strix, 261.
Stromateidæ, 111.
Stromateus, 112.
Stud-fish, 85.
Sturgeon, 33, 34.
Sturnella, 279.
Sucker, 43, 46.
Sucker-mouthed Buffalo, 44.
Sucking-fish, 102.
Sula, 225.
Sulidæ, 225.
Sulphur-bottom Whale, 334.
Summer Red-bird, 291.
Summer Warbler, 301.
Sunapee Trout, 81.
Sun-fish, 113, 116-119, 171.
Surf-bird, 250.
Surf Duck, 234.
Surf Whiting. 144.
Surmullet, 141.
Surnia, 263.
Swallow, 291, 292.
Swamp Sparrow, 289.
Swan. 235.
Sweet Sucker, 46.
Swell-fish, 169, 170.
Swell-toad, 170.
Swift (reptile), 203.
Swift (bird). 270.
Swingle-tail, 17.
Sword-fish, 103.
Sword-Grampus, 331.
Sylvania, 305.
Sylviidæ, 310.
Symphemia, 248.
Symphurus, 168.
Synaptomys, 320.
Synguathidæ, 95.
Synentognathi, 26, 91.
Synodontidæ, 75.
Synodus, 75.
Syrnium, 262.

Tachycineta, 292.
Tachysurus, 39.
Tadpole. 174.
Tailor Herring, 72.
Talpidæ, 326.
Tamias, 323.
Tanager, 291.

Tanagridæ, 291.
Tantalus, 237.
Tarpon, 71.
Tarpum, 71.
Tattler, 248.
Tauridea, 148.
Tautog, 146.
Tawny Thrush, 312.
Taxidea, 340.
Teal, 230.
Teeter-tail, 248.
Teidæ, 201.
Teleocephali, 37.
Teleostei, 26, 37.
Teleostomi, 25.
Tell-tale, 247.
Telmatodytes, 308.
Tennessee Warbler, 299.
Tenpounder, 70.
Tern, 221, 222.
Terrapin, 209.
Testudinata, 203.
Testudinidæ, 211.
Tetraodontidæ, 169.
Tetraonidæ, 251.
Tetrapturus, 104.
Thalaretos, 339.
Thalasseus, 221.
Thalassochelys, 205.
Thimble-eye, 107.
Thistle-bird, 285.
Thomomys, 319.
Thrasher, 307.
Thread-fish, 109.
Thread Herring, 73.
Thresher, 17.
Thrush, 311, 312.
Thryomanes, 308.
Thryothorus, 308.
Thunder-pumper, 144.
Thunder Snake, 196.
Thymallus, 79.
Tiger Shark, 16.
Tigoma, 67.
Tinnunculus, 260.
Tiny Perch, 113.
Tip-up, 248.
Titlark, 306.
Titmouse. 310.
Toad, 182.
Toad-fish, 154, 155.
Tobacco-box, 21.
Togue, 80.
Tom-cod, 163.

Tongue-fish, 168.
Toothed Herring, 69.
Top-Minnow, 86, 87.
Torpedinidæ, 21.
Torpedo, 22.
Tortoise, 204.
Tortoise-shell Turtle, 205.
Totanus, 247.
Towhee, 289.
Toxicophis, 199.
Trachinotus, 110.
Trachurops, 108.
Trachurus, 108.
Trachystomata, 175.
Tree Frog, 183.
Tree Sparrow, 284, 283.
Tree Toad, 184.
Trichiuridæ, 104.
Trichiurus, 104.
Trigger-fish, 168, 169.
Triglidæ, 152.
Triglops, 151.
Triglopsis, 150.
Tringa, 246.
Trionychidæ, 205.
Triple-tail, 138.
Trochilidæ, 270.
Trochilus, 270.
Troglodytes, 308.
Troglodytidæ, 306.
Tropidoclonium, 192.
Tropidonotus, 194.
Trout, 79, 80.
Trout Perch, 82.
True Fishes, 25.
Trumpeter Swan, 235.
Trumpet-fish, 97.
Trunk-back, 204.
Trygon, 22.
Tryngites, 248.
Tubinares, 222.
Tufted Titmouse, 310.
Tullibee, 78.
Tunicata, 5.
Tunny, 106.
Turbot, 166.
Turdidæ, 311.
Turdus, 312.
Turkey, 253.
Turkey Buzzard, 256.
Turn-stone, 250.
Tursiops, 330.
Turtle, 203.
Turtle Dove, 254.

Tylosurus, 92.
Tympanuchus, 252.
Typhlichthys, 83.
Tyrannidæ, 273.
Tyrannus, 274.

Ulocentra, 124.
Umbra, 87.
Umbridæ, 87.
Umbrula, 144.
Ungulata, 335.
Upland Sandpiper, 248.
Upsilonphorus, 156, 359.
Uranidea, 148.
Uranoscopidæ, 156.
Uria, 217.
Urinator, 216.
Urinatoridæ, 216.
Urochordata, 6.
Urocyon, 342.
Urodela, 175.
Urophycis, 162.
Ursidæ, 339.
Ursus, 339.

Vaillantia, 124.
Veery, 312.
Vertebrata, 5.
Vesperides, 328.
Vesper Sparrow, 286.
Vespertilio, 328.
Vespertilionidæ, 327.
Vesperugo, 328.
Vesperus, 328.
Vireo, 295, 296.
Vireonidæ, 294.
Vireosylva, 295.
Virginia, 191.
Vomer, 109.
Vulpes, 342.
Vulture, 255.

Wagtail, 306.
Wall-eye, 135.
Wapiti, 336.
Warbler, 296, 310.
Warbling Vireo, 296.
War-mouth, 115.
Water Dog, 175.
Water Hare, 317.
Water Shrew, 325.
Water Snake, 194.

Water Thrush, 304.
Water Turkey, 225.
Water Wagtail, 304.
Water Witch, 215.
Wax-wing, 293.
Weak-fish, 142.
Weasel, 341.
Whale, 333, 334.
Wharf Rat, 321.
Whipparee, 22.
Whippoorwill, 269.
Whiskey Jack, 277.
Whistling Swan, 235.
White Bass, 137.
White-bellied Swallow, 292.
White Cat-fish, 39, 40.
White-crowned Sparrow, 288.
White-eyed Vireo, 296.
White-fish, 77.
White Perch, 137, 144.
White Rabbit, 317.
White Shark, 19.
White Sturgeon, 34.
White Sucker, 46, 173.
White-tailed Deer, 336.
White-throated Sparrow, 288.
White Whale, 331.
White-wing Blackbird, 291.
Whiting, 143, 144, 163.
Widgeon, 231.
Wild Cat, 343.
Wild Goose, 235.
Willet, 248.
Willow Grouse, 252.
Wilson's Thrush, 312.
Window-pane, 166.
Winter Flounder, 167.
Winter Wren, 308.
Wolf, 342.
Wolf-fish, 159.
Wolverine, 340.
Woodchuck, 323.
Woodcock, 245.
Wood Duck, 231.
Wood-Frog, 185.
Wood Ibis, 237.
Woodland Caribou, 336.
Woodpeckers, 266, 267.
Wood Pewee, 275.
Wood Rat, 321.

Wood Thrush, 312.
Wood Tortoise, 210.
Worm-eating Warbler, 298.
Worm Snake, 191.
Wrasse, 145.
Wren, 306, 308.
Wry-mouth, 159.

Xanthocephalus, 279.
Xema, 221.
Xenopterygii, 155.
Xenotis, 118.
Xerobates, 211.
Xiphias, 103.
Xiphiidæ, 103.
Xystroplites, 119.

Yellow Bass, 137.
Yellow-bellied Terrapin, 209.
Yellow-bellied Woodpecker, 268.
Yellow-bird, 285.
Yellow-breasted Chat, 305.
Yellow Cat-fish, 40.
Yellow-hammer, 268.
Yellow-headed Blackbird, 279.
Yellow-legs, 248.
Yellow Mackerel, 109.
Yellow Perch, 134.
Yellow-rumped Warbler, 302.
Yellow-shanks, 247.
Yellow-tail, 143.
Yellow-throat, 305.
Yellow-throated Vireo, 296.
Yellow-throated Warbler, 303.
Yphantes, 279.

Zapodidæ, 318.
Zapus, 318.
Zenaidura, 254.
Ziphiidæ, 332.
Ziphius, 332.
Zoarces, 160.
Zonotrichia, 287.
Zygonectes, 86.

www.ingramcontent.com/pod-product-compliance
Lightning Source LLC
Chambersburg PA
CBHW021357210326
41599CB00011B/917